AF360883

Antimicrobial Resistance in Veterinary Medicine and Public Health

Antimicrobial Resistance in Veterinary Medicine and Public Health

Editors

Paola Roncada
Bruno Tilocca

MDPI • Basel • Beijing • Wuhan • Barcelona • Belgrade • Manchester • Tokyo • Cluj • Tianjin

Editors

Paola Roncada
University Magna Græcia of Catanzaro
Italy

Bruno Tilocca
University Magna Græcia of Catanzaro
Italy

Editorial Office
MDPI
St. Alban-Anlage 66
4052 Basel, Switzerland

This is a reprint of articles from the Special Issue published online in the open access journal *Animals* (ISSN 2076-2615) (available at: https://www.mdpi.com/journal/animals/special_issues/antimicrobial_resistance_veterinary_public_health).

For citation purposes, cite each article independently as indicated on the article page online and as indicated below:

LastName, A.A.; LastName, B.B.; LastName, C.C. Article Title. *Journal Name* **Year**, *Volume Number*, Page Range.

ISBN 978-3-0365-6342-8 (Hbk)
ISBN 978-3-0365-6343-5 (PDF)

Contents

Editorial

Antimicrobial Resistance in Veterinary Medicine and Public Health

Paola Roncada * and Bruno Tilocca

Department of Health Sciences, University Magna Græcia of Catanzaro, 88100 Catanzaro, Italy
* Correspondence: roncada@unicz.it

Editorial

Animal productions, and populations, have been rapidly expanding over the last decades, forcing the ever-closer coexistence of human beings and domestic animals on our yet "narrow" planet. In addition to the damage to crops, natural resources, and loss of animal biodiversity, the increased contact occurring between humans and animals greatly supports the transmission of zoonotic diseases. This, in turn, enables the onset and the rapid diffusion of antimicrobial resistance (AMR) traits across the three sectors of life (human, animal, and the environment). In this view, veterinary medicine and the surrounding environment play an important part in the public health threat, in line with the One-Health concept [1].

Major clinically relevant bacteria for virulence and antibiotic resistance traits are the so-called ESKAPE pathogens, an acronym referring to *Enterococcus faecium*, *Staphylococcus aureus*, *Klebsiella pneumoniae*, *Acinetobacter baumanii*, *Pseudomonas aeruginosa*, and *Enterobacter* spp. Antibiotics misuse (e.g., following improper prescription, overuse in the agricultural field and usage as growth promoting agents) results in a selective pressure stimulating the bacterial adaptation mechanisms which, unavoidably, include the increasing trend to chromosomal mutations and/or a higher tendency to acquire exogenous nucleic acids (i.e., free DNA, plasmids, etc.) seeking for more favourable (adapted) phenotypes. Antibiotic-resistant bacteria are subsequently selected, and their clonal expansion amplifies the AMR traits and their diffusion, both vertically to daughter cells and horizontally to further recipient cells [1,2].

The co-occurrence of mixed microbial communities in the same ecological niches favours the massive diffusion of the genetically determined resistance traits at both intra- and inter-specific levels. Evidence of interkingdom transmission of the antimicrobial traits has also been documented, underlining the impact of this phenomenon across boundaries and its potential invalidating effect on human and veterinary antibiotic-based therapies. In this light, further efforts are desirable in understanding the diverse facets of antimicrobial resistance, ranging from assessing the geographic diffusion of diverse AMR traits to how these dynamic fluxes evolve in time and space to reach the molecular mechanisms (novel and past) employed for the inactivation of antibiotic therapies [2].

This Special Issue harbours 11 published research articles summarizing studies on diverse aspects of the antimicrobial resistance phenomenon, providing significant levels of innovation and knowledge suggestive of further research routes. All the contributions of this Special Issue emphasize the need for the cautious assessment of the circulation of AMR traits in the clinical and veterinary fields, including the assessment of the contributions by microorganisms of food and environmental origin in the dissemination of antimicrobial resistance traits across the three sectors of life. Kwon et al. investigated extended-spectrum cephalosporin (ESC)-resistant *Salmonella* spp. of chicken origin to assess the transferability to humans of beta-lactamase gene-harbouring plasmid in vitro and in vivo. Warningly the study provides evidence on the ease of dissemination of the AMR traits between bacteria (*Salmonella–Escherichia*) in the case of coexistence, as it is common in microbial community

Citation: Roncada, P.; Tilocca, B. Antimicrobial Resistance in Veterinary Medicine and Public Health. *Animals* **2022**, *12*, 3253. https://doi.org/10.3390/ani12233253

Received: 31 October 2022
Accepted: 18 November 2022
Published: 23 November 2022

Publisher's Note: MDPI stays neutral with regard to jurisdictional claims in published maps and institutional affiliations.

settings, even in the absence of antibiotic-mediated selective pressure, providing a clear glimpse on the role of the food chain in the dissemination of the AMR from a One-Health perspective [3]. Another study by Kwon and colleagues [4] assessed the prevalence of *Campylobacter* spp. over the whole chicken production process. The study highlights the hot spot of AMR dissemination in the chicken meat production process, underlining that resistance to fluoroquinolones was the most frequently observed form of resistance in their sample.

Mtemisika and colleagues [5] surveyed the presence of resistant *Escherichia coli* in pigs and poultry populations reared in Mwanza, Tanzania, according to standard operating procedures and international guidelines. This study highlights that although different *E. coli* phenotypes are harboured in the animal's intestine, AMR traits have been recorded in both host types.

Piras et al. [6] investigated the persistence of *S. aureus* strains as means of biofilm production, a very well-known burden in food, farm and clinical contexts. The study detected mechanisms related to the control of catabolites, the production of proteins with moonlighting activities and the detoxification of compounds with antimicrobial activities, suggesting potential biomarkers/metabolic routes to be targeted to prevent and/or mitigate this phenomenon. Another study by Piras and co-workers [7] focused on the effective resistance traits of the milk-associated microbial community, focusing on the microbial protein repertoire. Proteins are indeed effective actors in the resistance mechanisms, and their study provides a detailed glimpse into the microbial dynamics and the metabolic influence that milk microbiota members exert on each other.

Overton et al. [8] employed a metabolomic approach to investigate the effect of multiple antibiotic molecules on *Salmonella typhimurium* cultured from various hosts. Interestingly, the authors highlight a significant modulation in the metabolic profiles, suggesting metabolomics as an innovative approach for the quick evaluation of resistance traits against multiple antibiotics.

The study of Abd El-Aziz et al. [9] provides the first survey into the virulence traits, antimicrobial and biocide resistance and epidemiological typing of *Streptococcus uberis* isolated from bovine clinical mastitis in dairy farms of diverse hygienic interventions in Egypt. The survey describes the prevalence of *S. uberis* infections and the genes involved in its virulence and antimicrobial resistance in relation to the hygienic conditions of the dairy farm, underlining the importance of a combined intervention while facing the dissemination of pathogens and their resistance traits.

A further study by Abd El-Aziz and colleagues [10] investigated, for the first time, class 1 integrons and associated gene cassettes among pan-drug-resistant (PDR), extensively drug-resistant (XDR) and multidrug-resistant (MDR) *Campylobacter spp*. Although no pan-susceptible isolates were found, the study describes the detection of multiple resistance traits, in addition to providing the very first identification of a putative phage tail tape measure protein which is indicative of a possible *Campylobacter*–bacteriophage interaction, resulting in the consequent spread of the resistance trait via horizontal gene transfer.

Kang and colleagues [11] focused on environmental mastitis-causing enterococci as an emerging cause of nosocomial infections of relevance due to their antimicrobial resistance traits. The survey underlined a dramatically high proportion of the bulk tank milk-derived enterococci as the vector of resistance traits against single- and multi-antimicrobial drugs.

Zou and collaborators [12] described the prevalence and molecular features of extraintestinal pathogenic *Escherichia coli* (ExPEC). In addition, the study observes the presence of the plasmid-borne colistin resistance gene along with an additional carbapenemase gene in some isolates. Altogether, their findings suggest that healthy chickens can serve as a potential reservoir for multidrug-resistant ExPEC isolates.

Interestingly, this Special Issue also includes a study by Nazish et al. [13] on antinematode resistance of *Haemonchus contortus*. It suggests the use of *Comamonas* spp. and *Pseudomonas weihenstephanesis* as biological control agents to be employed as alternatives to synthetic anthelmintic compounds, showing a significant mortality rate against the parasite with little-to-no selective pressure triggering the onset and diffusion of the antimicrobial resistance traits.

Altogether, the above research articles deal with several aspects of antimicrobial resistance, providing an enlightening view and suggestive approaches for future research lines aimed at tackling antimicrobial resistance in the diverse aspects of the life sciences. Both expert scientists and readers approaching this fascinating field can benefit from the cutting-edge analysis of an impressive range of data from diverse sample types, and we hope you enjoy it.

Author Contributions: B.T. and P.R. writing original draft preparation review and editing. All authors have read and agreed to the published version of the manuscript.

Institutional Review Board Statement: Not applicable.

Conflicts of Interest: The authors declare no conflict of interest.

References

1. Murugaiyan, J.; Kumar, P.A.; Rao, G.S.; Iskandar, K.; Hawser, S.; Hays, J.P.; Mohsen, Y.; Adukkadukkam, S.; Awuah, W.A.; Jose, R.A.M.; et al. Progress in Alternative Strategies to Combat Antimicrobial Resistance: Focus on Antibiotics. *Antibiotics* **2022**, *11*, 200. [CrossRef] [PubMed]
2. Palma, E.; Tilocca, B.; Roncada, P. Antimicrobial Resistance in Veterinary Medicine: An Overview. *Int. J. Mol. Sci.* **2020**, *21*, 1914. [CrossRef] [PubMed]
3. Kwon, B.-R.; Wei, B.; Cha, S.-Y.; Shang, K.; Zhang, J.-F.; Jang, H.-K.; Kang, M. Characterization of Extended-Spectrum Cephalosporin (ESC) Resistance in *Salmonella* Isolated from Chicken and Identification of High Frequency Transfer of bla_{CMY-2} Gene Harboring Plasmid In Vitro and In Vivo. *Animals* **2021**, *11*, 1778. [CrossRef] [PubMed]
4. Kwon, B.-R.; Wei, B.; Cha, S.-Y.; Shang, K.; Zhang, J.-F.; Kang, M.; Jang, H.-K. Longitudinal Study of the Distribution of Antimicrobial-Resistant *Campylobacter* Isolates from an Integrated Broiler Chicken Operation. *Animals* **2021**, *11*, 246. [CrossRef] [PubMed]
5. Mtemisika, C.I.; Nyawale, H.; Benju, R.J.; Genchwere, J.M.; Silago, V.; Mushi, M.F.; Mwanga, J.; Konje, E.; Mirambo, M.M.; Mshana, S.E. Epidemiological Cut-Off Values and Multidrug Resistance of *Escherichia coli* Isolated from Domesticated Poultry and Pigs Reared in Mwanza, Tanzania: A Cross-Section Study. *Animals* **2022**, *12*, 835. [CrossRef] [PubMed]
6. Piras, C.; Di Ciccio, P.; Soggiu, A.; Greco, V.; Tilocca, B.; Costanzo, N.; Ceniti, C.; Urbani, A.; Bonizzi, L.; Ianieri, A.; et al. *S. aureus* Biofilm Protein Expression Linked to Antimicrobial Resistance: A Proteomic Study. *Animals* **2021**, *11*, 966. [CrossRef] [PubMed]
7. Piras, C.; Greco, V.; Gugliandolo, E.; Soggiu, A.; Tilocca, B.; Bonizzi, L.; Zecconi, A.; Cramer, R.; Britti, D.; Urbani, A.; et al. Raw Cow Milk Bacterial Consortium as Bioindicator of Circulating Anti-Microbial Resistance (AMR). *Animals* **2020**, *10*, 2378. [CrossRef] [PubMed]
8. Overton, J.M.; Linke, L.; Magnuson, R.; Broeckling, C.D.; Rao, S. Metabolomic Profiles of Multidrug-Resistant *Salmonella* Typhimurium from Humans, Bovine, and Porcine Hosts. *Animals* **2022**, *12*, 1518. [CrossRef] [PubMed]
9. El-Aziz, N.A.; Ammar, A.; El Damaty, H.; Elkader, R.A.; Saad, H.; El-Kazzaz, W.; Khalifa, E. Environmental *Streptococcus uberis* Associated with Clinical Mastitis in Dairy Cows: Virulence Traits, Antimicrobial and Biocide Resistance, and Epidemiological Typing. *Animals* **2021**, *11*, 1849. [CrossRef] [PubMed]
10. El-Aziz, N.K.A.; Ammar, A.M.; Hamdy, M.M.; Gobouri, A.A.; Azab, E.; Sewid, A.H. First Report of *aacC5-aadA7Δ4* Gene Cassette Array and Phage Tail Tape Measure Protein on Class 1 Integrons of *Campylobacter* Species Isolated from Animal and Human Sources in Egypt. *Animals* **2020**, *10*, 2067. [CrossRef] [PubMed]
11. Kang, H.; Yoon, S.; Kim, K.; Lee, Y. Characteristics of High-Level Aminoglycoside-Resistant *Enterococcus faecalis* Isolated from Bulk Tank Milk in Korea. *Animals* **2021**, *11*, 1724. [CrossRef] [PubMed]
12. Zou, M.; Ma, P.-P.; Liu, W.-S.; Liang, X.; Li, X.-Y.; Li, Y.-Z.; Liu, B.-T. Prevalence and Antibiotic Resistance Characteristics of Extraintestinal Pathogenic *Escherichia coli* among Healthy Chickens from Farms and Live Poultry Markets in China. *Animals* **2021**, *11*, 1112. [CrossRef] [PubMed]
13. Nazish, A.; Fozia; Khattak, B.; Khan, T.A.; Ahmad, I.; Ullah, R.; Bari, A.; Asmari, M.; Mahmood, H.; Sohaib, M.; et al. Antinematode Activity of Abomasum Bacterial Culture Filtrates against *Haemonchus contortus* in Small Ruminants. *Animals* **2021**, *11*, 1843. [CrossRef]

Article

First Report of *aacC5-aadA7Δ4* Gene Cassette Array and Phage Tail Tape Measure Protein on Class 1 Integrons of *Campylobacter* Species Isolated from Animal and Human Sources in Egypt

Norhan K. Abd El-Aziz [1,*], **Ahmed M. Ammar** [1], **Mona M. Hamdy** [1], **Adil A. Gobouri** [2], **Ehab Azab** [3] **and Alaa H. Sewid** [1]

[1] Department of Microbiology, Faculty of Veterinary Medicine, Zagazig University, Zagazig 44511, Egypt; prof.ahmedammar_2000@yahoo.com (A.M.A.); dr.mona_micro@yahoo.com (M.M.H.); veterinarian.alaa.sweed@gmail.com (A.H.S.)

[2] Department of Chemistry, College of Science, Taif University, P.O. Box 11099, Taif 21944, Saudi Arabia; a.gobouri@tu.edu.sa

[3] Department of Biotechnology, College of Science, Taif University, P.O. Box 11099, Taif 21944, Saudi Arabia; e.azab@tu.edu.sa

* Correspondence: norhan_vet@hotmail.com; Tel.: +20-1226369943

Received: 2 October 2020; Accepted: 2 November 2020; Published: 8 November 2020

Simple Summary: *Campylobacter* species are among the major causes of bacterial foodborne infections. Here, we investigate, for the first time, class 1 integrons and associated gene cassettes among pan drug-resistant (PDR), extensively drug-resistant (XDR), and multidrug-resistant (MDR) *Campylobacter* species isolated from livestock animals and humans in Egypt. Our results revealed alarming PDR (2.55%) and inordinate XDR (68.94%) and MDR (28.5%) *Campylobacter* isolates. None of the examined isolates were pan-susceptible. The existence of a novel gene cassette array, namely *aacC5-aadA7Δ4* and a putative phage tail tape measure protein on class 1 integrons of *Campylobacter* species is the most highlighted novelty of the current study. Evidence from this study showed the possibility of *Campylobacter*–bacteriophage interactions as well as treatment failure in animals and humans due to horizontal gene transfer mediated by class 1 integrons.

Abstract: *Campylobacter* species are common commensals in the gastrointestinal tract of livestock animals; thus, animal-to-human transmission occurs frequently. We investigated for the first time, class 1 integrons and associated gene cassettes among pan drug-resistant (PDR), extensively drug-resistant (XDR), and multidrug-resistant (MDR) *Campylobacter* species isolated from livestock animals and humans in Egypt. *Campylobacter* species were detected in 58.11% of the analyzed chicken samples represented as 67.53% *Campylobacter jejuni* (*C. jejuni*) and 32.47% *Campylobacter coli* (*C. coli*). *C. jejuni* isolates were reported in 51.42%, 74.28%, and 66.67% of examined minced meat, raw milk, and human stool samples, respectively. Variable antimicrobial resistance phenotypes; PDR (2.55%), XDR (68.94%), and MDR (28.5%) campylobacters were reported. Molecular analysis revealed that 97.36% of examined campylobacters were *integrase* gene-positive; all harbored the class 1 integrons, except one possessed an empty integron structure. DNA sequence analysis revealed the predominance of *aadA* (81.08%) and *dfrA* (67.56%) alleles accounting for resistance to aminoglycosides and trimethoprim, respectively. This is the first report of *aacC5-aadA7Δ4* gene cassette array and a putative phage tail tape measure protein on class 1 integrons of *Campylobacter* isolates. Evidence from this study showed the possibility of *Campylobacter*–bacteriophage interactions and treatment failure in animals and humans due to horizontal gene transfer mediated by class 1 integrons.

Keywords: *Campylobacter* species; class 1 integrons; extensively drug-resistance; pan drug-resistance; gene cassette arrays

1. Introduction

Thermophilic *Campylobacter* species, particularly *Campylobacter jejuni* (*C. jejuni*) and *Campylobacter coli* (*C. coli*) pose veterinary and public health concerns due to their zoonotic potential, the enormous range of reservoir hosts, and persistence in the environment [1]. Consumption of contaminated food, especially poultry products, unpasteurized milk, and undercooked meat, as well as water, is a risk factor for *C. jejuni* and *C. coli* infections [2,3].

Most *Campylobacter* infections are self-limiting and require no therapeutic intervention other than supportive and rehydration therapy. However, prompt antimicrobial treatment is employed in immunocompromised individuals, patients whose symptoms are severe or persistent, intense or prolonged enteritis, cases of bacteremia, and those with extraintestinal infections [4]. Attention to the resistance of campylobacters has been launched due to the indiscriminate abuse of antibiotics [5]. Clinical, veterinary, and environmental surveys have shown that bacteria harboring integrons are frequently associated with the multidrug-resistant (MDR) phenotype [6]. However, the extensively drug-resistant (XDR) and pan drug-resistant (PDR) bacteria are epidemiologically significant not only due to their resistance to multiple antimicrobial agents but also to their ominous prospect of being resistant to almost all or all approved antimicrobial agents [7,8].

The integron is a site-specific recombination system capable of integrating mobile gene cassettes, which can be expressed and disseminated via horizontal gene transfer [9,10]. Class 1 integron includes two conserved segments (CSs), denoted as 5′- and 3′-CSs, flanking a gene cassette. An *int1* gene encoding an integrase enzyme is located within the 5′-CS and is responsible for the recombination of a gene cassette [11]. The 3′-CS possesses *qacEΔ1* and *sul1* genes encoding resistance to quaternary ammonium compounds and sulfonamide, respectively. Integrons can incorporate and express more than one gene cassette conferring resistance to multiple antimicrobial classes such as beta-lactams, aminoglycosides, trimethoprim, chloramphenicol, fosfomycin, macrolides, lincosamides, rifampicin, and quinolones [11].

Previous studies reported class 1 integrons carried aminoglycoside (*aadA* and *aacA4*) and trimethoprim (*dfr1* and *dfr9*) resistance gene cassettes in both *C. jejuni* and *C. coli* isolated from chicken house environment [12], poultry, and human sources [13–16] without showing the antimicrobial resistance profiles of the isolates. However, these genetic elements were not detected in an XDR *C. jejuni* CCARM 13,322 isolate recovered from a human case of diarrhea associated with international travel [17]. Hence, this study reports, for the first time, class 1 integrons and associated gene cassettes in thermophilic *Campylobacter* species isolated from livestock animals and humans showing variable antimicrobial resistance phenotypes.

2. Materials and Methods

2.1. Samples

A total of 550 samples comprising chickens and chicken products (*n* = 265), meat and meat products (*n* = 160), milk and milk products (*n* = 95) as well as human stools (*n* = 30) were collected during the period from January 2018 to December 2019. Samples of animal origins were obtained from various retail outlets, Zagazig city, Sharkia Governorate, Egypt. Human samples were collected from patients being affected by diarrhea and gastroenteritis, attending various private laboratories located in Zagazig city. The samples were transported immediately in an icebox to the bacteriology laboratory for further analysis. The animal study was approved by the Animal Welfare and Research Ethics Committee, Faculty of Veterinary Medicine, Zagazig University. The human study was conducted following the Ethics of the World Medical Association (Declaration of Helsinki) and was approved by the research ethics committee of the Faculty of Medicine, Zagazig University (ApprovalNoZU-IRB#2056-18-05-2019). The patients participating in the research study provided written informed consent.

2.2. Bacteriological Analysis and Molecular Identification

Isolation of *Campylobacter* species was performed under microaerobic conditions according to the protocol established by Vandepitte et al. [18]. Samples were enriched in Preston *Campylobacter* selective enrichment broth (Oxoid, Cambridge, UK) at 42 °C for 48 h. The enrichment broth was plated onto modified charcoal cefoperazone deoxycholate agar (mCCDA; Oxoid, Cambridge, UK) then transferred onto Columbia agar (Oxoid, Cambridge, UK) plates supplemented with 5% sterile defibrinated horse blood. Presumptive *Campylobacter* colonies were confirmed by oxidase, catalase, hippurate, and indoxyl acetate hydrolyses biochemical tests, in addition to testing their susceptibilities to nalidixic acid and cephalothin antimicrobials (30 mg/disc, each) [19]. The bacterial DNA was extracted using a QIAamp DNA Mini kit (Qiagen GmbH, Hilden, Germany) according to the manufacturer's instructions. Polymerase chain reaction (PCR) amplifications of the *23S rRNA* gene of *Campylobacter* species [20] in addition to *mapA* and *ceuE* genes of *C. jejuni* and *C. coli*, respectively, [21] were applied using oligonucleotide primers listed in Table S1.

2.3. Antimicrobial Susceptibility Testing

The antimicrobial susceptibilities of *Campylobacter* isolates were tested on Mueller–Hinton agar media (Oxoid-CM0337B, Cambridge, UK) supplemented with 5% sterile defibrinated horse blood under microaerobic conditions using the disc diffusion method [22] following the guidelines of the Clinical and Laboratory Standards Institute (CLSI) [23]. A panel of 25 standard antimicrobial discs (Oxoid, Cambridge, UK) within different 14 antimicrobial categories were examined including penicillins [ampicillin (AM; 10 µg) and amoxicillin (AX; 25 µg)], penicillin combinations [ampicillin-sulbactam (SAM; 20/10 µg) and amoxycillin-clavulanic acid (AMC; 20/10 µg)], cephalosporines [cephalothin (KF; 30 µg), cefoxitin (FOX; 30 µg), cefoperazone (CEP; 75 µg) and cefepime (FEP; 30 µg)], carbapenemes [meropenem (MEM; 10 µg)], monobactams [azetronam (ATM; 30 µg)], aminoglycosides [streptomycin (S; 10 µg), tobramycin (TOB; 10 µg), gentamycin (CN; 10 µg) and amikacin (AK; 30 µg)], macrolides [erythromycin (E; 15 µg), azithromycin (AZM; 15 µg) and clarithromycin (CLR; 15 µg)], quinolones [nalidixic acid (NA; 30 µg) and ciprofloxacin (CIP; 5 µg)], sulfonamides [sulfamethoxazole-trimethoprim (SXT; 23.75/1.25 µg)], amphenicols [chloramphenicol (C; 30 µg)], polypeptides [colistin (CT; 10 µg)], oxazolidones [lenzolid (LNZ; 30 µg)], lincosamides [clindamycin (DA; 2 µg)] and tetracyclines [doxycycline (DO; 30 µg)]. The interpretive criteria of CLSI (for most antimicrobials) [23] or the European Committee for Antimicrobial Susceptibility Testing (EUCAST) (for macrolides) were followed to classify *Campylobacter* isolates as susceptible, intermediate, or resistant [24].

The multiple antimicrobial resistance (MAR) indices were calculated as previously reported [25]. Pan drug-resistance (resistance to all antimicrobial agents), extensive drug-resistance (resistance to all classes of antimicrobial agents except 2 or fewer), and multidrug-resistance (resistance to three or more classes of antimicrobial agents) were determined as reported elsewhere [26].

2.4. PCR Amplification of Class 1 Integrons and Associated Gene Cassettes

Campylobacter isolates exhibited variable antimicrobial resistance profiles (PDR, XDR, and MDR) were subjected to DNA extraction, using the QIAamp DNA Mini kit (Qiagen, Gmbh, Hilden, Germany) following the manufacturer's recommendations. The isolates were screened for possession of the *integrase* gene as well as class 1 integrons using intI1 and hep primer sets, respectively [27,28] (Table S1). Isolates containing class 1 integrons were screened for the existence of contiguous resistance gene cassettes inserted in 5′ and 3′ conserved regions using 3′CS and 5′CS-targeted primers [29] (Table S1). The DNA of *C. jejuni* ATCC 33560 and sterile saline were included in all PCR assays as positive and negative controls, respectively.

2.5. Characterization of Gene Cassettes Arrays by DNA Sequencing

One of each amplified PCR product of repetitive distinct close size was selected, purified by PureLink PCR purification kit (Qiagen, Valencia, Spain) and sequenced using Big Dye Terminator V3.1 cycle sequencing kit (Perkin-Elmer Gmbh, Rodgau, Germany) in an Applied Biosystems 3130 genetic analyzer (California, USA). The resulting sequences were assembled using the SeqMan program within the Laser gene suite version 7 (DNAstar, Inc., Madison, WI, USA), then compared with the sequences in the GenBank database using the Basic Local Alignment Search Tool (http://www.ncbi.nlm.nih.gov/BLAST). The best BLAST hits on our query nucleotide sequences were selected based on the highest identity in the GenBank database. Alignment of the nucleotide sequences was performed using ClustalW sequence alignments (http://www.ebi.ac.uk/clustalw), then translation into amino acid sequences was performed using the ExPASy Translate Tool (http://us.expasy.org/, Swiss Institute of Bioinformatics SIB, Geneva, Switzerland). The novel complete gene cassette array (*aacC5-aadA7Δ4)* generated here assigned a new in number (in 1983) using the Integron Database INTEGRALL (http://integrall.bio.ua.pt/).

2.6. Bioinformatics and Statistical Analysis

Statistical Package for Social Sciences software (SPSS; v. 25, IBM, Armonk, NY, United States) was used for statistical analysis of data. Chi-squared test was used to determine if there were significant differences in the occurrence of antimicrobial resistance among different hosts (i.e., cattle, chicken, human) and between the two *Campylobacter* species being studied (*C. jejuni* and *C. coli*). *P* value was considered significant if <0.05. The overall distribution of the resistance phenotypes in *Campylobacter* isolates was visualized using a heat map. The clustering pattern of the isolates and the antimicrobial resistance phenotypes were determined by the hierarchical clustering dendrogram [30]. To predict the correlation among integron patterns and antimicrobial resistance phenotypes, correlation analyses were done on the raw data after conversion to a binary outcome (1 = variable presence, 0 = variable absence). The significance of the correlation was estimated at a significance level of 0.05. The variables ampicillin, amoxicillin, cephalothin, erythromycin, and sulfamethoxazole-trimethoprim were excluded from the analyses as they were identical among all isolates under study. The correlation analyses and visualization were done using R packages *corrplot*, *heatmaply*, *hmisc*, and *ggpubr* [31–33]. To estimate the similarities among *Campylobacter* isolates concerning various analyzed hosts (*n* = 3), the binary distances were calculated based on the presence or absence of certain integron patterns. This analysis was done using the functions *dist* and *hlcust* in the R environment.

2.7. Nucleotide Sequence Accession Numbers

DNA sequences generated in this study were submitted to GenBank and assigned the accession numbers of MT612446-MT612453.

3. Results

3.1. Prevalence of Campylobacter Species in Livestock Animals and Humans

As shown in Table 1, the overall occurrence rate of *Campylobacter* species was 42.72% (235/550), which significantly (*p* < 0.05) differed between species being 71.48% (168/235) for *C. jejuni* and 28.51% (67/235) for *C. coli*. Out of 265 samples of chicken origin, 154 (58.11%) *Campylobacter* isolates were detected, represented as 67.53% *C. jejuni* and 32.47% *C. coli*. The higher prevalence of *C. jejuni* was detected in chicken organs (61.54%), followed by cloacal swabs (57.14%) and chicken muscles (48.00%), while the isolation rate of *C. coli* from these sources was close to 30%, each. Moreover, *C. jejuni* were isolated from 18 of 35 (51.42%) minced meat samples, 26 of 35 (74.28%) raw milk, and 20 of 30 (66.67%) human stool samples, while *C. coli* were recorded by lower percentages. On the other hand, processed food products including chicken and meat luncheon, chicken and meat beef, smoked meat, canned milk, and canned and raw cheese were free from *Campylobacter* contamination. *Campylobacter* isolates yielded characteristic small, shiny, round, and gray colonies on mCCDA agar and no hemolysis on Columbia blood agar. All isolates were positive for oxidase, catalase, and nitrate reduction testing and exhibited sensitivity to nalidixic acid and resistance to cephalothin. *C. jejuni* isolates could hydrolyze indoxyl acetate and hippurate, while *C. coli* were indoxyl acetate-positive and hippurate-negative. *Campylobacter* isolates were further confirmed by PCR-based detection of the genus (*23S rRNA*) and species-specific (*mapA* for *C. jejuni* and *ceuE* for *C. coli*) genes.

Table 1. Prevalence of *Campylobacter* species isolated from animal and human sources.

Source (No.)	Sample Type (No.)	Overall Prevalence of *Campylobacter* Isolates No. (%)	*Campylobacter* Species No. (%)		*p* Value
			C. jejuni	*C. coli*	
Chickens and chicken products (265)	Cloacal swab (70)	60 (85.71)	40 (57.14)	20 (28.57)	0.001
	Breast muscle (25)	18 (72.00)	14 (56.00)	4 (16.00)	0.003
	Thigh chicken muscle (25)	16 (64.00)	10 (40.00)	6 (24.00)	0.225
	Liver (25)	25 (100.00)	16 (64.00)	9 (36.00)	0.048
	Spleen (20)	18 (90.00)	14 (70.00)	4 (20.00)	0.001
	Intestine (20)	17 (85.00)	10 (50.00)	7 (35.00)	0.337
	Chicken beef (40)	0 (0.00)	0 (0.00)	0 (0.00)	NE
	Chicken luncheon (40)	0 (0.00)	0 (0.00)	0 (0.00)	NE
Meat and meat products (160)	Minced meat (35)	28 (80.00)	18 (51.43)	10 (28.57)	0.05
	Smoked meat (45)	0 (0.00)	0 (0.00)	0 (0.00)	NE
	Meat beef (40)	0 (0.00)	0 (0.00)	0 (0.00)	NE
	Meat luncheon (40)	0 (0.00)	0 (0.00)	0 (0.00)	NE
Dairy products (95)	Raw milk (35)	29 (82.86)	26 (74.29)	3 (8.57)	<0.0001
	Canned milk (20)	0 (0.00)	0 (0.00)	0 (0.00)	NE
	Canned cheese (20)	0 (0.00)	0 (0.00)	0 (0.00)	NE
	Raw cheese (20)	0 (0.00)	0 (0.00)	0 (0.00)	NE
Human (30)	Stool (30)	24 (80.00)	20 (66.67)	4 (13.33)	<0.0001
Total	550	235 (42.73)	168 (71.49)	67 (28.51)	<0.0001

NE, not estimated, *p* values < 0.05 are statistically significant.

3.2. Antimicrobial Resistance Profiles

The in vitro antimicrobial susceptibilities of 235 *Campylobacter* isolates comprising 168 *C. jejuni* and 67 *C. coli* against 25 antimicrobial agents are summarized in Table 2. The results revealed that all *Campylobacter* isolates originating from animal and human sources were resistant to amoxicillin, ampicillin, erythromycin, cephalothin, and sulfamethoxazole-trimethoprim (100%, each). Moreover, high levels of resistance were recorded for clarithromycin (100% and 97%), clindamycin (96.4% and 95.5%), nalidixic acid (90.5% and 86.6%), amoxycillin-clavulanic acid (89.3% and 80.6%), cefepime (88.1% and 83.6%), doxycycline (86.3% and 86.5%), colistin (83.9% and 88%) and chloramphenicol (83.3% and 80.6%) for *C. jejuni* and *C. coli* isolates, respectively. On the other hand, lower resistance rates were reported for amikacin (21.4% and 20.9%) and cefoxitin (26.8% and 43.2%) against *C. jejuni* and *C. coli* isolates, respectively. Of note, *C. jejuni* and *C. coli* were resistant to meropenem with alarming percentages (19.6% and 32.8%, respectively). Statistical analysis revealed significant differences in the resistance of *Campylobacter* species isolated from different sources to the most tested antimicrobials ($p < 0.05$) except for ampicillin-sulbactam that showed non-significant variation ($p > 0.05$). However, non-significant differences ($p > 0.05$) were reported between resistance of *C. jejuni* and *C. coli* to almost half of the examined antimicrobial agents.

Table 2. Antimicrobial resistance pattern of *Campylobacter* species isolated from different sources.

| AMA | Number of *C. jejuni* Isolates | | | | | | | Number of *C. coli* Isolates | | | | | | | *p* Value | |
| | Chickens (*n* = 104) | | | Raw Milk (26) | Minced Meat (18) | Human Stool (20) | Total (168) | Chickens (*n* = 50) | | | Raw Milk (3) | Minced Meat (10) | Human Stool (4) | Total (67) | Various Hosts | *Campylobacter* Species |
	Muscle (24)	Internal Organs (40)	Cloacal Swabs (40)					Muscle (10)	Internal Organs (20)	Cloacal Swabs (20)						
AX	24 (100.00)	40 (100.00)	40 (100.00)	26 (100.00)	18 (100.00)	20 (100.00)	168 (100.00)	10 (100.00)	20 (100.00)	20 (100.00)	3 (100.00)	10 (100.00)	4 (100.00)	67 (100.00)	NA	NA
AM	24 (100.00)	40 (100.00)	40 (100.00)	26 (100.00)	18 (100.00)	20 (100.00)	168 (100.00)	10 (100.00)	20 (100.00)	20 (100.00)	3 (100.00)	10 (100.00)	4 (100.00)	67 (100.00)	NA	NA
SAM	15 (62.50)	27 (67.50)	20 (50.00)	11 (42.31)	10 (55.55)	10 (50.00)	93 (55.36)	0.00 (0.00)	11 (55.00)	16 (80.00)	3 (100.00)	1 (10.00)	0.00 (0.00)	31 (46.27)	0.1022	0.2
AMC	24 (100.00)	34 (85.00)	34 (85.00)	26 (100.00)	17 (94.44)	15 (75.00)	150 (89.29)	7 (70.00)	20 (100.00)	18 (90.00)	3 (100.00)	10 (100.00)	0.00 (0.00)	58 (86.57)	<0.0001	0.55
KF	24 (100.00)	40 (100.00)	40 (100.00)	26 (100.00)	18 (100.00)	20 (100.00)	168 (100.00)	10 (100.00)	20 (100.00)	20 (100.00)	3 (100.00)	10 (100.00)	4 (100.00)	67 (100.00)	NA	NA
FOX	8 (33.33)	4 (10.00)	18 (45.00)	17 (65.38)	12 (66.67)	9 (45.00)	68 (40.48)	0.00 (0.00)	4 (20.00)	10 (50.00)	0.00 (0.00)	0.00 (0.00)	4 (100.00)	18 (26.87)	0.002	0.05
CEP	14 (58.33)	34 (85.00)	19 (47.50)	26 (100.00)	12 (66.67)	18 (90.00)	123 (73.21)	10 (100.00)	20 (100.00)	12 (60.00)	3 (100.00)	9 (90.00)	4 (100.00)	58 (86.57)	0.0068	0.02
FEP	16 (66.67)	38 (95.00)	39 (97.50)	17 (65.38)	10 (55.56)	20 (100.00)	140 (83.33)	8 (80.00)	19 (95.00)	20 (100.00)	3 (100.00)	6 (60.00)	4 (100.00)	60 (89.55)	<0.0001	0.2
MEM	1 (4.17)	2 (5.00)	20 (50.00)	0.00 (0.00)	6 (33.33)	4 (20.00)	33 (19.64)	0.00 (0.00)	14 (70.00)	8 (40.00)	0.00 (0.00)	0.00 (0.00)	0.00 (0.00)	22 (32.84)	0.0124	0.03
ATM	14 (58.33)	39 (97.50)	36 (90.00)	26 (100.00)	17 (94.44)	10 (50.00)	142 (84.52)	0.00 (0.00)	14 (70.00)	12 (60.00)	3 (100.00)	10 (100.00)	4 (100.00)	43 (64.18)	<0.0001	0.0006
S	22 (91.67)	16 (40.00)	34 (85.00)	12 (46.15)	7 (38.89)	14 (70.00)	105 (62.50)	4 (40.00)	17 (85.00)	18 (90.00)	0.00 (0.00)	2 (20.00)	3 (75.00)	44 (65.67)	<0.0001	0.6
TOB	22 (91.67)	16 (40.00)	30 (75.00)	8 (30.77)	6 (33.33)	13 (65.00)	95 (56.55)	4 (40.00)	17 (85.00)	17 (85.00)	0.00 (0.00)	0.00 (0.00)	4 (100.00)	42 (62.69)	<0.0001	0.38
CN	24 (100.00)	24 (60.00)	36 (90.00)	14 (53.85)	2 (11.11)	9 (45.00)	109 (64.88)	6 (60.00)	17 (85.00)	14 (70.00)	0.00 (0.00)	5 (50.00)	4 (100.00)	46 (68.66)	<0.0001	0.58

Table 2. *Cont.*

| AMA | Number of *C. jejuni* Isolates | | | | | | | | | | | | | | p Value | |
| | Chickens (n = 104) | | | Raw Milk (26) | Minced Meat (18) | Human Stool (20) | Total (168) | Chickens (n = 50) | | | Raw Milk (3) | Minced Meat (10) | Human Stool (4) | Total (67) | Various Hosts | *Campylobacter* Species |
	Muscle (24)	Internal Organs (40)	Cloacal Swabs (40)					Muscle (10)	Internal Organs (20)	Cloacal Swabs (20)						
AK	18 (75.00)	0.00 (0.00)	14 (35.00)	0.00 (0.00)	6 (33.33)	6 (30.00)	44 (26.19)	6 (60.00)	3 (15.00)	6 (30.00)	0.00 (0.00)	00.00 (0.00)	4 (100.00)	19 (28.36)	0.0032	0.73
E	24 (100.00)	40 (100.00)	40 (100.00)	26 (100.00)	18 (100.00)	20 (100.00)	168 (100.00)	10 (100.00)	20 (100.00)	20 (100.00)	3 (100.00)	10 (100.00)	4 (100.00)	67 (100.00)	NA	NA
AZM	15 (62.50)	33 (82.50)	32 (80.00)	26 (100.00)	18 (100.00)	20 (100.00)	144 (85.71)	4 (40.00)	20 (100.00)	14 (70.00)	3 (100.00)	10 (100.00)	4 (100.00)	55 (82.09)	<0.0001	0.48
CLR	24 (100.00)	40 (100.00)	6 (15.00)	26 (100.00)	18 (100.00)	20 (100.00)	134 (79.76)	10 (100.00)	20 (100.00)	2 (10.00)	3 (100.00)	10 (100.00)	0.00 (0.00)	45 (67.16)	<0.0001	0.04
CIP	24 (100.00)	16 (40.00)	32 (80.00)	26 (100.00)	1 (5.56)	11 (55.00)	110 (65.48)	6 (60.00)	9 (45.00)	16 (80.00)	0.00 (0.00)	9 (90.00)	0.00 (0.00)	40 (59.70)	<0.0001	0.4
NA	24 (100.00)	33 (82.50)	40 (100.00)	26 (100.00)	17 (94.44)	10 (50.00)	150 (89.29)	6 (60.00)	20 (100.00)	18 (90.00)	0.00 (0.00)	10 (100.00)	4 (100.00)	58 (86.57)	<0.0001	0.55
SXT	24 (100.00)	40 (100.00)	40 (100.00)	26 (100.00)	18 (100.00)	20 (100.00)	168 (100.00)	10 (100.00)	20 (100.00)	20 (100.00)	3 (100.00)	10 (100.00)	4 (100.00)	67 (100.00)	<0.0001	0.005
C	24 (100.00)	26 (65.00)	32 (80.00)	26 (100.00)	17 (94.44)	16 (80.00)	141 (83.93)	6 (60.00)	16 (80.00)	16 (80.00)	3 (100.00)	5 (50.00)	0.00 (0.00)	46 (68.66)	0.0461	0.0088
CT	14 (58.33)	28 (70.00)	34 (85.00)	26 (100.00)	18 (100.00)	11 (55.00)	131 (77.98)	6 (60.00)	16 (80.00)	18 (90.00)	3 (100.00)	10 (100.00)	4 (100.00)	57 (85.07)	<0.0001	0.21
LNZ	24 (100.00)	39 (97.50)	28 (70.00)	26 (100.00)	14 (77.78)	13 (65.00)	144 (85.71)	10 (100.00)	9 (45.00)	18 (90.00)	0.00 (0.00)	10 (100.00)	0.00 (0.00)	47 (70.15)	0.0012	0.005
DA	23 (95.83)	40 (100.00)	39 (97.50)	26 (100.00)	18 (100.00)	20 (100.00)	166 (98.81)	10 (100.00)	20 (100.00)	20 (100.00)	3 (100.00)	10 (100.00)	0.00 (0.00)	63 (94.03)	<0.0001	0.03
DO	24 (100.00)	37 (92.50)	24 (60.00)	26 (100.00)	18 (100.00)	20 (100.00)	149 (88.69)	10 (100.00)	20 (100.00)	18 (90.00)	3 (100.00)	10 (100.00)	0.00 (0.00)	61 (91.04)	0.0103	0.59

Values represent number of *Campylobacter* isolates (%), p values were calculated based on Chi-squared test; p values < 0.05 are statistically significant; p values < 0.01 are highly significant; AMA, antimicrobial agent; AX, amoxicillin; AM, ampicillin; SAM, ampicillin-sulbactam; AMC, amoxycillin-clavulanic acid; KF, cephalothin; FOX, cefoxitin; CEP, cefoperazone; FEP, cefepime; MEM, meropenem; ATM, aztreonam; S, streptomycin; TOB, tobramycin; CN, gentamicin; AK, amikacin; E, erythromycin; AZM, azithromycin; CLR, clarithromycin; CIP, ciprofloxacin; NA, nalidixic acid; SXT, sulfamethoxazole-trimethoprim; C, chloramphenicol; CT, colistin; LNZ, lenzolid; DA, clindamycin; DO, doxycycline; NA, non-applicable.

As shown in Figure 1 and Table S2, the antibiogram analysis revealed that *Campylobacter* isolates showed resistance to 11–25 antimicrobial agents with MAR indices ranged from 0.44 to 1.00 and demonstrated 93 distinct resistance patterns. The antibiotype 55 was the most prevalent among the analyzed isolates (*n* = 8; 3.40%) (Table S2).

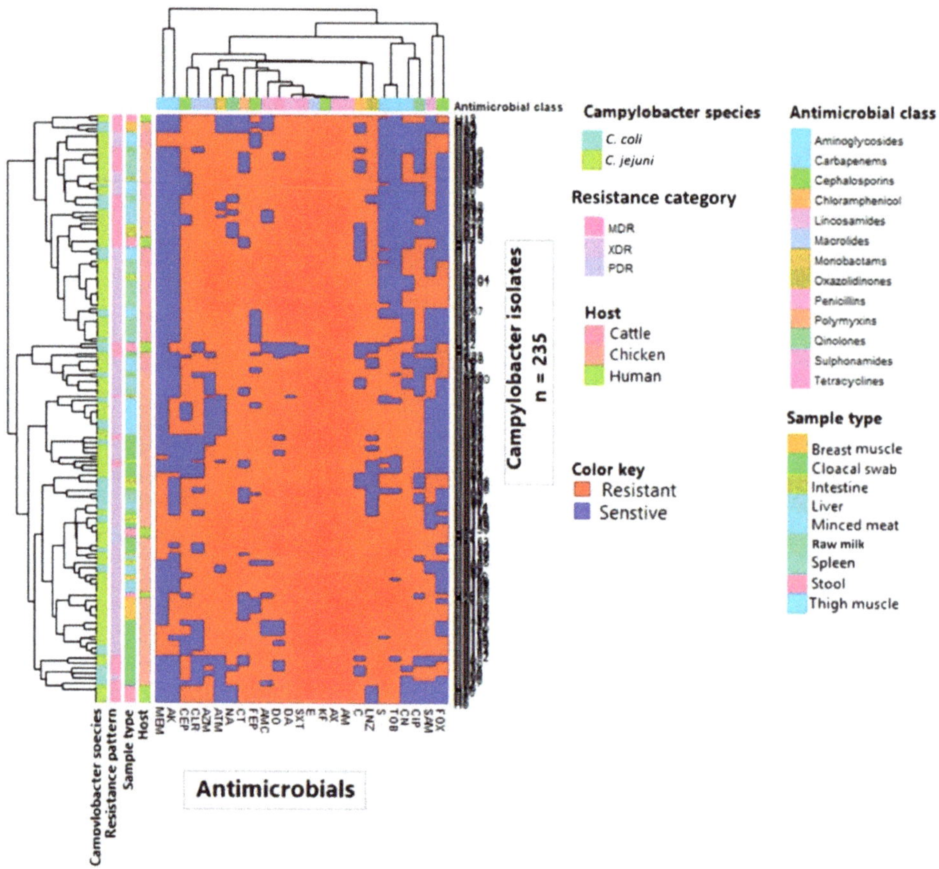

Figure 1. Overall distribution and clustering of *Campylobacter* isolates (*n* = 235) under study and the patterns of their antimicrobial resistance. Different *Campylobacter* species, hosts, sample types, antimicrobial classes, and resistance categories are shown for each isolate as color codes. The heat map represents the hierarchical clustering of the isolates and the antimicrobial classes.

The PDR, XDR, and MDR patterns were reported among the analyzed isolates (Table 3 and Figures S1 and S2). In total, 2.55% (6/235) of *Campylobacter* isolates exhibited PDR patterns being resistant to all tested antimicrobial agents. The XDR profiles were extremely increased among analyzed isolates with a percentage of 68.94% (162/235). However, 28.5% (67/235) of the isolates showed MDR patterns. None of the examined *Campylobacter* isolates was pan-susceptible. Regarding the isolation source, most *C. jejuni* (68.26%) and *C. coli* (74%) isolates originated from chicken samples were XDR. All *C. jejuni* and *C. coli* isolates recovered from raw milk showed XDR and MDR patterns, respectively. Moreover, all *C. coli* isolated from human stool and minced meat exhibited MDR and XDR profiles, respectively, while 88.9% of *C. jejuni* originated from the minced meat were XDR. The PDR *Campylobacter* isolates originated from chicken cloacal swabs (*n* = 2) and human stool (*n* = 4).

Table 3. Occurrence of MDR, XDR and PDR categories in *Campylobacter* isolates from different sources.

Resistance Category	Resistance to Antimicrobial Class (*n* = 14)	Resistance to Antimicrobial Agent (*n* = 25)	No. of Resistant *Campylobacter* Isolates (Source)	
			C. jejuni (*n* = 168)	*C. coli* (*n* = 67)
MDR (*n* = 67)	7	11	0	2 (chicken muscle)
		12	0	1 (chicken muscle)
		14	0	1 (human stool)
		15	0	3 (human stool)
	8	13	0	1 (chicken muscle)
	10	13	2 (cloacal swab)	0
		14	1 (minced meat)	0
		15	8 (chicken organ)	0
		16	2 (cloacal swab), 9 (human stool)	3 (raw milk)
		17	5 (human stool), 1 (cloacal swab)	2 (cloacal swab)
	11	15	1 (minced meat)	2 (cloacal swab)
		16	4 (cloacal swab), 2 (chicken organ)	0
		17	8 (chicken organ), 2 (cloacal swab)	3 (chicken organ)
		18	1 (cloacal swab)	0
		19	1 (chicken muscle)	2 (cloacal swab)
XDR (*n* = 162)	12	16	2 (cloacal swab)	0
		17	3 (minced meat)	4 (minced meat)
		18	2 (chicken muscle), 5 (chicken organ), 4 (cloacal swab)	4 (minced meat), 2 (chicken muscle), 2 (cloacal swab)
		19	5 (chicken muscle), 5 (chicken organ), 2 (cloacal swab), 4 (minced meat)	1 (minced meat), 1 (chicken organ), 2 (cloacal swab)
		20	5 (raw milk), 2 (chicken muscle), 2 (cloacal swab), 2 (chicken organ), 2 (human stool)	6 (chicken organ), 4 (chicken muscle)
		21	2 (chicken muscle)	4 (chicken organ)
		22	4 (chicken muscle)	0
		23	3 (chicken muscle)	0
	13	18	1 (minced meat)	1 (minced meat), 2 (cloacal swab)
		19	9 (raw milk), 1 (chicken muscle), 3 (chicken organ)	1 (cloacal swab)
		20	2 (minced meat), 5 (raw milk)	1 (cloacal swab)
		21	6 (raw milk), 3 (chicken muscle), 6 (cloacal swab)	3 (chicken organ)
		22	1 (raw milk), 5 (chicken organ), 2 (cloacal swab)	0
		23	1 (chicken muscle)	0
	14	18	1 (cloacal swab)	0
		19	1 (cloacal swab)	0
		20	6 (minced meat)	0
		21	1 (cloacal swab)	0
		22	3 (cloacal swab)	0
		23	2 (cloacal organ)	3 (chicken organ), 4 (cloacal swab)
		24	2 (cloacal swab)	2 (cloacal swab)
PDR (*n* = 6)	14	25	4 (human stool), 2 (cloacal swab)	0

MDR, multiple drug-resistance; XDR, extensively drug-resistance; PDR, pan drug-resistance.

3.3. Screening for Class 1 Integrons and Characterization of Associated Gene Cassettes in Campylobacter Isolates

Thirty-eight *Campylobacter* isolates (28 *C. jejuni*, and 10 *C. coli*) categorized as MDR (*n* = 5), XDR (*n* = 31), and PDR (*n* = 2) representing all sample origins and being resistant to at least 15 antimicrobial agents were screened for the possession of class 1 integrons using PCR assay. Overall, 37 of 38 (97.36%) examined isolates were positive for the *integrase* gene *(intI1)*, all harbored class 1 integrons carrying gene cassettes of varying sizes ranging from 349 to 2600 bp. Only one *C. jejuni* isolate (code No. 16) possessed an empty integron structure with no gene cassettes inserted between its conserved segments (Table 4). Eight repetitive distinct gene cassettes were selected among integron positive isolates for DNA sequencing. Other gene cassette arrays were identified according to their PCR product sizes based on relevant previously published data.

Table 4. Antimicrobial resistance patterns and gene cassette arrays carried by class 1 integrons in MDR, XDR and PDR *Campylobacter* isolates recovered from different sources ($n = 38$).

Isolate No.	Code No.	*Campylobacter* Species	Source	Antimicrobial Resistance Pattern	Resistance to Antimicrobial Agents	Resistance to Antimicrobial Classes	Antimicrobial Resistance Type	intI1/Class 1 Integron	Resistance Gene(s) in Class 1 Integron (Size in bp)
1	CK105	*C. coli*	Chicken intestine	AX, AM, SAM, AMC, KF, FOX, CEP, FEP, MEM, ATM, S, TOB, CN, E, AZM, CLR, NA, SXT, CT, DA, DO	21	12	XDR	+/+	*aadA2* (995)
2	CK134	*C. jejuni*	Chicken cloacal swab	AX, AM, SAM, AMC, KF, FOX, FEP, MEM, ATM, TOB, CN, E, AZM, CIP, NA, SXT, C, CT, LNZ, DA, DO	21	14	XDR	+/+	*dfrA15* (738)
3	CK110	*C. jejuni*	Chicken thigh muscle	AX, AM, SAM, AMC, KF, FEP, ATM, S, TOB, CN, E, AZM, CLR, CIP, NA, SXT, C, CT, LNZ, DA, DO	21	13	XDR	+/+	*dfrA15* (738); *oxa1-aadA1a* (2000)
4	H12	*C. jejuni*	Human stool	AX, AM, KF, FOX, CEP, FEP, S, CN, E, AZM, CLR, CIP, SXT, C, LNZ, DA, DO	17	10	MDR	+/+	*dfrA17ab-aadA5* (1513); *dfrA12-gcu-aadA2* (1864)
5	H19	*C. jejuni*	Human stool	AX, AM, SAM, KF, FOX, FEP, ATM, S, TOB, CN, AK, E, AZM, CLR, NA, SXT, C, LNZ, DA, DO	20	12	XDR	+/+	*dfrA15* (738); *dfrA17ab-aadA5* (1513); *oxa1-aadA1a* (2000)
6	CK139	*C. jejuni*	Chicken breast muscle	AX,AM, SAM, AMC, KF, FOX, CEP, FEP, MEM, ATM, S, CN, TOB, E, AZM, CLR, CIP, NA, SXT, C, LNZ, DA, DO	23	13	XDR	+/+	*dfrA12-gcu-aadA2* (1864)
7	CK104	*C. jejuni*	Chicken intestine	AX, AM, SAM, AMC, KF, FOX, CEP, FEP, ATM, E, AZM, CLR, CIP, NA, SXT, C, CT, LNZ, DA, DO	20	12	XDR	+/+	Phage tail TMP (349)
8	CK147	*C. jejuni*	Chicken Liver	AX, AM, AMC, SAM, KF, FEP, CEP, FOX, MEM, ATM, S, TOB, CN, E, AZM, CLR, NA, SXT, C, CT, LNZ, DA, DO	23	14	XDR	+/+	*dfrA1-gcuC* (1177)
9	C41	*C. jejuni*	Raw milk	AX, AM, SAM, AMC, KF, FOX, CEP, FEP, ATM, CN, E, AZM, CLR, CIP, NA, SXT, C, CT, LNZ, DA, DO	21	13	XDR	+/+	*aacA4-cmlA4* (2600)
10	CK143	*C. coli*	Chicken cloacal swab	AX, AM, SAM, AMC, KF, FOX, FEP, MEM, ATM, S, TOB, CN, AK, E, AZM, CIP, NA, SXT, C, CT, LNZ, DA, DO	23	14	XDR	+/+	*dfrA15* (738); *dfrA17ab-aadA5* (1513)

Table 4. *Cont.*

Isolate No.	Code No.	*Campylobacter* Species	Source	Antimicrobial Resistance Pattern	Resistance to Antimicrobial Agents	Resistance to Antimicrobial Classes	Antimicrobial Resistance Type	intI1/Class 1 Integron	Resistance Gene(s) in Class 1 Integron (Size in bp)
11	H1	*C. coli*	Human stool	AX, AM, KF, FOX, CEP, FEP, ATM, TOB, CN, AK, E, AZM, NA, SXT, CT	15	8	MDR	+/+	*dfrA15* (700); *dfrA1ah-gcuCΔ* (1146)
12	C52	*C. jejuni*	Raw milk	AX, AM, SAM, AMC, KF, FOX, CEP, FEP, ATM, S, CN, E, AZM, CLR, CIP, NA, SXT, C, CT, LNZ, DA, DO	22	13	XDR	+/+	*dfrA17ab-aadA5* (1513); *dfrA12-gcu-aadA2* (1864)
13	C8	*C. jejuni*	Minced meat	AX, AM, AMC, SAM, KF, FEP, ATM, S, E, AZM, CLR, NA, SXT, C, CT, LNZ, DA, DO	18	12	XDR	+/+	*dfrA17-gcu5-aadA5* (1900)
14	H5	*C. jejuni*	Human stool	AX, AM, AMC, KF, CEP, FEP, S, TOB, E, AZM, CLR, SXT, C, CT, DA, DO	15	10	MDR	+/+	Phage tail TMP (349)
15	CK54	*C. coli*	Chicken cloacal swab	AX, AM, AMC, KF, FEP, ATM, S, TOB, CN, E, CIP, NA, SXT, C, CT, LNZ, DA, DO	18	13	XDR	+/+	*aadA22* (947); *dfrA17ab-aadA5* (1513)
16	CK80	*C. jejuni*	Chicken breast muscle	AX, AM, SAM, AMC, KF, CEP, FEP, TOB, CN, E, AZM, CLR, CIP, NA, SXT, C, CT, LNZ, DO	19	11	MDR	-/-	ND
17	CK125	*C. jejuni*	Chicken Liver	AX, AM, SAM, AMC, KF, FEP, CEP, ATM, S, TOB, CN, E, AZM, CLR, CIP, NA, SXT, C, CT, LNZ, DA, DO	22	13	XDR	+/+	*dfrA1-gcuC* (1177); *dfrA12-gcu-aadA2* (1864)
18	C12	*C. coli*	Minced meat	AX, AM, AMC, KF, CEP, FEP, ATM, S, CN, E, AZM, CLR, CIP, NA, SXT, CT, LNZ, DA, DO	19	12	XDR	+/+	*dfrA15* (700); *dfrA1-gcuC* (1177); *dfrA17ab-aadA5* (1513)
19	CK78	*C. jejuni*	Chicken cloacal swab	AX, AM, AMC, KF, FEP, MEM, ATM, S, CN, E, AZM, CIP, NA, SXT, C, CT, LNZ, DA, DO	19	14	XDR	+/+	*dfrA15* (738); *aadA22* (947); *dfrA1ah-gcuCΔ* (1146)
20	C46	*C. jejuni*	Raw milk	AX, AM, AMC, KF, FOX, CEP, FEP, ATM, S, CN, E, AZM, CLR, CIP, NA, SXT, C, CT, LNZ, DA, DO	21	13	XDR	+/+	*aadA2* (995); *bla pse-1* (1200)
21	CK109	*C. jejuni*	Chicken breast muscle	AX, AM, SAM, AMC, KF, CEP, FEP, ATM, S, CN, E, AZM, CLR, CIP, NA, SXT, C, CT, LNZ, DA, DO	21	13	XDR	+/+	*dfrA15* (700); *dfrA1-aadA1a* (1600)

Table 4. *Cont.*

Isolate No.	Code No.	*Campylobacter* Species	Source	Antimicrobial Resistance Pattern	Resistance to Antimicrobial Agents	Resistance to Antimicrobial Classes	Antimicrobial Resistance Type	intI1/Class 1 Integron	Resistance Gene(s) in Class 1 Integron (Size in bp)
22	CK153	*C. jejuni*	Chicken cloacal swab	AX, AM, SAM, AMC, KF, FOX, CEP, FEP, MEM, ATM, S, TOB, CN, AK, E, AZM, CLR, CIP, NA, SXT, C, CT, LNZ, DA, DO	25	14	PDR	+/+	*dfrA15* (738); *dfrA17-gcu5-aadA5* (1900)
23	CK50	*C. jejuni*	Chicken liver	AX, AM, SAM, AMC, KF, FEP, CEP, ATM, S, TOB, E, CLR, NA, SXT, CT, LNZ, DA, DO	18	12	XDR	+/+	*dfrA15* (738); *bla pse-1* (1200); *dfrA1-aadA1a* (1600)
24	CK117	*C. jejuni*	Chicken cloacal swab	AX, AM, SAM, KF, CEP, FEP, MEM, ATM, S, TOB, CN, AK, E, AZM, CIP, NA, SXT, C, CT, LNZ, DA	21	13	XDR	+/+	*aadA2* (995); *dfrA17-gcu5-aadA5* (1900)
25	H21	*C. jejuni*	Human stool	AX, AM, SAM, AMC, KF, FOX, CEP, FEP, MEM, ATM, S, TOB, CN, AK, E, AZM, CLR, CIP, NA, SXT, C, CT, LNZ, DA, DO	25	14	PDR	+/+	*aadA22* (947); *aacA4-cmlA4* (2600)
26	C32	*C. jejuni*	Raw milk	AX, AM, AMC, KF, CEP, ATM, S, CN, E, AZM, CLR, CIP, NA, SXT, C, CT, LNZ, DA, DO	19	13	XDR	+/+	*dfrA15* (700); *aadA2* (995); *bla pse-1* (1200); **aacC5-aadA7Δ4 (1799)**
27	CK44	*C. jejuni*	Chicken thigh muscle	AX, AM, AMC, KF, S, TOB, CN, AK, E, CLR, CIP, NA, SXT, C, CT, LNZ, DA, DO	18	12	XDR	+/+	*aadA2* (1000)
28	H6	*C. jejuni*	Human stool	AX, AM, AMC, KF,CEP, FEP, S, TOB, E, AZM, CLR, SXT, C, CT, DA, DO	16	10	MDR	+/+	*aadA2* (995)
29	CK112	*C. coli*	Chicken Liver	AX, AM, AMC, KF, FEP, CEP, MEM, ATM, S, TOB, CN, E, AZM, CLR, CIP, NA, SXT, C, CT, DA, DO	21	13	XDR	+/+	**dfrA1-gcuC (1177)**
30	C17	*C. jejuni*	Minced meat	AX, AM, AMC, KF, FOX, CEP, FEP, ATM, S, TOB, E, AZM, CLR, NA, SXT, C, CT, DA, DO	19	12	XDR	+/+	**dfrA15 (738)**; *aadA2* (1000)
31	CK106	*C. coli*	Chicken intestine	AX, AM, SAM, AMC, KF, FOX, CEP, FEP, MEM, ATM, S, TOB, CN, E, AZM, CLR, NA, SXT, CT, DA, DO	21	12	XDR	+/+	**Phage tail TMP (349); aadA22 (947)**
32	CK99	*C. coli*	Chicken cloacal swab	AX, AM, SAM, AMC, KF, FOX, CEP, FEP, ATM, S, CN, E, NA, SXT, C, CT, LNZ, DA, DO	19	13	XDR	+/+	*dfrA15* (738); *aacC5-aadA7Δ4* (1799)

Table 4. *Cont.*

Isolate No.	Code No.	*Campylobacter* Species	Source	Antimicrobial Resistance Pattern	Resistance to Antimicrobial Agents	Resistance to Antimicrobial Classes	Antimicrobial Resistance Type	intI1/Class 1 Integron	Resistance Gene(s) in Class 1 Integron (Size in bp)
33	C53	*C. jejuni*	Raw milk	AX, AM, SAM, AMC, KF, FOX, CEP, FEP, ATM, S, CN, E, AZM, CLR, CIP, NA, SXT, C, CT, LNZ, DA, DO	22	13	XDR	+/+	*aadA2* (995)
34	C4	*C. coli*	Minced meat	AX, AM, AMC, KF, CEP, ATM, S, E, AZM, CLR, CIP, NA, SXT, C, CT, LNZ, DA, DO	18	13	XDR	+/+	*aadA22* (947)
35	C21	*C. jejuni*	Minced meat	AX, AM, AMC, KF, FOX, CEP, ATM, S, TOB, CN, E, AZM, CLR, NA, SXT, C, CT, LNZ, DA, DO	20	13	XDR	+/+	*oxa1-aadA1a* (2000)
36	CK46	*C. jejuni*	Chicken cloacal swab	AX, AM, AMC, KF, FEP, ATM, S, TOB, CN, E, AZM, CIP, NA, SXT, C, CT, DA, DO	18	12	XDR	+/+	Phage tail TMP (349); *dfrA1ah-gcuCΔ* (1146); *oxa1-aadA1a* (2000)
37	CK151	*C. coli*	Chicken cloacal swab	AX, AM, SAM, AMC, KF, FOX, FEP, MEM, ATM, S, TOB, CN, AK, E, AZM, CLR, CIP, NA, SXT, C, CT, LNZ, DA, DO	24	14	XDR	+/+	*dfrA15* (738); *aadA22* (947); *bla pse-1* (1200)
38	CK136	*C. jejuni*	Chicken breast muscle	AX, AM, SAM, AMC, KF, FOX, CEP, FEP, ATM, S, CN, TOB, AK, E, AZM, CLR, CIP, NA, SXT, C, LNZ, DA, DO	23	12	XDR	+/+	***dfrA17ab-aadA5*** **(1513)**

AX, amoxicillin; AM, ampicillin; SAM, ampicillin-sulbactam; AMC, amoxycillin-clavulanic acid; KF, cephalothin; FOX, cefoxitin; CEP, cefoperazone; FEP, cefepime; MEM, meropenem; ATM, aztreonam; S, Streptomycin; TOB, tobramycin; CN, gentamicin; AK, amikacin; E, erythromycin; AZM, azithromycin; CLR, clarithromycin; NA, nalidixic acid; CIP, ciprofloxacin; SXT, sulfamethoxazole-trimethoprim; C, chloramphenicol; CT, colistin; LNZ, lenzolid; DA, clindamycin; DO, doxycycline; MDR, multiple drug-resistance; XDR, extensively drug-resistance; PDR, pan drug-resistance; TMP, tape measure protein; ND, not detected; CK, chicken; H, human; C, cattle; +, positive; -, negative. Bold gene cassette arrays were subjected to DNA sequencing, deposited in the GenBank database and assigned their accession numbers as depicted in the Material and Methods section.

As shown in Table 4, 16 gene cassette arrays were identified among class1 integron-positive isolates. DNA sequence analysis revealed the predominance of *aadA* alleles (*aadA1a*, *aadA2*, *aadA5*, *aadA7Δ4*, and *aadA22*) in 30 out of 37 (81.08%) analyzed isolates, accounting resistance for aminoglycosides, particularly streptomycin. Other frequent gene cassettes reported herein were *dfrA* (25/37; 67.56%) alleles (*dfrA1*, *dfrA1*, *dfrA12*, *dfrA15*, and *dfrA17*), conferring resistance to the trimethoprim antimicrobial agent. Despite the high frequency of resistant *Campylobacter* isolates to B-lactams, the *bla pse-1* and *oxa1* gene cassettes were detected in only four isolates (4/37; 10.81% each). Likewise, the gene cassette *aacA4-cmlA4* conferring resistance to chloramphenicol was found in only two *Campylobacter* isolates.

The most striking finding in the current study is the exclusive existence of a novel gene cassette array namely *aacC5-aadA7Δ4* (In number in 1983) as a first report according to the INTEGRALL database in only two *Campylobacter* isolates (code Nos. 26 and 32). This conferred resistance to aminoglycosides in particular gentamicin and streptomycin, but not tobramycin, amikacin, nor kanamycin.

Two gene cassette arrays were reported within class 1 integrons of *Campylobacter* species, each one harbored triple genes. The *dfrA17-gcu5-aadA5* (1900 bp) integron-borne cassette array existed in two XDR (code Nos. 13 and 24) and one PDR (code No. 22) *Campylobacter* isolates and *dfrA12-gcu-aadA2* (1864 bp) gene cassette incorporated in four XDR *Campylobacter* isolates (code Nos. 4, 6, 12 and 17), both cassettes conferred resistance to aminoglycosides and trimethoprim antimicrobial agents (Table 4).

Of interest, all gene cassettes reported here were linked to antimicrobial resistance except one, whose product is a putative phage tail tape measure protein (349 bp; accession number MT612449). It was reported in three XDR *Campylobacter* isolates (code Nos. 7, 31, and 36) originated from chicken and one MDR *C. jejuni* isolate (code No. 14) of human origin, thus facilitates DNA transit to the cell cytoplasm during infection. To our knowledge, this is the first report of a putative phage tail protein associated with class 1 integrons in *Campylobacter* species.

3.4. Correlation between Class 1 Integrons and Antimicrobial Resistance Phenotypes in Campylobacter Isolates

As depicted in Figure 2 and Table S2, PCR results and DNA sequence analysis were consistent with certain antimicrobial susceptibility phenotypes. It was noted that the existence of *aadA* and *aacC5* genes positively correlated ($r = 0.09$–0.18) with streptomycin resistance. Moreover, class 1 integron-positive isolates carrying *aacC5-aadA7Δ4* and *aacA4-cmlA4* cassette arrays showed positive correlations with resistance to gentamicin ($r = 0.11$) and tobramycin ($r = 0.17$), respectively. The presence of *bla pse-1* and *oxa1* genes non-significantly ($p > 0.05$) associated with resistance to amoxicillin-clavulanate ($r = 0.12$ each) and cefoperazone ($r = 0.21$ and 0.01, respectively). However, both genes did not confer resistance to ampicillin-sulbactam ($r = -0.07$ and -0.25, respectively), cefoxitin ($r = -0.21$ and -0.04, respectively) or cefepime ($r = -0.16$ each).

The clustering pattern of class 1 integron-positive *Campylobacter* isolates is illustrated in Figure 2. The two variables (gene cassette arrays and antimicrobial resistance phenotypes) produced two distinct clusters (A and B). Notably, the *aacA4-cmlA4* gene cassette gathered with tobramycin in cluster A. While, the *aadA* genes and *bla pse-1* and *oxa1-aadA1* cassette arrays, which confer resistance to streptomycin and amoxicillin-clavulanate and cefoperazone, respectively, were clustered together in cluster B.

3.5. Cluster Analysis of Gene Cassette Arrays in Campylobacter Isolates from Human and Animal Populations

The dendrogram analysis (Figure 3) of class 1 integron-positive isolates ($n = 38$) simplified the existence of gene cassettes across livestock animals and humans. Three clusters were noticed in our dataset (A, B and, C). A close relatedness was observed among certain *Campylobacter* isolates of different sources. As exemplified, a *Campylobacter* isolate of the chicken source (code CK153) was closer to another one of human origin (code H19), both were gathered in cluster A. In addition, two *Campylobacter* isolates of human (H21) and chicken (CK143) sources clustered closely together in cluster B. Regarding the cluster C, several isolates of the three populations (cattle, chicken and human) clustered together.

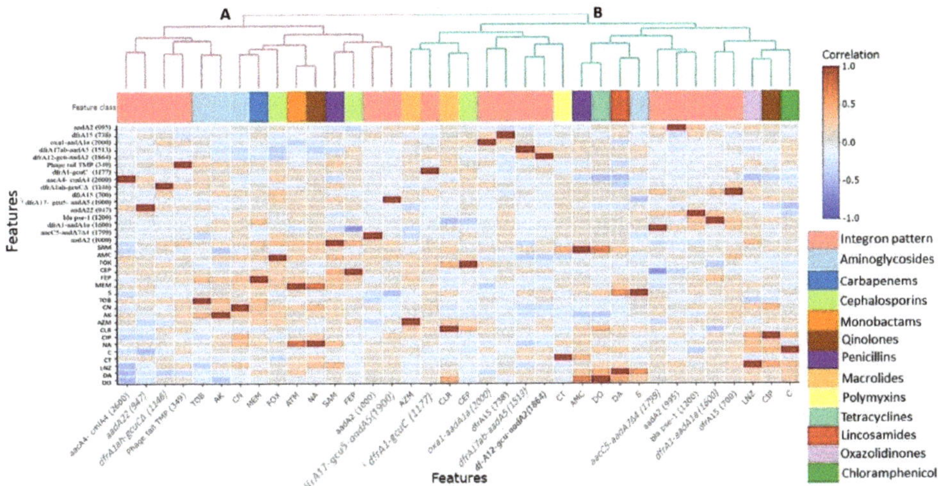

Figure 2. Pairwise correlation (*R*) among different antimicrobial resistance phenotypes and integron gene cassette arrays. Red and blue colors indicate positive and negative correlations, respectively. The color key refers to the correlation coefficient (*R*). The darker colors imply stronger positive or negative correlations. The hierarchical clustering of the variables is shown as a dendrogram illustrating different clusters with different colors and letters (e.g., A and B). Variables that are identical among all strains are excluded, and thus not shown in this figure. Classes of antimicrobials are color-coded below the dendrogram.

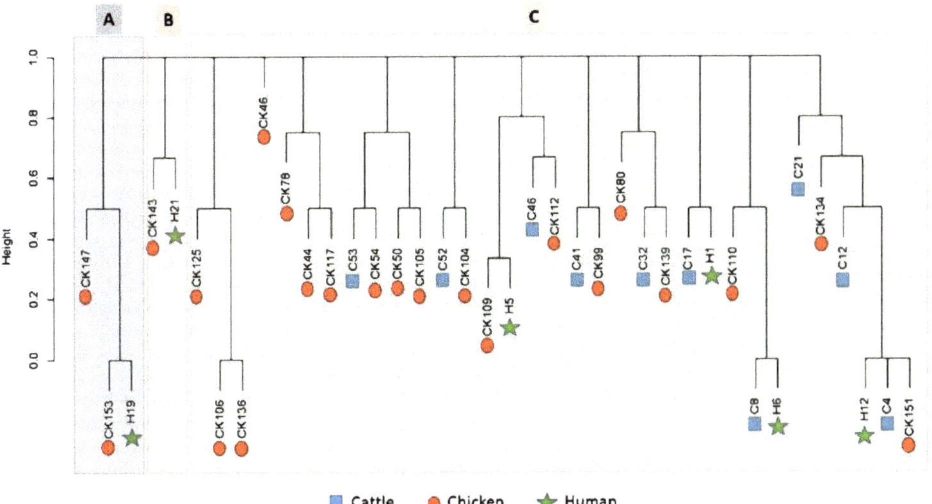

Figure 3. Dendrogram showing the binary distances among different *Campylobacter* isolates based on the integron patterns. The isolates are categorized based on their host, which are shown as different colors and symbols. The *X*-axis refers to the binary distance scale. Isolates codes are illustrated in Table 4.

4. Discussion

Thermophilic campylobacters as *C. jejuni* and *C. coli* are associated with infections in humans due to the consumption of undercooked meat, particularly poultry, and unpasteurized milk [2,3]. Currently, increasing resistance to major antibiotics in use among campylobacters is an emerging problem [34]. This is the first report to provide insights into the carriage of class 1 integrons by PDR, XDR, and MDR *Campylobacter* species isolated from livestock animals and humans in Egypt. In this study, the overall occurrence rate of *C. jejuni* (71.48%) was higher than that of *C. coli* (28.51%). Most *Campylobacter* isolates were detected in chicken samples (58.11%) represented as 67.53% *C. jejuni* and 32.47% *C. coli*, while previous studies reported varying rates of *Campylobacter* prevalence in chickens ranging from 24% to 62% [35,36].

Raw milk acts as a second main source of campylobacters [37]. The consumption of unpasteurized milk and milk products has been implicated in infections of 23% of human cases with campylobacteriosis in Egypt [38]. We reported a *C. jejuni* prevalence rate of 74.29% in raw milk samples, which have been previously documented with a lower prevalence rate (34%) [39]. Cross-contamination with *Campylobacter* species could occur during slaughter and milking of cattle. Herein, the prevalence of *C. jejuni* in fresh meat was 51.43%, which was lower than that reported in a previous study in Ethiopia (72%) [40]. Of note, milk and meat products were free from *Campylobacter* species, which was consistent with a previous study in India [41].

Poultry, milk, and meat are a reservoir for campylobacters; therefore, food processing with poor sanitation is an important source of transmission leading to increase the risk of human exposure, especially those in contact with food-producing animals. A higher prevalence of *C. jejuni* in the human stool (66.67%) was detected in this study when compared with previous studies conducted in Egypt with prevalence rates of 27.5% [35], 16.66% [42], and 4.07% [43].

Campylobacteriosis does not usually require antibiotic treatment; however, in some cases, antibiotics may be administered. Ciprofloxacin and erythromycin are considered drugs of choice for treating *Campylobacter* infections in humans [44]. However, the unregulated use of antibiotics in human and veterinary medicine, resulting in increasing their resistance [45].

Increasing resistance to the major antibiotics in use among campylobacters is an emerging problem [34]. Therefore, an investigation of the resistance rates and mechanisms is essential to prevent the spread of antibiotic-resistant campylobacters in livestock animals and humans. Herein, we provided better insight into different drug resistance patterns (*n* = 93-pattern) as well as an alarming increase of PDR, XDR, and MDR categories, while testing 25 antimicrobials among 14 antimicrobial categories.

Resistance to three or more antimicrobial classes (MDR) is a worldwide disturbing situation in *C. jejuni* [46]. In this study, 28.5% of *Campylobacter* isolates originating from livestock animals and humans exhibited MDR pattern. This level of resistance was less than that reported previously [47], while the resistance profile reaching 11 to 19 drugs is worrying compared with previous studies that recorded resistance to five to six [48] or three to four antimicrobial agents [49].

Resistance to all classes of antimicrobial agents except two or fewer is defined as XDR [26]. Of interest, this is the first report of XDR *Campylobacter* isolates (68.94%) among livestock animals and humans. In a previous study, an XDR *C. jejuni* CCARM 13,322 was recovered from a human case of diarrhea associated with international travel [17]. Moreover, 2.55% of *Campylobacter* isolates showed PDR (resistant to all antimicrobial agents among 14 categories), which was not reported in any previous study yet.

Class 1 integron is the main cause of multiple antibiotic resistance gene cassettes transmission in Gram-negative bacteria causing multidrug resistance [9,50]. Till date, few reports detected class 1 integrons in both *C. jejuni* and *C. coli* isolated from chicken house environment [12] and human sources [13,14] without showing the antimicrobial resistance profiles of isolates. However, no reports could detect class 1 integrons in PDR or XDR *Campylobacter* isolates [17].

In this study, the interested report of *integrase* gene and class 1 integrons (97.36% each) in *C. jejuni* and *C. coli* of chicken, cattle, and human origins representing MDR, XDR, and PDR patterns were documented for the first time, at least in Egypt. These integrons were associated with gene cassettes of different sizes ranging from 349 to 2600 bp. Considering, the previously published data, all *Campylobacter* isolates originating from a chicken house environment harbored a single cassette in the integron with 900 bp amplicon [12]. Additionally, those isolates originating from human and poultry sources had gene cassettes of molecular weights ranging from 300 to 1.4 kb [14].

Class 1 integrons detected in this study associated with 16 resistance gene cassettes. The most frequently reported were trimethoprim (*dfrA1*, *dfrA1*, *dfrA12*, *dfrA15*, and *dfrA17*) and aminoglycoside (*aadA1a*, *aadA2*, *aadA5*, *aadA22*, and *aadA7Δ4*) resistance gene cassette arrays. Consistently, previous studies detected class 1 integrons associated with aminoglycoside resistance genes (*aadA2* and *aacA4*) in both *C. jejuni* and *C. coli* resulting from the sequencing of 1000 bp, and 900 bp amplicons, respectively [12,14]. Moreover, the trimethoprim resistance gene cassettes (*dfr1* and *dfr9*) were carried by class 1 integrons in clinical isolates of *C. jejuni* following the sequencing of 399 bp, and 254 bp amplicons, respectively [13,15].

The most surprising points in the current study are the carriage of a novel gene cassette array, namely *aacC5-aadA7Δ4* (in number in 1983) as a first report according to the INTEGRALL database. In addition, an unusual phage tail tape measure protein gene cassette was harbored by four *Campylobacter* isolates as a first record in campylobacters worldwide. It is widely assumed that *Campylobacter*–bacteriophage interactions may play a role in horizontal gene transfer. In accordance with previous reports, bacteriophages cause genomic instability in *C. jejuni* and mediate interstrain transfer of large DNA fragments [51,52].

Herein, the correlation between resistance to certain antimicrobials and the corresponding gene cassettes was shown similar to other studies as *Campylobacter* isolates harboring aminoglycoside resistance gene, *aacA4*, conferred higher tobramycin MICs but slightly increased resistance to gentamicin [12]. In addition, high-level resistance to trimethoprim in *C. jejuni* was associated with the acquisition of *dfr* genes [15].

Previous studies showed that the *dfr* cassette is mostly associated with the *aadA* gene cassette [53]. DNA sequence analysis of a 1513 bp amplicon of class 1 integron revealed *dfrA17ab-aadA5* gene cassette in seven XDR *Campylobacter* isolates. Moreover, *dfrA12-gcu-aadA2* gene cassette array was detected in four analyzed isolates. According to our results, the *dfrA1-aadA1* gene cassette has been reported in MDR *Citrobacter* species with 1600 bp fragment size [54]. In addition, a previous study detected these gene cassettes of 2000 bp in MDR *Salmonella* isolates [55].

5. Conclusions

This is the first report, at least in Egypt, that showed the prevalence of PDR and XDR *Campylobacter* species in livestock animals and humans. Moreover, we demonstrated the existence of class 1 integrons and associated gene cassettes in analyzed isolates, which confer antimicrobial resistance and the possibility of *Campylobacter*–bacteriophage interactions by the carriage of an unusual phage tail protein as a first report.

Supplementary Materials: The following are available online at http://www.mdpi.com/2076-2615/10/11/2067/s1, Table S1. Oligonucleotide primer sequences used in the study, Table S2: Antibiotypes, MAR-indices, and detection of PDR, XDR, and MDR *Campylobacter* isolates originated from different sources. Figure S1: Distribution and clustering of studied *C. jejuni* isolates (n = 168) and the patterns of their antimicrobial resistance. Different hosts, sample types, antimicrobial classes, and resistance categories are shown for each isolate as color codes. The heat map represents the hierarchical clustering of the isolates and the antimicrobial classes, Figure S2: Distribution and clustering of studied *C. coli* isolates (n = 67) and the patterns of their antimicrobial resistance. Different hosts, sample types, antimicrobial classes, and resistance categories are shown for each isolate as color codes. The heat map represents the hierarchical clustering of the isolates and the antimicrobial classes.

Author Contributions: N.K.A.E.-A. and A.M.A. contributed equally in the conception and design of the study. M.M.H. performed the classical microbiological techniques. A.A.G. and E.A. conceived the study and participated in the design. A.A.G., E.A., A.H.S. and M.M.H. participated in acquisition of data and analysis and interpretation of data. N.K.A.E.-A. carried out the sequence analysis and participated in the data analysis. N.K.A.E.-A. and A.H.S. wrote the initial draft of the manuscript and revised it critically for important intellectual content. A.A.G.

and E.A. are responsible for the project administration and funding resources. All authors revised the manuscript and gave the final approval of the version to be published. All authors have read and agreed to the published version of the manuscript.

Funding: This research received no external funding.

Acknowledgments: We thank Taif University researchers supporting project number TURSP-2020/13, Taif University, Taif, Saudi Arabia. We thank the staff members of Faculty of Veterinary Medicine, Zagazig University, Egypt for their technical support during this study.

Conflicts of Interest: The authors declare no conflict of interest.

References

1. Moore, J.E.; Corcoran, D.; Dooley, J.S.; Fanning, S.; Lucey, B.; Matsuda, M.; McDowell, D.A.; Mégraud, F.; Millar, B.C.; O'Mahony, R. Campylobacter. *Vet. Res.* **2005**, *36*, 351–382. [CrossRef]

2. Friedman, C.R.; Hoekstra, R.M.; Samuel, M.; Marcus, R.; Bender, J.; Shiferaw, B.; Reddy, S.; Ahuja, S.D.; Helfrick, D.L.; Hardnett, F. Risk factors for sporadic *Campylobacter* infection in the United States: A case-control study in FoodNet sites. *Clin. Infect. Dis.* **2004**, *38*, S285–S296. [CrossRef]

3. Neimann, J.; Engberg, J.; Mølbak, K.; Wegener, H.C. A case–control study of risk factors for sporadic *Campylobacter* infections in Denmark. *Epidemiol. Infect.* **2003**, *130*, 353–366. [CrossRef]

4. European Food Safety Authority; European Centre for Disease Prevention and Control. EU Summary Report on antimicrobial resistance in zoonotic and indicator bacteria from humans, animals and food in 2013. *EFSA J.* **2015**, *13*, 4036.

5. Luangtongkum, T.; Jeon, B.; Han, J.; Plummer, P.; Logue, C.M.; Zhang, Q. Antibiotic resistance in Campylobacter: Emergence, transmission and persistence. *Future Microbiol.* **2009**, *4*, 189–200. [CrossRef]

6. Laroche, E.; Pawlak, B.; Berthe, T.; Skurnik, D.; Petit, F. Occurrence of antibiotic resistance and class 1, 2 and 3 integrons in *Escherichia coli* isolated from a densely populated estuary (Seine, France). *FEMS Microbiol. Ecol.* **2009**, *68*, 118–130. [CrossRef]

7. Brink, A.; Feldman, C.; Richards, G.; Moolman, J.; Senekal, M. Emergence of extensive drug resistance (XDR) among Gram-negative bacilli in South Africa looms nearer. *S. Afr. Med. J.* **2008**, *98*, 586–592.

8. Park, Y.K.; Peck, K.R.; Cheong, H.S.; Chung, D.-R.; Song, J.-H.; Ko, K.S. Extreme drug resistance in *Acinetobacter baumannii* infections in intensive care units, South Korea. *Emerg. Infect. Dis.* **2009**, *15*, 1325. [CrossRef]

9. Mazel, D. Integrons: Agents of bacterial evolution. *Nat. Rev. Microbiol.* **2006**, *4*, 608–620. [CrossRef]

10. Labbate, M.; Case, R.J.; Stokes, H.W. The integron/gene cassette system: An active player in bacterial adaptation. In *Horizontal Gene Transfer*; Springer: Berlin/Heidelberg, Germany, 2009; pp. 103–125.

11. Recchia, G.D.; Hall, R.M. Gene cassettes: A new class of mobile element. *Microbiology* **1995**, *141*, 3015–3027. [CrossRef] [PubMed]

12. Lee, M.D.; Sanchez, S.; Zimmer, M.; Idris, U.; Berrang, M.E.; McDermott, P.F. Class 1 integron-associated tobramycin-gentamicin resistance in *Campylobacter jejuni* isolated from the broiler chicken house environment. *Antimicrob. Agents Chemother.* **2002**, *46*, 3660–3664. [CrossRef]

13. Gibreel, A.; Sköld, O. An integron cassette carrying *dfr1* with 90-bp repeat sequences located on the chromosome of trimethoprim-resistant isolates of *Campylobacter jejuni*. *Microb. Drug Resist.* **2000**, *6*, 91–98. [CrossRef] [PubMed]

14. O'Halloran, F.; Lucey, B.; Cryan, B.; Buckley, T.; Fanning, S. Molecular characterization of class 1 integrons from Irish thermophilic *Campylobacter* spp. *J. Antimicrob. Chemother.* **2004**, *53*, 952–957. [CrossRef]

15. Gibreel, A.; Sköld, O. High-level resistance to trimethoprim in clinical isolates of *Campylobacter jejuni* by acquisition of foreign genes (*dfr1* and *dfr9*) expressing drug-insensitive dihydrofolate reductases. *Antimicrob. Agents Chemother.* **1998**, *42*, 3059–3064. [CrossRef] [PubMed]

16. Chang, Y.-C.; Tien, N.; Yang, J.-S.; Lu, C.-C.; Tsai, F.-J.; Huang, T.-J.; Wang, I.-K. Class 1 integrons and plasmid-mediated multiple resistance genes of the *Campylobacter* species from pediatric patient of a university hospital in Taiwan. *Gut Pathog.* **2017**, *9*, 1–8. [CrossRef] [PubMed]

17. Shin, E.; Hong, H.; Oh, Y.; Lee, Y. First report and molecular characterization of a *Campylobacter jejuni* isolate with extensive drug resistance from a travel-associated human case. *Antimicrob. Agents Chemother.* **2015**, *59*, 6670–6672. [CrossRef]

18. Vandepitte, J.; Verhaegen, J.; Engbaek, K.; Rohner, P.; Piot, P.; Heuck, C.; Heuck, C. *Basic Laboratory Procedures in Clinical Bacteriology*; World Health Organization: Geneva, Switzerland, 2003.

19. Quinn, P.J.; Markey, B.; Carter, M.E. *Campylobacter* spp. In *Clinical Veterinary Microbiology*, 1st ed.; Wolfe Publishing: London, UK, 1994; pp. 268–272.

20. Wang, G.; Clark, C.G.; Taylor, T.M.; Pucknell, C.; Barton, C.; Price, L.; Woodward, D.L.; Rodgers, F.G. Colony multiplex PCR assay for identification and differentiation of *Campylobacter jejuni, C. coli, C. lari, C. upsaliensis, and C. fetus* subsp. *fetus*. *J. Clin. Microbiol.* **2002**, *40*, 4744–4747. [CrossRef]

21. Shin, E.; Lee, Y. Comparison of three different methods for *Campylobacter* isolation from porcine intestines. *J. Microbiol. Biotechnol.* **2009**, *19*, 647–650.

22. Bauer, A.; Kirby, W.; Sherris, J.; Turck, M. Antibiotic susceptibility testing by a standardized single disc method. *Am. J. Clin. Pathol.* **1966**, *45*, 493–496. [CrossRef]

23. Clinical and Laboratory Standards Institute (CLSI). *Performance Standards for Antimicrobial Susceptibility Testing*; Twenty-First Informational Supplement, CLSI document M100-S21; Clinical and Laboratory Standards Institute: Wayne, PA, USA, 2011.

24. European Committee on Antimicrobial Susceptibility Testing. Breakpoint tables for interpretation of mics and zone diameters. Version 7.1, Valid from 2017-03-10. Available online: http://www.eucast.org/fileadmin/src/media/PDFs/EUCAST_files/Breakpoint_tables/v_7.1_Breakpoint_Tables.pdf (accessed on 20 August 2019).

25. Tambekar, D.; Dhanorkar, D.; Gulhane, S.; Khandelwal, V.; Dudhane, M. Antibacterial susceptibility of some urinary tract pathogens to commonly used antibiotics. *Afr. J. Biotechnol.* **2006**, *5*, 1562–1565.

26. Magiorakos, A.P.; Srinivasan, A.; Carey, R.B.; Carmeli, Y.; Falagas, M.E.; Giske, C.G.; Harbarth, S.; Hindler, J.F.; Kahlmeter, G.; Olsson-Liljequist, B.; et al. Multidrug-resistant, extensively drug-resistant and pandrug-resistant bacteria: An international expert proposal for interim standard definitions for acquired resistance. *Clin. Microbiol. Infect.* **2012**, *18*, 268–281. [CrossRef]

27. Kashif, J.; Buriro, R.; Memon, J.; Yaqoob, M.; Soomro, J.; Dongxue, D.; Jinhu, H.; Liping, W. Detection of class 1 and 2 integrons, β-lactamase genes and molecular characterization of sulfonamide resistance in *Escherichia coli* isolates recovered from poultry in China. *Pak. Vet. J.* **2013**, *33*, 321–324.

28. White, P.A.; McIver, C.J.; Deng, Y.-M.; Rawlinson, W.D. Characterisation of two new gene cassettes, aadA5 and dfrA17. *FEMS Microbiol. Lett.* **2000**, *182*, 265–269. [CrossRef]

29. Sow, A.G.; Wane, A.A.; Diallo, M.H.; Boye, C.S.-B.; Aïdara-Kane, A. Genotypic characterization of antibiotic-resistant *Salmonella* Enteritidis isolates in Dakar, Senegal. *J. Infect. Dev. Ctries.* **2007**, *1*, 284–288.

30. Kolde, R. Package 'pheatmap': Pretty Heat Map. 2018. Available online: https://rdrr.io/cran/pheatmap/ (accessed on 12 May 2020).

31. Friendly, M. Corrgrams. *Am. Stat.* **2002**, *56*, 316–324. [CrossRef]

32. Galili, T.; O'Callaghan, A.; Sidi, J.; Sievert, C. Heatmaply: An R package for creating interactive cluster heatmaps for online publishing. *Bioinformatics* **2018**, *34*, 1600–1602. [CrossRef]

33. Harrell, F.E., Jr. Package 'Hmisc'. Available online: https://cran.r-project.org/web/packages/Hmisc/Hmisc.pdf (accessed on 20 May 2020).

34. Mshana, S.; Joloba, M.; Kakooza, A.; Kaddu-Mulindwa, D. *Campylobacter* spp among Children with acute diarrhea attending Mulago hospital in Kampala-Uganda. *Afr. Health Sci.* **2009**, *9*, 201.

35. Abushahba, M.F.; Ahmed, S.O.; Ibrahim, A.A.; Mosa, H.A. Prevalence of zoonotic species of *Campylobacter* in broiler chicken and humans in Assiut governorate, Egypt. *Approaches Poult. Dairy Vet. Sci* **2018**, *3*, 1–9. [CrossRef]

36. Jouahri, M.; Asehraou, A.; Karib, H.; Hakkou, A.; Touhami, M. Prevalence and control of thermo-tolerant *Campylobacter* species in raw poultry meat in Morocco. *MESO Prvi Hrvat. Časopis Mesu* **2007**, *9*, 262–267.

37. Leedom, J.M. Milk of nonhuman origin and infectious diseases in humans. *Clin. Infect. Dis.* **2006**, *43*, 610–615. [CrossRef] [PubMed]

38. Jun, W.; Guo, Y.C.; Ning, L. Prevalence and risk assessment of *Campylobacter jejuni* in chicken in China. *Biomed. Environ. Sci.* **2013**, *26*, 243–248.

39. Wicker, C.; Giordano, M.; Rouger, S.; Sorin, M.-L.; Arbault, P. *Campylobacter* detection in food using an ELISA based method. *Int. J. Med. Microbiol.* **2001**, *291*, 1–12.

40. Woldemariam, T.; Asrat, D.; Zewde, G. Prevalence of thermophilic *Campylobacter* species in carcasses from sheep and goats in an abattoir in Debre Zeit area, Ethiopia. *Ethiop. J. Health Dev.* **2009**, *23*. [CrossRef]

41. Singh, H.; Rathore, R.S.; Singh, S.; Cheema, P.S. Comparative analysis of cultural isolation and PCR based assay for detection of *Campylobacter jejuni* in food and faecal samples. *Braz. J. Microbiol.* **2009**, *42*, 181–186. [CrossRef]

42. Hassanain, N.A. Antimicrobial resistant *Campylobacter jejuni* isolated from humans and animals in Egypt. *Glob. Vet.* **2011**, *6*, 195–200.

43. El-Hamid, M.I.; El-Aziz, A.; Khairy, N.; Samir, M.; Remela, A.; Etab, M.; El-Naenaeey, E.-s.Y.; Bendary, M.M.; Mosbah, R.A. Genetic diversity of *Campylobacter jejuni* isolated from Avian and human sources in Egypt. *Front. Microbial.* **2019**, *10*, 2353. [CrossRef] [PubMed]

44. Ghunaim, H.; Behnke, J.M.; Aigha, I.; Sharma, A.; Doiphode, S.H.; Deshmukh, A.; Abu-Madi, M.M. Analysis of resistance to antimicrobials and presence of virulence/stress response genes in *Campylobacter* isolates from patients with severe diarrhoea. *PLoS ONE* **2015**, *10*, e0119268. [CrossRef] [PubMed]

45. De Vries, S.P.; Vurayai, M.; Holmes, M.; Gupta, S.; Bateman, M.; Goldfarb, D.; Maskell, D.J.; Matsheka, M.I.; Grant, A.J. Phylogenetic analyses and antimicrobial resistance profiles of *Campylobacter* spp. from diarrhoeal patients and chickens in Botswana. *PLoS ONE* **2018**, *13*. [CrossRef]

46. Mdegela, R.; Nonga, H.; Ngowi, H.; Kazwala, R. Prevalence of thermophilic *Campylobacter* infections in humans, chickens and crows in Morogoro, Tanzania. *J. Vet. Med. Ser. B* **2006**, *53*, 116–121. [CrossRef]

47. Bouhamed, R.; Bouayad, L.; Messad, S.; Zenia, S.; Naïm, M.; Hamdi, T.-M. Sources of contamination, prevalence, and antimicrobial resistance of thermophilic *Campylobacter* isolated from turkeys. *Vet. World* **2018**, *11*, 1074. [CrossRef]

48. Marotta, F.; Garofolo, G.; di Marcantonio, L.; Di Serafino, G.; Neri, D.; Romantini, R.; Sacchini, L.; Alessiani, A.; Di Donato, G.; Nuvoloni, R. Antimicrobial resistance genotypes and phenotypes of *Campylobacter jejuni* isolated in Italy from humans, birds from wild and urban habitats, and poultry. *PLoS ONE* **2019**, *14*, e0223804. [CrossRef] [PubMed]

49. Di Giannatale, E.; Calistri, P.; Di Donato, G.; Decastelli, L.; Goffredo, E.; Adriano, D.; Mancini, M.E.; Galleggiante, A.; Neri, D.; Antoci, S. Thermotolerant *Campylobacter* spp. in chicken and bovine meat in Italy: Prevalence, level of contamination and molecular characterization of isolates. *PLoS ONE* **2019**, *14*, e0225957. [CrossRef]

50. Hall, M.A.L.-v.; Blok, H.E.M.; Donders, T.R.; Paauw, A.; Fluit, A.C.; Verhoef, J. Multidrug resistance among *Enterobacteriaceae* is strongly associated with the presence of integrons and is independent of species or isolate origin. *J. Infec. Dis.* **2003**, *187*, 251–259.

51. Scott, A.E.; Timms, A.R.; Connerton, P.L.; Carrillo, C.L.; Radzum, K.A.; Connerton, I.F. Genome dynamics of *Campylobacter jejuni* in response to bacteriophage predation. *PLoS Pathog.* **2007**, *3*. [CrossRef]

52. Scott, A.E.; Timms, A.R.; Connerton, P.L.; El-Shibiny, A.; Connerton, I.F. Bacteriophage influence *Campylobacter jejuni* types populating broiler chickens. *Environ. Microbiol.* **2007**, *9*, 2341–2353. [CrossRef]

53. Yan, H.; Li, L.; Zong, M.; Alam, M.J.; Shinoda, S.; Shi, L. Occurrence and characteristics of Class 1 and 2 integrons in clinical bacterial isolates from patients in south China. *J. Health Sci.* **2010**, *56*, 442–450. [CrossRef]

54. Lorestani, R.C.; Akya, A.; Elahi, A.; Hamzavi, Y. Gene cassettes of class 1 integron-associated with antimicrobial resistance in isolates of *Citrobacter* spp. with multidrug resistance. *Iran. J. Microbiol.* **2018**, *10*, 22.

55. Zhao, X.; Yang, J.; Zhang, B.; Sun, S.; Chang, W. Characterization of integrons and resistance genes in *Salmonella* isolates from farm animals in Shandong Province, China. *Front. Microbiol.* **2017**, *8*, 1300. [CrossRef]

Article

Longitudinal Study of the Distribution of Antimicrobial-Resistant *Campylobacter* Isolates from an Integrated Broiler Chicken Operation

Bo-Ram Kwon [1,†], Bai Wei [1,†], Se-Yeoun Cha [1,†], Ke Shang [1], Jun-Feng Zhang [1], Min Kang [1,2,*] and Hyung-Kwan Jang [1,2,*]

[1] Department of Veterinary Infectious Diseases and Avian Diseases, College of Veterinary Medicine and Center for Poultry Diseases Control, Jeonbuk National University, Iksan 54596, Korea; aniscikwon@gmail.com (B.-R.K.); weibai116@hotmail.com (B.W.); kshmnk@hanmail.net (S.-Y.C.); shangke0624@gmail.com (K.S.); jfzhang018@gmail.com (J.-F.Z.)

[2] Bio Disease Control (BIOD) Co., Ltd., Iksan 54596, Korea

[*] Correspondence: minkang@jbnu.ac.kr (M.K.); hkjang@jbnu.ac.kr (H.-K.J.); Tel.: +82-63-850-0690 (M.K.); +82-63-850-0945 (H.-K.J.); Fax: +82-858-0686 (M.K.); +82-858-9155 (H.-K.J.)

[†] These authors contributed equally to this work.

Simple Summary: Investigation of *Campylobacter* prevalence throughout the entire chicken production process from farms to retail meat is still limited. In this study, we examined the prevalence and antimicrobial susceptibility of *Campylobacter* in 10 production lines from one of the largest integrated poultry production companies in Korea. The prevalence of *Campylobacter* in breeder farm, hatchery, broiler farm, slaughterhouse, and retail meat products was 50.0%, 0%, 3.3%, 13.4%, and 68.4%, respectively. Resistance to fluoroquinolones was the most frequently observed, and 16 isolates from breeder farm were resistant to both azithromycin and ciprofloxacin. Diverse pulsed-field gel electrophoresis genotypes were presented with discontinuous patterns along the whole production chain. Thirty percent of *Campylobacter*-free flocks became positive after slaughtering. An identical genotype was simultaneously detected from both breeder farm and retail meat, even from different production lines. This study reveals that antimicrobial-resistant *Campylobacter* contamination can occur at all stages of the chicken supply chain. In particular, the breeder farm and slaughterhouse should be the main control points, as they are the potential stages at which antimicrobial-resistant *Campylobacter* could spread to retail meat products by horizontal transmission.

Citation: Kwon, B.-R.; Wei, B.; Cha, S.-Y.; Shang, K.; Zhang, J.-F.; Kang, M.; Jang, H.-K. Longitudinal Study of the Distribution of Antimicrobial-Resistant *Campylobacter* Isolates from an Integrated Broiler Chicken Operation. *Animals* **2021**, *11*, 246. https://doi.org/10.3390/ani11020246

Received: 6 November 2020
Accepted: 15 January 2021
Published: 20 January 2021

Abstract: The aim of this study was to analyze the prevalence, antimicrobial resistance, and genetic diversity of *Campylobacter* isolates that were obtained from whole chicken production stages in Korea. A total of 1348 samples were collected from 10 production lines. The prevalence of *Campylobacter* in breeder farm, broiler farm, slaughterhouse, and retail meat products was 50.0%, 3.3%, 13.4%, and 68.4%, respectively, and *Campylobacter* was not detected at the hatchery stage. Resistance to quinolones/fluoroquinolones was the most prevalent at all stages. Among the multidrug-resistant isolates, 16 isolates (19.8%) from breeder farm were resistant to both azithromycin and ciprofloxacin. A total of 182 isolates were subdivided into 82 pulsed-field gel electrophoresis (PFGE) genotypes with 100% similarity. Diverse genotypes were presented with discontinuous patterns along the whole production chain. Thirty percent of *Campylobacter*-free flocks became positive after slaughtering. An identical genotype was simultaneously detected from both breeder farm and retail meat, even from different production lines. This study reveals that antimicrobial-resistant *Campylobacter* contamination can occur at all stages of the chicken supply chain. In particular, the breeder farm and slaughterhouse should be the main control points, as they are the potential stages at which antimicrobial-resistant *Campylobacter* could spread to retail meat products by horizontal transmission.

Keywords: *Campylobacter*; whole-chicken production chain; antimicrobial resistance; longitudinal study; PFGE; genetic diversity

1. Introduction

Campylobacter spp. are a leading cause of food-borne diarrheal illnesses globally, and *Campylobacter* infection is among the most frequently reported causes of gastroenteritis in humans worldwide [1]. Poultry and poultry products, particularly contaminated chicken products, are considered to be major sources of human infection [2]. *Campylobacter* can be isolated at all stages of the chicken supply chain from farms to retail meat products [3]. *Campylobacter* usually colonizes from the third week of age after the beginning of the rearing period and, once colonized, *Campylobacter* will rapidly reach high numbers in flocks and the farm environment [4]. Many studies have found that *Campylobacter* is rarely detected in day-old chicks, possibly due to the protection that is offered by maternal antibodies [2]. According to a previous study, vertical transmission from parent flocks to their progeny still remains unknown [5]; meanwhile, several suspected horizontal transmission sources or vectors, including the poultry house environment, small animals on the farm, flies, and rodents, have been identified as major factors of flock colonization [4]. Various contamination factors specifically exist in slaughterhouses; for example, direct contact between carcasses can frequently induce cross-contamination during defeathering and evisceration, and contact with common surfaces, such as rubber fingers, conveyor belts, and cutting tables, is also a main reason for cross-contamination and the presentation of various colonies of *Campylobacter* [6,7]. Some isolates seem to survive in the slaughter equipment and during processing [7]. The persistence of *Campylobacter* in the equipment may lead to the contamination of *Campylobacter*-negative flocks that are slaughtered after *Campylobacter*-positive flocks [8]. Numerous studies concluded that the most effective measures should aim at reducing the prevalence of *Campylobacter*-positive flocks and the level of contamination of *Campylobacter* on broiler carcasses [9].

When *Campylobacter* infection requires antimicrobial treatment, macrolides and fluoroquinolones are normally considered to be first and second antimicrobials of choice, respectively [10]. However, the recent emergence of resistance to these antimicrobial groups and multidrug-resistant *Campylobacter* isolates has been observed within the food supply chain [3]. The use of enrofloxacin in poultry was banned in the US in 2005 because increased levels of fluoroquinolone resistance have raised public health concern. In Korea, the use of antimicrobial agents as growth promoters was withdrawn in 2011, but antimicrobial agents, including quinolone, macrolides, tetracyclines, and penicillin, are still widely used in the conventional chicken industry for treating diseases [11]. Despite the policy against antimicrobial usage, persistent use of antimicrobial agents may induce the development of resistance and affect other properties, such as the ability to colonize an animal host or persist in the farm or food processing environment [12].

Most studies monitoring the prevalence and antimicrobial resistance of *Campylobacter* have focused on slaughterhouse and retail meat [13–16]. However, investigation covering the whole chicken production stage from farms to retail meat are still limited [17]. Tracing the distribution of *Campylobacter* longitudinally in whole chicken production stages would help to identify the relatedness of transmission to subsequent stages and determine the mode of transmission between the vertical and horizontal routes. We examined the most prevalent contamination spot and antimicrobial susceptibility of *Campylobacter* in 10 production lines from integrated chicken operation. Given that resistant *Campylobacter* strains could be directly transmitted to the people who had direct contact with the contaminated breeder chicken [18], we included breeder farm as the beginning of production stage and observed the prevalence of antimicrobial resistant *Campylobacter*, especially those that are resistant to fluoroquinolone and/or azithromycin; both of which are used widely in human. In addition, pulsed-field gel electrophoresis (PFGE), which is widely regarded as the gold standard for tracing outbreaks [7], was performed for molecular typing of *Campylobacter* isolates in order to clarify the transmission routes and epidemiological relationships among isolates of the same species.

2. Materials and Methods

2.1. Description of Production Company and Farms

From August 2015 to August 2016, 10 chicken production lines (lines 1 to 10), which belonged to one of the largest integrated poultry production company in South Korea, were chronologically investigated from breeder farms to retail meat products. The breeder farms were located in various provinces. The size of breeder farm flocks varied from 16,000 to 50,000 chickens. Every broiler hatching egg produced by these farms was transported to hatchery assigned to same integrated company. Newly hatched chicks were then again transported to, and reared at, broiler farms, which contained an average of 70,000–100,000 broilers and three to five separate flocks, until their slaughter age of 30 days. Finally, chickens from all broiler farms in this study were gathered and slaughtered in one processing plant. All of the breeder and broiler farms in this study used ampicillin, florfenicol, and tetracycline for disease treatment.

2.2. Sampling and Isolation

All of the animals used were commercially raised and reared in conventional chicken farms under the supervision of the local veterinary authorities; in particular, sampling was performed in accordance with the relevant guidelines (Guide for the Care and Use of Laboratory Animals 2014, Korean Ministry of Food and Drug Safety) and regulations (Korean Council on Animal Care and Korean Animal Protection Law, 2015; Article 23) for experiments with livestock animals in farm. No chickens were killed for this study, and sampling was carried by a veterinarian according to the standard protocols and with prior consent of the farmer/manager of the facilities. Furthermore, written informed consent was obtained from the owners for the participation of their animals in this study.

In order to increase the relatedness between samples, the sampling procedure was implemented in an orderly manner from breeder farms to retail meat, and sampling was limited to one cycle—from breeder farm to production as retail meat; furthermore, the samples were acquired as evenly as possible in order to avoid oversampling at a specific time or region. From 10 chicken production lines, a total of 1348 samples from all stages, such as breeder farm, hatchery, broiler farm, slaughterhouse, and retail meat, were collected (Supplementary Table S1). Cloacal swab samples (n = 110) from 28–65-week-old chickens and litter samples (n = 66) were collected from breeder farms (n = 176). In hatcheries, cloacal swab samples were collected from newly hatched chicks (n = 165). All of the cloacal swab samples that were collected from breeder farms and hatcheries were separately pooled from five chickens to one. As for broiler farms, two flocks per farm were sampled three times within a 30-day period (chickens at 1–14 d, 15–24 d, and >25 d of age) during one rearing cycle. Broiler farm sampling was repeated during a second, separate production cycle (n = 720). The cloacal swab samples (n = 300) were randomly collected from 25 chickens in the entire area of the flock. In detail, a flock was divided equally into five sectors, and five cloacal samples were obtained from each sector and then pooled into one sample, making five samples in total for a flock. Environmental samples of feed (n = 120), litter (n = 180), and water (n = 120) were uniformly collected from equally divided sectors of the flock, and each sample from the same sector was pooled into one sample (Supplementary Table S1). The samples from slaughterhouse (n = 230) were collected at the beginning of each sampling day during the slaughtering of the first batch of broilers chickens; different production lines were sampled at different sampling dates. From lairage, five cloacal swab samples from five different chicken were collected, which were then pooled into one sample (n = 50). Furthermore, environmental samples in slaughterhouse were collected by aseptically swabbing on the surface of each slaughtering site; they were also pooled into one sample (n = 180). Retail meat samples (n = 57) were collected from the meat that was purchased from retail markets in Jeonbuk province, South Korea. All of the samples were placed into plastic bags and boxes and then transported in a box with ice to the laboratory where they were analyzed immediately.

Pooled cloacal swab samples and environmental swab samples were pre-enriched in Bolton broth (Oxoid Ltd., Basingstoke, UK) that was supplemented with cefoperazone, vancomycin, trimethoprim, and cycloheximide (Oxoid). Fresh samples (1 g (or mL)) of feed, litter, and water were separately mixed with 9 mL (1:9 dilution) of Bolton broth. Subsequently, these samples were incubated in a microaerophilic environment of 10% CO_2, 5% O_2, and 85% N_2 at 42 °C for 48 h for enrichment. Each retail meat sample was aseptically rinsed with 100 mL of buffered peptone water (Difco, Sparks, MD, USA) in sterile plastic bags [19]. From rinsed meat, 10 mL of rinse solution was added to 10 mL of 2× Bolton broth. Next, the samples were incubated, as above. After enrichment for 48 h, a loop full of each sample was streaked onto a plate of modified charcoal cefoperazone deoxycholate agar (mCCDA) that was prepared with *Campylobacter* blood-free selective agar base (Oxoid) supplemented with a CCDA selective supplement containing cefoperazone and amphotericin (Oxoid). After incubation, the plates were examined for typical colonies, which are generally small, gray, shiny, and drop-like in shape. At least three presumptive *Campylobacter* colonies from each selective agar plate were further cultured on 5% sheep blood agar plates (Komed, Seongnam, South Korea) microaerobically at 42 °C for 48 h. Presumptive *Campylobacter* isolates were confirmed by polymerase chain reaction (PCR) assay, as described previously [20]. After identifying each isolate, *Campylobacter* isolates were stored in brain heart infusion broth (Oxoid) with 20% glycerol at −70 °C.

2.3. Antimicrobial Susceptibility Testing

The susceptibility of all *Campylobacter* isolates to 11 antimicrobial agents was determined by agar dilution method and using Sensititre susceptibility plates (TREK Diagnostic Systems, Incheon, Korea). The standard agar dilution method, as described by the Clinical Laboratory Standards Institute [21], was followed in order to confirm the susceptibility to two antimicrobial agents, namely enrofloxacin (ENR; Daesung Microbiological, Uiwang, Korea) and ampicillin (AMP; Sigma-Aldrich, St. Louis, MO, USA). Mueller–Hinton agar (Oxoid) plates supplemented with 5% lysed sheep blood (Oxoid) and antimicrobial agents at concentrations of 0.125–128 μg/mL for ENR and 8–128 μg/mL for AMP in two-fold serial dilutions were used. Plates were inoculated with 1-mm-diameter inoculating pins and incubated at 42 °C for 24 h under microaerobic conditions. The rest of the nine antimicrobial agents were tested by Sensititre susceptibility plates containing azithromycin (AZM; 0.015–64 μg/mL), erythromycin (ERY; 0.03–64 μg/mL), telithromycin (TEL; 0.015–8 μg/mL), nalidixic acid (NAL; 4–64 μg/mL), ciprofloxacin (CIP; 0.015–64 μg/mL), clindamycin (CLI; 0.03–16 μg/mL), gentamicin (GEN; 0.12–32 μg/mL), florfenicol (FFN; 0.03–64 μg/mL), and tetracycline (TET; 0.06–64 μg/mL). The plates were incubated under microaerobic conditions at 42 °C for 24 h. The results were evaluated according to the interpretation criteria of the National Antimicrobial Resistance Monitoring System [22]. We used the breakpoints for *Enterobacteriaceae* from the Clinical and Laboratory Standards Institute criteria, as no enrofloxacin and ampicillin breakpoints are available for *Campylobacter* [23]. *Campylobacter jejuni* ATCC 33560 was used as a quality control isolate. Multidrug resistant (MDR) isolates were those with resistance to two or more classes of antimicrobials.

2.4. Pulsed-Field Gel Electrophoresis (PFGE)

The isolates of *C. coli* and *C. jejuni* were genotyped while using PFGE according to protocols from the Centers for Disease Control and Prevention available on PulseNet. Genomic DNA (extraction using 1% sodium dodecylsulfate and 1-mg/mL proteinase K, Biosesang, Seoul, Korea) of *Campylobacter* isolates was digested with *Sma*I (Thermo Fisher Scientific, Inchon, Korea), and *Xba*I-digested DNA from *Salmonella* Braenderup H9812 was used as the standard size. The PFGE results were analyzed using BioNumerics (version 6.6 for Windows, Kortrijk, Belgium). Dice coefficients were calculated based on a pairwise comparison of the PFGE types of the isolates. The isolates were defined as closely related based on molecular typing when their PFGE patterns had dice coefficients with

100% similarity level. Dice coefficients, with an optimization of 2.0% and a band position tolerance of 1.5%, were applied.

2.5. Statistical Analysis

The prevalence of *Campylobacter* spp. between different production stages was compared with the chi-square test. The statistical significance of the differences in resistance to all antimicrobials between *Campylobacter* spp. was also tested while using chi-square test. Differences were considered to be statistically significant at *p* values less than 0.05.

3. Results

3.1. Distribution of Campylobacter spp. along the Chicken Production Chain

The prevalence of *Campylobacter* in breeder, broiler farm, slaughterhouse, and retail meat products was 50.0% (88/176), 3.3% (24/720), 13.5% (31/230), and 68.4% (39/57), respectively (Figure 1), which indicated the highest prevalence in retail meat products ($p < 0.05$). *Campylobacter* was not detected in samples that were acquired from the hatchery stage. The distribution of *Campylobacter* species from the chicken production stage is shown in Table 1. Overall, 182 isolates (13.5%) out of 1348 samples were positive for *Campylobacter*, either *C. coli* (80 isolates, 44%) or *C. jejuni* (102 isolates, 56%). Except in breeder farms and retail meat products, *C. jejuni* was more prevalent than *C. coli* at all other stages. Each production line showed various distribution patterns of *Campylobacter* isolates. Lines 2 and 8 were positive for *Campylobacter* at the breeder farm and retail meat product stages, but not at other stages. Regarding lines 5, 6, and 7, *Campylobacter* spp. were isolated from every stage of the chicken supply chain, except at the hatchery stage.

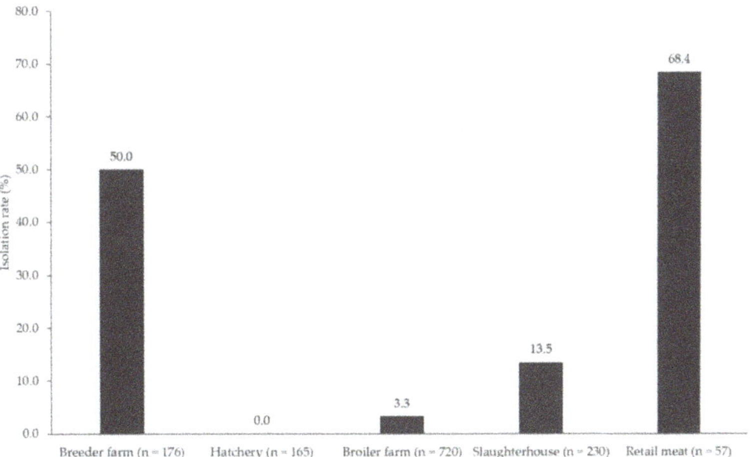

Figure 1. Prevalence of *Campylobacter* isolated from the chicken production chain. (n = total number of samples from each production stage).

3.2. Antimicrobial Susceptibility

Table 2 presents the results of antimicrobial susceptibility testing performed on the 182 isolates. Resistance to CIP and ENR was the most common (170/182, 93.4%), followed by resistance to NAL (161/182, 88.5%), AMP (133/182, 73.1%), and TET (103/182, 56.6%). Resistance to CLI, GEN, and FFN was only found in 1.6%, 4.4%, and 0.5% samples, respectively. *Campylobacter* resistance to macrolides, such as AZM and ERY, was only noted in isolates that were derived from breeder farms, with resistance rates of 9.9% and 8.8%, respectively. All of the *C. coli* isolates were resistant to at least one antimicrobial tested in this study. The resistance rate for antimicrobials was statistically ($p < 0.05$) higher in *C. coli* than in *C. jejuni* for CIP, ENR, and TET.

Table 1. Distribution of *Campylobacter coli* and *C. jejuni* isolated from different lines along the chicken production chain.

| Line | C. coli (80/182, 44.0%) | | | | | | | C. jejuni (102/182, 56.0%) | | | | | | |
| | Breeder n/(%) | Hatchery n/(%) | Broiler n/(%) | | | SlaughterHouse n/(%) | Retail Meat n/(%) | Breeder n/(%) | Hatchery n/(%) | Broiler n/(%) | | | SlaughterHouse n/(%) | Retail Meat n/(%) |
			1 d–14 d	15 d–24 d	>25 d					1 d–14 d	15 d–24 d	>25 d		
1	4/32 (12.5)	0/20 (0.0)	0/24 (0.0)	0/24 (0.0)	0/24 (0.0)	5/34 14.7) [c]	3/3 (100.0)	8/32 (25.0)	0/20 (0.0)	0/24 (0.0)	0/24 (0.0)	0/24 (0.0)	0/34 (0.0)	0/3 (0.0)
2	1/16 (6.3)	0/20 (0.0)	0/24 (0.0)	0/24 (0.0)	0/24 (0.0)	0/44 (0.0)	8/14 (57.1)	7/16 (43.8)	0/20 (0.0)	0/24 (0.0)	0/24 (0.0)	0/24 (0.0)	0/44 (0.0)	2/14 (14.3)
3	10/24 (41.7)	0/10 (0.0)	0/24 (0.0)	0/24 (0.0)	0/24 (0.0)	0/34 (0.0)	4/7 (57.1)	10/24 (41.7)	0/10 (0.0)	0/24 (0.0)	0/24 (0.0)	0/24 (0.0)	5/34 (14.7) [c]	3/7 (42.9)
4	6/16 (37.5)	0/10 (0.0)	0/24 (0.0)	0/24 (0.0)	0/24 (0.0)	0/34 (0.0)	0/3 (0.0)	3/16 (18.8)	0/10 (0.0)	0/24 (0.0)	0/24 (0.0)	0/24 (0.0)	1/34 (2.9) [c]	0/3 (0.0)
5	4/32 (12.5)	0/20 (0.0)	0/24 (0.0)	0/24 (0.0)	4/24 (16.7) [a]	2/5 (40.0) [b]	1/3 (33.3)	8/32 (25.0)	0/20 (0.0)	0/24 (0.0)	0/24 (0.0)	1/24 (4.7) [a]	0/5 (0.0)	0/3 (0.0)
6	8/16 (50.0)	0/25 (0.0)	0/24 (0.0)	0/24 (0.0)	0/24 (0.0)	0/5 (0.0)	0/6 (0.0)	1/16 (6.3)	0/25 (0.0)	6/24 (25.0) [a]	0/24 (0.0)	0/24 (0.0)	5/5 (100)	2/6 (33.3)
7	4/24 (16.7)	0/20 (0.0)	0/24 (0.0)	0/24 (0.0)	0/24 (0.0)	0/20 (0.0)	5/9 (55.6)	4/24 (16.7)	0/20 (0.0)	1/24 (4.7) [b]	0/24 (0.0)	0/24 (0.0)	3/20 (15.0) [b]	4/9 (44.4)
8	8/16 (50.0)	0/20 (0.0)	0/24 (0.0)	0/24 (0.0)	0/24 (0.0)	0/10 (0.0)	0/6 (0.0)	2/16 (12.5)	0/20 (0.0)	0/24 (0.0)	0/24 (0.0)	0/24 (0.0)	0/10 (0.0)	1/6 (16.7)
9	-	0/10 (0.0)	0/24 (0.0)	0/24 (0.0)	0/24 (0.0)	0/39 (0.0)	3/3 (100.0)	-	0/10 (0.0)	0/24 (0.0)	3/24 (12.5) [b]	0/24 (0.0)	10/39 (25.6) [d]	0/3 (0.0)
10	-	0/10 (0.0)	0/24 (0.0)	0/24 (0.0)	0/24 (0.0)	0/5 (0.0)	0/3 (0.0)	-	0/10 (0.0)	4/24 (16.7) [a]	0/24 (0.0)	5/24 (20.8) [b]	0/5 (0.0)	3/3 (100.0)
Total	45/176 (25.6)	0/165 (0.0)	0/240 (0.0)	0/240 (0.0)	4/240 (1.7)	7/230 (3.0)	24/57 (42.1)	43/176 (24.4)	0/165 (0.0)	11/240 (4.6)	3/240 (1.3)	6/240 (2.5)	24/230 (10.4)	15/57 (26.3)

[a] Isolates including cloacal swab and rearing materials (feed, litter, and water). [b] Isolates only from cloacal swab. [c] Isolates only from environmental sources. [d] Isolates including cloacal swab and environmental sources. All positive isolates from breeder farm were isolated from cloacal swab. - Sampling was not included.

Table 2. Antimicrobial resistance profiles of *Campylobacter* isolates from the chicken production chain.

| Antimicrobial Agent | Campylobacter spp. | C. coli | | | | | C. jejuni | | | | |
	Total (n = 182)	Breeder (n = 45)	Broiler (n = 4)	Slaughterhouse (n = 7)	Retail Meat (n = 24)	Total (n = 80)	Breeder (n = 43)	Broiler (n = 20)	Slaughterhouse (n = 24)	Retail Meat (n = 15)	Total (n = 102)
Azithromycin	18 (9.9%)	10 (22.2%)	0 (0.0%)	0 (0.0%)	0 (0.0%)	10 (12.5%)	8 (18.6%)	0 (0.0%)	0 (0.0%)	0 (0.0%)	8 (7.8%)
Erythromycin	16 (8.8%)	9 (20.0%)	0 (0.0%)	0 (0.0%)	0 (0.0%)	9 (11.3%)	7 (16.3%)	0 (0.0%)	0 (0.0%)	0 (0.0%)	7 (6.9%)
Telithromycin	11 (6.0%)	7 (15.6%)	0 (0.0%)	0 (0.0%)	0 (0.0%)	7 (8.8%)	4 (9.3%)	0 (0.0%)	0 (0.0%)	0 (0.0%)	4 (3.9%)
Nalidixic acid	161 (88.5%)	43 (95.6%)	4 (100.0%)	5 (71.4%)	19 (79.2%)	71 (88.8%)	43 (100.0%)	16 (80.0%)	16 (66.7%)	15 (100.0%)	90 (88.2%)
Ciprofloxacin	170 (93.4%)	45 (100.0%)	4 (100.0%)	7 (100.0%)	24 (100.0%)	80 (100%)	43 (100.0%)	16 (80.0%)	16 (66.7%)	15 (100.0%)	90 (88.2%)
Enrofloxacin	170 (93.4%)	45 (100.0%)	4 (100.0%)	7 (100.0%)	24 (100.0%)	80 (100%)	43 (100.0%)	16 (80.0%)	16 (66.7%)	15 (100.0%)	90 (88.2%)
Clindamycin	3 (1.6%)	0 (0.0%)	0 (0.0%)	0 (0.0%)	0 (0.0%)	0 (0.0%)	3 (7.0%)	0 (0.0%)	0 (0.0%)	0 (0.0%)	3 (3.0%)
Gentamicin	8 (4.4%)	6 (13.3%)	0 (0.0%)	0 (0.0%)	0 (0.0%)	6 (7.5%)	2 (4.7%)	0 (0.0%)	0 (0.0%)	0 (0.0%)	2 (2.0%)
Florfenicol	1 (0.5%)	0 (0.0%)	0 (0.0%)	0 (0.0%)	0 (0.0%)	0 (0.0%)	1 (2.3%)	0 (0.0%)	0 (0.0%)	0 (0.0%)	1 (1.0%)
Tetracycline	103 (56.6%)	29 (64.4%)	4 (100.0%)	7 (100.0%)	20 (83.3%)	60 (75.0%)	23 (53.5%)	6 (30.0%)	5 (20.8%)	9 (60.0%)	43 (42.2%)
Ampicillin	133 (73.1%)	35 (77.8%)	4 (100.0%)	5 (71.4%)	18 (75.0%)	62 (77.5%)	39 (90.7%)	12 (60.0%)	8 (33.3%)	12 (80.0%)	71 (69.6%)

Isolates that were resistant to more than two antimicrobial classes were defined as MDR isolates; 57.5% (46/80) of *C. coli* and in 34.3% (35/102) of *C. jejuni* were identified as MDR isolates (Table 3). The most common multidrug resistance pattern in *Campylobacter* spp. was the resistance to quinolones/fluoroquinolones (NAL, CIP, ENR), tetracyclines (TET), and penicillin (AMP). This pattern was observed at all stages of the chicken supply chain. Furthermore, 19.8% (16/81) of MDR isolates were resistant to both AZM and CIP, and they were only detected in samples from breeder farms.

3.3. Pulsed-Field Gel Electrophoresis Profiles

After analyzing the PFGE results, the 182 isolates were subdivided into 86 PFGE types with 100% similarity (Table 4). Two predominant types (types 6 and 10) of *C. coli* were associated with six isolates and three predominant types (types 17, 19, and 20) of *C. jejuni* were with nine, 10, and eight isolates, respectively. Most types of *C. coli* (24, 54.5%) and *C. jejuni* (24, 57.1%) were shared with one isolate. There was genotype diversity of the isolates for both *C. coli* and *C. jejuni* in the poultry production chain, with the highest diversity being detected at the breeder stage. The breeder farms carried a large variety of PFGE types, with 30 and 24 types of *C. coli* (Supplementary Figure S1) and *C. jejuni*, respectively (Supplementary Figure S2). The cross-contamination of *C. coli* and *C. jejuni* isolates was common among breeder farms of different production lines. Herein, PFGE type 27 of *C. coli* and PFGE type 9 of *C. jejuni* in a breeder farm were simultaneously found in production lines 4 and 8, which indicated a high frequency of cross-contamination between the two production lines. Moreover, PFGE type 27 of *C. coli* was found in three different production lines (lines 3, 4, and 8), and PFGE types 8 and 21 were found in two different production lines (lines 6 and 7 and lines 6 and 8, respectively). PFGE types 9, 19, and 36 of *C. jejuni* were found in two different production lines, which are lines 4 and 8, lines 1 and 3, and lines 1 and 2, respectively.

The same genotype (type 11 of *C. jejuni*) was simultaneously detected from both breeder farm and retail meat, even from different production lines. PFGE type 19 of *C. jejuni* was found in four different production lines (lines 1, 3, 4, and 5), providing the evidence of contamination across the farm stage, including at the breeder and broiler farm stages. Furthermore, serious cross-contamination between different production lines was found at the slaughterhouse stage. PFGE types 16 and 18 of *C. jejuni* were first found in lines 6 and 9 in slaughterhouses and they were later recovered from retail chicken meat in line 2 and lines 6 and 8, respectively. Some of the new PFGE types (*C. coli* from lines 1, 2, 3, 5, 7, and 9; *C. jejuni* from lines 2, 3, 6, 7, 8, and 10) were recovered from retail meat products but were not detected in previous stages of the same production line. Only one PFGE type (type 19 of *C. jejuni*) continuously existed from breeder farm to slaughterhouse, even in different lines, but the rest of the types did not persist across different stages until the final product. Some PFGE types, such as types 8, 11, 16, 18, and 36 of *C. jejuni*, were sparsely detected from different production lines and stages. Fifteen out of 44 types of *C. coli* and 21 types out of 42 types of *C. jejuni* were considered to be non-MDR isolates. Twelve PFGE types out of 86 were MDR isolates, including those non-sensitive to azithromycin and ciprofloxacin, as identified using human therapeutic treatment.

Table 3. Antimicrobial resistance patterns of *Campylobacter coli* and *C. jejuni* isolates from the chicken production chain.

No. of Antimicrobial Agents	Antimicrobial Resistance Pattern	n [a] (%)	No. of *C. coli* in Each Stage				No. of *C. jejuni* in Each Stage			
			Breeder	Broiler	Slaughterhouse	Retail Meat	Breeder	Broiler	Slaughterhouse	Retail Meat
	Susceptible	7 (3.8)							7	
1	AMP	5 (2.7)						4	1	
2	CIP+ENR	2 (1.1)	1			1				
3	NAL+CIP+ENR	27 (14.8)	9				4	8	5	1
3	CIP+ENR+AMP	2 (1.1)	1			1				
4	NAL+CIP+ENR+AMP	44 (24.2)	12			2	16	2	7	5
4	NAL+CIP+ENR+TET	14 (7.7)			3	5	1		3	2
4	CIP+ENR+TET+AMP	5 (2.7)			2	3				
5	NAL+CIP+ENR+TET+AMP	54 (29.7)	12	4	1	11	12	6	2	6
5	NAL+CIP+ENR+GEN+AMP	1 (0.5)	1							
6	NAL+CIP+ENR+GEN+TET+AMP	4 (2.2)	3				1			
6	AZM+NAL+CIP+ENR+TET+AMP	1 (0.5)					1			
6	NAL+CIP+ENR+FFN+TET+AMP	1 (0.5)					1			
7	AZM+ERY+NAL+CIP+ENR+TET+AMP	4 (2.2)	1				3			
7	AZM+NAL+CIP+ENR+GEN+TET+AMP	1 (0.5)	1							
8	AZM+ERY+NAL+CIP+ENR+GEN+TET+AMP	2 (1.1)	1				1			
8	AZM+ERY+TEL+NAL+CIP+ENR+TET+AMP	6 (3.3)	6							
8	AZM+ERY+NAL+CIP+ENR+CLI+TET+AMP	2 (1.1)					2			

AZM: Azithromycin, ERY: Erythromycin, TEL: Telithromycin, NAL: Nalidixic acid, CIP: Ciprofloxacin, ENR: Enrofloxacin, CLI: Clindamycin, GEN: Gentamicin, FFN: Florfenicol, TET: Tetracycline, AMP: Ampicillin. [a] Number of *Campylobacter* spp.

Table 4. Pulsed-field gel electrophoresis type of multi-drug resistant (MDR) *Campylobacter coli* and *C. jejuni* isolates from different lines along the chicken production chain.

Line	MDR *C. coli*				MDR *C. jejuni*			
	Breeder	Broiler	Slaughterhouse	Retail Meat	Breeder	Broiler	Slaughterhouse	Retail Meat
1	12[a],17[a],29[a],38		25[c], 30[c]	20, 32	19[a], 27[b], 33[a], 36, 37[b], 41[a]			
2	3			6[c], 7, 13[a]	2, 5, 10, 12, 36			16[a], 25
3	2, 11[a], 18[b], 27[a], 42[a], 43[a], 44			5, 19[b,c]	3[a], 19[b,c], 42		38[a]	15, 40[a]
4	27[a], 28[a], 37[a]				9, 28[b], 32[a]		19[a]	
5	22[b],26	23, 24	33	34	13, 26[b], 29[b,c], 30[a]	19[a]		
6	4[a], 8, 14[b], 15[b], 21, 39[c], 40[c]				11[a]	35, 36	3[a], 4[a], 16[a]	18[a]
7	1, 8, 9[a], 41			10[c], 34[a]	24[a], 34[a], 39[a]	22[a]	8[a]	6, 11, 21[a], 31
8	16, 21, 27[a], 31[b], 35[a], 36[a]				1, 9			18[a]
9	-			10	-	22[a], 23[a]	14, 18, 20[a]	
10	-				-	17[a]		7[c], 8[a]

[a] Non-MDR strain. [b] Including non-susceptible to azithromycin and/or ciprofloxacin. [c] Including non-MDR strain. - Sampling was not included.

4. Discussion

Campylobacter is the most common gastroenteritis-causing pathogen worldwide. Foodborne transmission accounts for most cases of *Campylobacter* infection, and up to 80% of *Campylobacter* infections can be attributed to the consumption of poultry, particularly the consumption of contaminated chicken meat [24]. This study shows that monitoring the distribution of antimicrobial resistant *Campylobacter* and its resistance patterns and tracing the route of transmission from comprehensive longitudinal sampling in the whole production stages are important for better understanding the occurrence resistant *Campylobacter* contamination.

A previous study on *Campylobacter* emergence suggested that *Campylobacter* contamination is due to vertical and horizontal transfer in broiler farms [4]. In the present study, all of the production stages, except hatchery, were contaminated with *Campylobacter* (Figure 1). The finding of a hatchery being *Campylobacter*-negative, despite a *Campylobacter*-positive parent flock, indicates that vertical transmission is not a major infectious route as it was in previous studies [5,25]. Furthermore, horizontal transmission comes across as a major potential source of flock infection via feed, litter, water, footwear, and chicken sheds [4].

The implementation of strict biosecurity practices was considered to be effective method to prevent or delay *Campylobacter* colonization in broiler chickens during the rearing period. In addition, low prevalence of *Campylobacter* isolates from broiler farms in this study could be due to a short rearing time of about 30 days before slaughter [26]. This result was consistent with the report that identified slaughter age as a risk factor for *Campylobacter* colonization in broiler chickens and suggested that reducing the rearing period of broiler chicken would decrease the prevalence of *Campylobacter* [27]. However, when compared with the low prevalence of *Campylobacter* in broiler farms and slaughterhouses, the isolation rate rapidly increased in retail meat samples in this study. These results were in accordance with the finding that suggested the possibility of contamination during slaughter [7,9,28]. Therefore, *Campylobacter* control in poultry faces many hurdles that need to be overcome and probably several strategies will have to be combined in order to achieve this goal. Although the best way to reduce *Campylobacter* contamination in chicken carcasses is to prevent colonization in the broiler house, an effective, suitable, and reliable strategy to eradicate this foodborne pathogen should focus not only on rearing farms, but also on the subsequent stages [29].

Most *Campylobacter* isolates (175/182) were resistant to at least one antimicrobial agent. Notably, extremely high resistance to nalidixic acid (88.5%), ciprofloxacin (93.4%), and enrofloxacin (93.4%) were found in this study, which is a finding that is consistent with previous studies [13,19,30]. In addition, 44.5% (81/182) of isolates showed multidrug resistance, and 16 isolates (16/81, 19.8%) were resistant to both azithromycin and ciprofloxacin. Extremely high resistance to fluoroquinolones and a steady increase in macrolide resistance would pose a serious public health threat of the transmission of such resistant *Campylobacter* through the chicken production stages [3]. Contrary to the high resistance to fluoroquinolones (>90%), tetracycline (56.6%), and ampicillin (73.1%), low resistance to gentamicin (4.4%) and florfenicol (0.5%) was identified in this study, which is consistent with the findings of a previous study [31]. Although these antimicrobials (gentamicin and chloramphenicol) are not the routine choice of treatment for human *Campylobacter* infection, increasing the resistance to the first-line antimicrobials and the decline in newly developed antimicrobials necessitated the monitoring of these alternative antimicrobial agents; this is because monitoring antimicrobial resistance is crucial in establishing the prevention and control measures in order to limit the dissemination of the resistant isolates. Thus, enhanced monitoring of *Campylobacter* resistance to these antimicrobials is required in order to better prevent infections that are caused by resistant pathogens and protect public health.

In general, *C. jejuni* was reported as the predominant *Campylobacter* species in poultry. However, our results showed a similar prevalence of *C. coli* and *C. jejuni*. Similar results, showing that the prevalence of *C. coli* was similar to that of *C. jejuni* or that *C. coli* showed an even higher prevalence than *C. jejuni* in poultry, have been reported in China, Thailand, and Reunion Island, among other places [17,32,33]. In addition, *C. coli* always showed higher antimicrobial resistance than *C. jejuni* and, accordingly, the choice of disinfectants and antimicrobials used in farms could be targeted at certain *Campylobacter* populations [32,34,35]. This study shows that *C. coli* demonstrated higher rates of antimicrobial resistance than *C. jejuni* in accordance with previous studies [32,34,35]. Our results suggest that the use of antimicrobial agents, such as ampicillin, florfenicol, and tetracycline, in farms may lead to favorably selected antimicrobial resistant *C. coli* being higher in prevalence than *C. jejuni*. It poses a potential public health threat and, thus, should be monitored in high priority in order to control the widespread of *C. coli*.

In this study, the genetic diversity among *Campylobacter* isolates and the presence of *Campylobacter* isolates along the chicken meat supply chain were evaluated. In contrast to other studies, the discontinuous appearance of *Campylobacter* and the diversity of PFGE types of isolates were mostly present along the entire chicken production process [7,36]. This result suggests that various contamination sources, such as wild animals, insects, farm staff, transport vehicles, and slaughtering environment, and equipment with *Campylobacter*-positive flocks, exacerbate the risk of bringing new resistant isolates into the chicken production stages [4]. In addition, genetic instability has been reported in *Campylobacter* isolates that are highly sensitive to environmental stress both in farms and slaughterhouse [37,38]. Furthermore, we have to acknowledge the limitations that are associated with our small sample sizes for slaughterhouse and retail meat, particularly when compared to the whole flock, which has about 15,000–20,000 broiler chickens; however, a larger number of samples was acquired for several sampling sites [39,40].

During the processing of poultry carcasses in slaughterhouse, cross-contamination between production lines seems to be relatively frequent. We found that PFGE type 16 of *C. jejuni* isolates from line 6 in the slaughterhouse was re-isolated from retail meat of line 2. Furthermore, PFGE type 18 of *C. jejuni* isolates from line 9 in the slaughterhouse was re-isolated from retail meat of lines 6 and 8. In addition, a *Campylobacter*-free flock could become positive after processing in the slaughterhouse. From our results, retail meat from 30% (3/10) flocks became *Campylobacter*-positive, even when these flocks (lines 2, 8, and 10) were negative at earlier stages. The primary source of contamination of *Campylobacter* for these *Campylobacter*-free flocks may be the *Campylobacter*-positive flocks that were slaughtered on previous days. These results suggested that some strains of *Campylobacter* form biofilms outside the host and may form a film on metal, glass, or rubber surfaces in the slaughterhouse; furthermore, *Campylobacter* can survive in the slaughter environment, even after cleaning with disinfectants [41–43]. Some surviving isolates could persist up to three weeks in the slaughterhouse environment, and these colonies could pose a high contamination risk to the following chicken flock [7,44]. These interventions at the slaughterhouse stage are an urgent requirement, as current interventions against *Campylobacter* contamination during poultry slaughter are not implemented in Asia [45].

We also noted the spread of the same genotype (PFGE type 11 of *C. jejuni*) that was isolated from breeder chicken and retail meat from different production lines. This result corroborated a previous study, which reported that the breeder chicken was the reservoir of *Campylobacter* with antimicrobial resistance and that these resistant *Campylobacter* may horizontally or vertically spread to retail meat along the chicken production stages [4,25]; our results are also in agreement with another study, stating that these resistant *Campylobacter* isolates could be directly transmitted to people who come into direct contact with the contaminated breeder chicken [18]. In agreement with previous studies, our results showed that *Campylobacter* isolates from the breeder chicken had higher antimicrobial resistance than those from broiler chicken; furthermore, the breeder chicken also had a higher prevalence of isolates with co-resistance to azithromycin and ciprofloxacin than broiler chicken [46]. With a high possibility to obtain antimicrobial treatment in its long life cycle, breeder chicken could accumulate MDR isolates and would be a persistently infected source for spreading the MDR isolates to the environment and downstream broiler chicken or retail meat via horizontal transmission [25,46]. We also noticed that multiple *Campylobacter* genotypes were shared between different breeder farms (types 8, 21, and 27 of *C. coli* and types 9, 19, and 36 of *C. jejuni*), despite the high biosecurity measures being implemented in breeder chicken farms in South Korea [47]. Moreover, one PFGE type (type 19 of *C. jejuni* from lines 1 and 3) was transmitted from the breeder farm to downstream production stages beyond production lines. The presence of the same genotypes at different production stages and in different lines highlights a common source from the same company that could be shared during the transport of birds, feeding, and veterinary visits, among other ways [4]. Based on these factors, the circulation of specific genotypes in an integrated production system could occur. This result was also supported by the fact that the long life cycle of breeder chicken increased the risk of pathogen exchange by increasing the number of encounters among breeder chicken farms [47]. Therefore, the breeder chicken cannot be excluded from the antimicrobial resistance monitoring program to limit and prevent the spread of resistant *Campylobacter*.

5. Conclusions

In conclusion, we found that the significant contamination of antimicrobial-resistant *Campylobacter* was prevalent at all production stages, except at the hatchery stage; moreover, the transmission of *Campylobacter* occurred by multiple routes and it induced a variety of genotypes. To our knowledge, this is the first report on the occurrence of antimicrobial-resistant *Campylobacter* investigated longitudinally from breeder farms to retail meat along the chicken supply chain in Korea. High prevalence of antimicrobial-resistant *Campylobacter* in breeder farms according to the bird age suggests that epidemiological investigations should include breeder farms, which could be a source of transmission of antimicrobial-resistant *Campylobacter*, including the antimicrobials that were used in human treatment, in the chicken supply chain. According to the PFGE results, new types were mainly introduced at farm and slaughterhouse stages with numerous factors that resulted in the accumulation of various genotypes. In particular, the slaughtering process may contaminate *Campylobacter*-negative flocks with various genotypes by the end of the process. These findings indicated that further studies are necessary in order to figure out the contamination factors or routes from rearing farm and slaughterhouse and develop interventions targeting slaughterhouse for improving food safety and public health.

Supplementary Materials: The following are available online at https://www.mdpi.com/2076-2615/11/2/246/s1, Table S1. Types and number of samples collected throughout the chicken production chain. Figure S1 A dendrogram of *Campylobacter coli* *Sma*I-PFGE patterns isolated from the chicken production chain and antimicrobial resistance. Isolate names are the following: S: cloacal swab sample; F: feed sample; from 189-C to 189-I: slaughter processing environment; w: age in weeks; d: age in days. Figure S2. A dendrogram of *Campylobacter jejuni* *Sma*I-PFGE patterns isolated from the chicken production chain and antimicrobial resistance. Isolate names are the following: S: cloacal swab sample; F: feed sample; W: water sample; L: litter sample; from 198-A to 198-H, 196-A, and from 194-E-1 to 194-J-1: slaughter processing environment; w: age in weeks; d: age in days.

Author Contributions: B.W., M.K., and H.-K.J. contributed to conception and design of experiments. S.-Y.C. and B.-R.K. contributed to acquisition, analysis, and interpretation of data. B.-R.K., B.W., S.-Y.C., J.-F.Z., K.S., M.K., and H.-K.J. drafted and/or revised the article. All authors have read and agreed to the published version of the manuscript.

Funding: This work was supported by Korea Institute of Planning and Evaluation for Technology in Food, Agriculture and Forestry (IPET) through Agriculture, Food and Rural Affairs Convergence Technologies Program for Educating Creative Global Leader (716002-7, 320005-4) and Animal Disease Management Technology Development Program (119059-2), funded by Ministry of Agriculture, Food and Rural Affairs (MAFRA). And this paper was supported by Basic Science Research Program through the National Research Foundation of Korea(NRF) funded by the Ministry of Education(No.2019R1I1A1A01060147). And this research was supported by "Research Base Construction Fund Support Program" funded by Jeonbuk National University in 2020.

Institutional Review Board Statement: The study was conducted according to the guidelines of Guide for the Care and Use of Laboratory Animals 2014, Korean Ministry of Food and Drug Safety and regulations of the Korean Council on Animal Care and Korean Animal Protection Law, 2015; Article 23 (Experiments with Animals). No chickens were killed and sampling was carried by a veterinarian with prior consent of the farmer/manager of the facilities.

Data Availability Statement: The data presented in this study are available from the corresponding author on reasonable request.

Conflicts of Interest: The authors declare no conflict of interest.

References

1. European Food Safety Authority (EFSA). The European Union Summary Report on Antimicrobial Resistance in zoonotic and indicator bacteria from humans, animals and food in 2017/2018. *EEFSA J.* **2020**, *18*, e06007. [CrossRef]
2. Young, K.T.; Davis, L.M.; DiRita, V.J. *Campylobacter jejuni*: Molecular biology and pathogenesis. *Nat. Rev. Microbiol.* **2007**, *5*, 665–679. [CrossRef] [PubMed]
3. Skarp, C.P.A.; Hanninen, M.L.; Rautelini, H.I.K. Campylobacteriosis: The role of poultry meat. *Clin. Microbiol. Infect.* **2016**, *22*, 103–109. [CrossRef] [PubMed]
4. Sahin, O.; Morishita, T.Y.; Zhang, Q. *Campylobacter* colonization in poultry: Sources of infection and modes of transmission. *Anim. Health Res. Rev.* **2002**, *3*, 95–105. [CrossRef] [PubMed]
5. Callicott, K.A.; Fridriksdottir, V.; Reiersen, J.; Lowman, R.; Bisaillon, J.R.; Gunnarsson, E.; Berndtson, E.; Hiett, K.L.; Needleman, D.S.; Stern, N.J. Lack of evidence for vertical transmission of *Campylobacter* spp. in chickens. *Appl. Environ. Microbiol.* **2006**, *72*, 5794–5798. [CrossRef] [PubMed]
6. Rosenquist, H.; Sommer, H.M.; Nielsen, N.L.; Christensen, B.B. The effect of slaughter operations on the contamination of chicken carcasses with thermotolerant *Campylobacter*. *Int. J. Food Microbiol.* **2006**, *108*, 226–232. [CrossRef]
7. Melero, B.; Juntunen, P.; Hanninen, M.L.; Jaime, I.; Rovira, J. Tracing *Campylobacter jejuni* strains along the poultry meat production chain from farm to retail by pulsed-field gel electrophoresis, and the antimicrobial resistance of isolates. *Food Microbiol.* **2012**, *32*, 124–128. [CrossRef]
8. Seliwiorstow, T.; Bare, J.; Berkvens, D.; Van Damme, I.; Uyttendaele, M.; De Zutter, L. Identification of risk factors for *Campylobacter* contamination levels on broiler carcasses during the slaughter process. *Int. J. Food Microbiol.* **2016**, *226*, 26–32. [CrossRef]
9. Roccato, A.; Mancin, M.; Barco, L.; Cibin, V.; Antonello, K.; Cocola, F.; Ricci, A. Usefulness of indicator bacteria as potential marker of *Campylobacter* contamination in broiler carcasses. *Int. J. Food Microbiol.* **2018**, *276*, 63–70. [CrossRef]
10. Lehtopolku, M.; Nakari, U.M.; Kotilainen, P.; Huovinen, P.; Siitonen, A.; Hakanen, A.J. Antimicrobial susceptibilies of multidrug-resistant *Campylobacter jejuni* and *C. coli* strains: In vitro activities of 20 antimicrobial agents. *Antimicrob. Agents Chemother.* **2010**, *54*, 1232–1236. [CrossRef]
11. Animal and Plant Quarantine Agency. *Establishment of Antimicrobial Resistance Surveillance System for Livestock, 2018*; Ministry of Agriculture, Food and Rural Affairs: Gimcheon and Sejong, Korea, 2019.
12. Aarestrup, F.M.; Wegener, H.C. The effects of antibiotic usage in food animals on the development of antimicrobial resistance of importance for humans in *Campylobacter* and *Escherichia coli*. *Microbes. Infect.* **1999**, *1*, 639–644. [CrossRef]
13. Han, K.; Jang, S.S.; Choo, E.; Heu, S.; Ryu, S. Prevalence, genetic diversity, and antibiotic resistance patterns of *Campylobacter jejuni* from retail raw chickens in Korea. *Int. J. Food Microbiol.* **2007**, *114*, 50–59. [CrossRef]
14. Kim, H.J.; Kim, J.H.; Kim, Y.I.; Choi, J.S.; Park, M.Y.; Nam, H.M.; Jung, S.C.; Kwon, J.W.; Lee, C.H.; Kim, Y.H.; et al. Prevalence and characterization of *Campylobacter* spp. isolated from domestic and imported poultry meat in Korea, 2004–2008. *Foodborne Pathog. Dis.* **2010**, *7*, 1203–1209. [CrossRef]
15. Wieczorek, K.; Denis, E.; Osek, J. Comparative analysis of antimicrobial resistance and genetic diversity of *Campylobacter* from broilers slaughtered in Poland. *Int. J. Food Microbiol.* **2015**, *210*, 24–32. [CrossRef] [PubMed]
16. Wurfel, S.F.R.; da Silva, W.P.; de Oliveira, M.G.; Kleinubing, N.R.; Lopes, G.V.; Gandra, E.A.; Dellagostin, O.A. Genetic diversity of *Campylobacter jejuni* and *Campylobacter coli* isolated from poultry meat products sold on the retail market in Southern Brazil. *Poult. Sci.* **2019**, *98*, 932–939. [CrossRef] [PubMed]
17. Prachantasena, S.; Charununtakorn, P.; Muangnoicharoen, S.; Hankla, L.; Techawal, N.; Chaveerach, P.; Tuitemwong, P.; Chokesajjawatee, N.; Williams, N.; Humphrey, T.; et al. Distribution and genetic profiles of *Campylobacter* in commercial broiler production from breeder to slaughter in Thailand. *PLoS ONE* **2016**, *11*, e0149585. [CrossRef] [PubMed]
18. Oh, J.Y.; Kwon, Y.K.; Wei, B.; Jang, H.K.; Lim, S.K.; Kim, C.H.; Jung, S.C.; Kang, M.S. Epidemiological relationships of *Campylobacter jejuni* strains isolated from humans and chickens in South Korea. *J. Microbiol.* **2017**, *55*, 13–20. [CrossRef] [PubMed]
19. Wei, B.; Cha, S.Y.; Yoon, R.H.; Kang, M.; Roh, J.H.; Seo, H.S.; Lee, J.A.; Jang, H.K. Prevalence and antimicrobial resistance of *Campylobacter* spp. isolated from retail chicken and duck meat in South Korea. *Food Control* **2016**, *62*, 63–68. [CrossRef]
20. Wei, B.; Cha, S.Y.; Kang, M.; Roh, J.H.; Seo, H.S.; Yoon, R.H.; Jang, H.K. Antimicrobial susceptibility profiles and molecular typing of *Campylobacter jejuni* and *Campylobacter coli* isolates from ducks in South Korea. *Appl. Environ. Microbiol.* **2014**, *80*, 7604–7610. [CrossRef]

21. Clinical and Laboratory Standards Institute (CLSI). *Methods for Antimicrobial Susceptibility Testing of Infrequently Isolated or Fastidious Bacteria Isolated from Animals*, 1st ed.; CLSI: Wayne, PA, USA, 2016.
22. National Antimicrobial Resistance Monitoring System (NARMS). *NARMS Retail Meat Annual Report, 2011*; Food and Drug Administration: Rockville, MD, USA, 2013.
23. Clinical and Laboratory Standards Institute (CLSI). *Performance Standards for Antimicrobial Disk and Dilution Susceptibility Tests for Bacteria Isolated from Animals*, 4th ed.; CLSI: Wayne, PA, USA, 2018.
24. Igwaran, A.; Okoh, A.I. Human campylobacteriosis: A public health concern of global importance. *Heliyon* **2019**, 5. [CrossRef]
25. Jacobs-Reitsma, W.F. *Campylobacter* bacteria in breeder flocks. *Avian Dis.* **1995**, *39*, 355–359. [CrossRef]
26. Ingresa-Capaccioni, S.; Jimenez-Trigos, E.; Marco-Jimenez, F.; Catala, P.; Vega, S.; Marin, C. *Campylobacter* epidemiology from breeders to their progeny in Eastern Spain. *Poult. Sci.* **2016**, *95*, 676–683. [CrossRef]
27. Bouwknegt, M.; van de Giessen, A.W.; Dam-Deisz, W.D.; Havelaar, A.H.; Nagelkerke, N.J.; Henken, A.M. Risk factors for the presence of *Campylobacter* spp. in Dutch broiler flocks. *Prev. Vet. Med.* **2004**, *62*, 35–49. [CrossRef]
28. Perez-Arnedo, I.; Gonzalez-Fandos, E. Prevalence of *Campylobacter* spp. in poultry in three Spanish farms, a slaughterhouse and a further processing plant. *Foods* **2019**, *8*, 111. [CrossRef]
29. Rasschaert, G.; De Zutter, L.; Herman, L.; Heyndrickx, M. *Campylobacter* contamination of broilers: The role of transport and slaughterhouse. *Int. J. Food Microbiol.* **2020**, *322*, 108564. [CrossRef] [PubMed]
30. Wieczorek, K.; Osek, J. Antimicrobial resistance mechanisms among *Campylobacter*. *BioMed Res. Int.* **2013**. [CrossRef] [PubMed]
31. Kassa, T.; Gebre-Selassie, S.; Asrat, D. Antimicrobial susceptibility patterns of thermotolerant *Campylobacter* strains isolated from food animals in Ethiopia. *Vet. Microbiol.* **2007**, *119*, 82–87. [CrossRef] [PubMed]
32. Ma, L.; Wang, Y.; Shen, J.; Zhang, Q.; Wu, C. Tracking *Campylobacter* contamination along a broiler chicken production chain from the farm level to retail in China. *Int. J. Food Microbiol.* **2014**, *181*, 77–84. [CrossRef] [PubMed]
33. Henry, I.; Reichardt, J.; Denis, M.; Cardinale, E. Prevalence and risk factors for *Campylobacter* spp. in chicken broiler flocks in Reunion Island (Indian Ocean).)*Prev. Vet. Med.* **2011**, *100*, 64–70. [CrossRef]
34. Gibreel, A.; Taylor, D.E. Macrolide resistance in *Campylobacter jejuni* and *Campylobacter coli*. *J. Antimicrob. Chemother.* **2006**, *58*, 243–255. [CrossRef]
35. Karki, A.B.; Marasini, D.; Oakey, C.K.; Mar, K.; Fakhr, M.K. *Campylobacter coli* from retail liver and meat products is more aerotolerant than *Campylobacter jejuni*. *Front. Microbiol.* **2018**, 9. [CrossRef]
36. Hansson, I.; Ederoth, M.; Andersson, L.; Vagsholm, I.; Olsson Engvall, E. Transmission of *Campylobacter* spp. to chickens during transport to slaughter. *J. Appl. Microbiol.* **2005**, *99*, 1149–1157. [CrossRef]
37. Manning, G.; Duim, B.; Wassenaar, T.; Wagenaar, J.A.; Ridley, A.; Newell, D.G. Evidence for a genetically stable strain of *Campylobacter jejuni*. *Appl. Environ. Microbiol.* **2001**, *67*, 1185–1189. [CrossRef]
38. Rivoal, K.; Ragimbeau, C.; Salvat, G.; Colin, P.; Ermel, G. Genomic diversity of *Campylobacter coli* and *Campylobacter jejuni* isolates recovered from free-range broiler farms and comparison with isolates of various origins. *Appl. Environ. Microbiol.* **2005**, *71*, 6216–6227. [CrossRef]
39. Bull, S.A.; Allen, V.M.; Domingue, G.; Jorgensen, F.; Frost, J.A.; Ure, R.; Whyte, R.; Tinker, D.; Corry, J.E.; Gillard-King, J.; et al. Sources of *Campylobacter* spp. colonizing housed broiler flocks during rearing. *Appl. Environ. Microbiol.* **2006**, *72*, 645–652. [CrossRef]
40. Vidal, A.B.; Colles, F.M.; Rodgers, J.D.; McCarthy, N.D.; Davies, R.H.; Maiden, M.C.J.; Clifton-Hadley, F.A. Genetic Diversity of *Campylobacter jejuni* and *Campylobacter coli* isolates from conventional broiler flocks and the impacts of sampling strategy and laboratory method. *Appl. Environ. Microbiol.* **2016**, *82*, 2347–2355. [CrossRef] [PubMed]
41. Peyrat, M.B.; Soumet, C.; Maris, P.; Sanders, P. Recovery of *Campylobacter jejuni* from surfaces of poultry slaughterhouses after cleaning and disinfection procedures: Analysis of a potential source of carcass contamination. *Int. J. Food Microbiol.* **2008**, *124*, 188–194. [CrossRef] [PubMed]
42. Brown, H.L.; Reuter, M.; Salt, L.J.; Cross, K.L.; Betts, R.P.; van Vliet, A.H. Chicken juice enhances surface attachment and biofilm formation of *Campylobacter jejuni*. *Appl. Environ. Microbiol.* **2014**, *80*, 7053–7060. [CrossRef] [PubMed]
43. Rossler, E.; Olivero, C.; Soto, L.P.; Frizzo, L.S.; Zimmermann, J.; Rosmini, M.R.; Sequeira, G.J.; Signorini, M.L.; Zbrun, M.V. Prevalence, genotypic diversity and detection of virulence genes in thermotolerant *Campylobacter* at different stages of the poultry meat supply chain. *Int. J. Food Microbiol.* **2020**, *326*, 108641. [CrossRef]
44. Garcia-Sanchez, L.; Melero, B.; Jaime, I.; Hanninen, M.L.; Rossi, M.; Rovira, J. *Campylobacter jejuni* survival in a poultry processing plant environment. *Food Microbiol.* **2017**, *65*, 185–192. [CrossRef]
45. Premarathne, J.; Satharasinghe, D.A.; Huat, J.T.Y.; Basri, D.F.; Rukayadi, Y.; Nakaguchi, Y.; Nishibuchi, M.; Radu, S. Impact of human *Campylobacter* infections in Southeast Asia: The contribution of the poultry sector. *Crit. Rev. Food Sci. Nutr.* **2017**, *57*, 3971–3986. [CrossRef] [PubMed]
46. Perez-Boto, D.; Garcia-Pena, F.J.; Abad-Moreno, J.C.; Echeita, M.A. Antimicrobial susceptibilities of *Campylobacter jejuni* and *Campylobacter coli* strains isolated from two early stages of poultry production. *Microb. Drug Resist.* **2013**, *19*, 323–330. [CrossRef]
47. Choi, S.W.; Ha, J.S.; Kim, B.Y.; Lee, D.H.; Park, J.K.; Youn, H.N.; Hong, Y.H.; Lee, S.B.; Lee, J.B.; Park, S.Y.; et al. Prevalence and characterization of *Salmonella* species in entire steps of a single integrated broiler supply chain in Korea. *Poult. Sci.* **2014**, *93*, 1251–1257. [CrossRef]

Article

Raw Cow Milk Bacterial Consortium as Bioindicator of Circulating Anti-Microbial Resistance (AMR)

Cristian Piras [1,*], Viviana Greco [2,3], Enrico Gugliandolo [4], Alessio Soggiu [5], Bruno Tilocca [1], Luigi Bonizzi [5], Alfonso Zecconi [5], Rainer Cramer [6], Domenico Britti [1], Andrea Urbani [2,3] and Paola Roncada [1,*]

[1] Department of Health Sciences, "Magna Græcia University" of Catanzaro, Campus Universitario "Salvatore Venuta" Viale Europa, 88100 Catanzaro, Italy; tilocca@unicz.it (B.T.); britti@unicz.it (D.B.)

[2] Department of Basic Biotechnological Sciences, Intensivological and Perioperative Clinics, Università Cattolica del Sacro Cuore, Largo Francesco Vito 1, 00168 Roma, Italy; viviana.greco@unicatt.it (V.G.); andrea.urbani@unicatt.it (A.U.)

[3] Dipartimento di Scienze di Laboratorio e Infettivologiche, Fondazione Policlinico Universitario Agostino Gemelli IRCCS, Largo A. Gemelli 8, 00168 Roma, Italy

[4] Department of Veterinary Sciences, University of Messina, 98168 Messina, Italy; egugliandolo@unime.it

[5] Surgical and Dental Sciences-One Health Unit, Department of Biomedical, University of Milano, Via Celoria 10, 20133 Milano, Italy; alessio.soggiu@unimi.it (A.S.); luigi.bonizzi@unimi.it (L.B.); alfonso.zecconi@unimi.it (A.Z.)

[6] Department of Chemistry, University of Reading, Reading RG6 6DX, UK; r.k.cramer@reading.ac.uk

* Correspondence: c.piras@unicz.it (C.P.); roncada@unicz.it (P.R.); Tel.: +39-096-1369-4236 (C.P.); +39-096-1369-4284 (P.R.)

Received: 20 November 2020; Accepted: 8 December 2020; Published: 11 December 2020

Simple Summary: Antimicrobials represent useful tools to fight bacterial infections that could harm human and animal health. Antimicrobial resistance occurs naturally or can be induced by the misuse of antibiotics. Its occurrence limits the efficiency of antibiotics and therefore the possibility to treat infections effectively. This can lead to an increasing severity of infectious diseases in humans and animals. Here, we describe the development of a workflow that provides a qualitative representation of the antimicrobial genes that are translated into proteins. Since proteins are ultimately the real effectors, the method herein described demonstrates that those genes are effectively enhancing antimicrobial resistance (AMR). The presented method is independent of any amplification step and provides useful information on the dynamics of the biochemical functions accomplished by the raw milk bacterial consortium.

Abstract: The environment, including animals and animal products, is colonized by bacterial species that are typical and specific of every different ecological niche. Natural and human-related ecological pressure promotes the selection and expression of genes related to antimicrobial resistance (AMR). These genes might be present in a bacterial consortium but might not necessarily be expressed. Their expression could be induced by the presence of antimicrobial compounds that could originate from a given ecological niche or from human activity. In this work, we applied (meta)proteomics analysis of bacterial compartment of raw milk in order to obtain a method that provides a measurement of circulating AMR involved proteins and gathers information about the whole bacterial composition. Results from milk analysis revealed the presence of 29 proteins/proteoforms linked to AMR. The detection of mainly β-lactamases suggests the possibility of using the milk microbiome as a bioindicator for the investigation of AMR. Moreover, it was possible to achieve a culture-free qualitative and functional analysis of raw milk bacterial consortia.

Keywords: proteomics; AMR; raw milk; microbiome; β-lactamase

1. Introduction

Bacteria are becoming more and more resistant to a greater number of antibiotics. Antimicrobial resistance (AMR) is a trait that can be horizontally transferred by previously resistant bacteria or can be generated by the occurrence of new mutations [1]. The complete panel of AMR genes present within a microbiome is defined as the "resistome". Moreover, bacteria can be intrinsically resistant to different classes of antibiotics conferring, to a given ecological niche, a certain level of resistance. The bacterial intrinsic resistome is defined as the entirety of elements contributing to antibiotic resistance regardless of previous exposure to antibiotics [2].

For example, soil microorganisms are carriers of resistance genes to many classes of antibiotics independently from human-derived antimicrobial pressure. The intrinsic resistome predates the clinical use of antibiotics posing the question whether AMR occurred earlier than the human antibiotics production and spread [3]. Naturally occurring AMR is related to the biological pressure of every ecological environment/niche that implicates the bacteria-bacteria competition or the bacteria-fungi competition. Penicillin was the first discovered antibiotic and is produced by the fungi of the genus *Penicillium*. Therefore, bacteria-fungi co-existence may have been the driver for the initial production and synthesis of the early forms of beta-lactamases.

Studies based on metagenomics and high-performance genome sequencing broadened the knowledge about bacterial genomes, leading to the discovery of a high concentration of AMR -related genes in many ecological niches. However, the detection of antimicrobial-related genes does not necessarily mean that those genes will be translated into proteins. Antimicrobial genes might be present within a bacterial consortium in the main genome or in the plasmids of the present species, but may remain silent, unless their expression is induced by the presence of antimicrobial compounds in the environment. The genes detected through next generation sequencing (NGS) methods may belong to bacterial species that are dead or unable to replicate. In order to have a deep knowledge of the composition and the biological functions of a microbial consortium, different investigation approaches need to be applied.

The study at protein level (protein expression level) is therefore necessary to assess the full functionality of a given microbial consortium. Mass spectrometry-based proteomics, and specifically metaproteomics, supported by the improved bioinformatic tools, allowed the detection of a high number of different proteins and proteoforms from different organisms in heterogeneous biological samples [4].

Metaproteomics studies represent a challenge for the computational resources because of the large dimension of the databases. Despite this challenge, we have already performed metaproteomics studies which proved to be efficient and reliable for the study of bacterial consortia of hard cheese [5] and of the gut microbiome of newborn mice [6]. The bacterial consortium of hard pasta cheese was enriched using an isoelectric precipitation of caseins to be discarded. The proteomics part was achieved using a bottom-up approach followed by the search against a database including all the bacterial protein sequences obtained from UNIPROT [7]. A similar approach was used to evaluate the diet-related shaping of the whole set of microorganisms present in the gut of newborn mice [6].

In case of raw, unpasteurized milk metaproteome, there are few challenges to overcome to successfully analyze the microbiome. First, unlike metagenomics, it is important to have a robust enrichment step because of the lack of amplification steps for proteins. Second, residual of both milk proteins and somatic cells (which include mainly leukocytes) proteins will be retained in the sample to be analyzed.

For the aforementioned reasons, the challenge of experimentally enriching the raw milk bacterial consortium was addressed with a rapid agitation step and the selective analysis of the bacterial proteins with a bottom-up proteomics approach coupled with database filtering. The main goal was to selectively investigate and demonstrate the expression of proteins related to AMR and, within the same experimental procedure, to evaluate the whole microbial composition up to the genre level.

2. Materials and Methods

2.1. Milk Sampling

Two bulk tank milk (BTM) samples were collected (each one in duplicate) at distance of 7 days in January 2018 from the official research facility for large animals of University of Milan "Azienda Agraria Didattico-Sperimentale "Angelo Menozzi"—Landriano (Pv)". The facility counts around 90 lactating cows. These two bulk milk samples were then used for two different extractions named extraction number 1 and extraction number 2.

For this step, 250 mL were taken from the top of the tank using a clean, sanitized dipper after the milk was agitated for 5–10 min as suggested [8]. One aliquot of both samples was delivered refrigerated to ARAL Laboratories for somatic cell count (SCC, 98,000 and 112,000 for the first and the second sample collected) that was performed by certified methods, currently applied by Italian Breeders Association (A.I.A.) on a Fossomatic FC (Foss DK) instrument. The second aliquot collected of each sample was kept at 4 °C and processed within 24 h for bacterial enrichment and metaproteomics analysis.

2.2. Bacterial Enrichment for Proteomics Analysis

For each sample, 160 mL of fresh milk were divided into sixteen 15 mL tubes (10 mL each tube) and horizontally placed over the plate of a FALC F320 stirrer for 10 min at 1600 rpm (Figure 1).

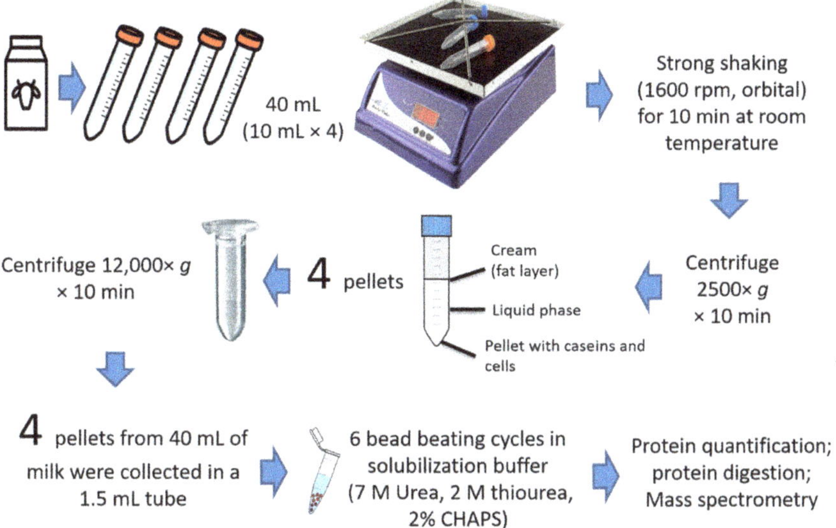

Figure 1. Workflow of the rapid bacterial enrichment method. The experimental phases of the whole procedure prior to the mass spectrometry analysis are described, starting with the separation of the bacterial fraction from the lipid fraction, to the collection and enrichment of the bacterial pellets.

After this step, the samples were kept in the same vials and centrifuged for 20 min at room temperature at 2500× g for cells and bacteria collection. A small red cellular pellet was visible in the bottom of the tube. The top layer (lipids) was removed with a spatula and the supernatant was discarded. Four pellets with the small amount of residual liquid were then gently mixed a pipette and merged in a 2 mL tube. This latter was centrifuged at 12,000× g at 4 °C for 20 min. Four of these obtained pellets (coming from 16 original 10 mL tubes) were then collected in one single 2 mL tube and centrifuged again at the same speed. The result is a cellular pellet collected from an original amount of 160 mL of raw milk. This method has been adapted from Brewster and Paul [9]. The supernatant was discarded, and the pellet was then solubilized with 300 μL of solubilization sample buffer (7M UREA, 2M Thiourea, 4% CHAPS). To ensure the complete disruption of the collected bacterial cells the samples were processed with 6 cycles of 1 min bead beating interspersed by a cycle of centrifuge (Figure 1). Bead beating steps were performed by adding to the sample the same amount (1:1 *v/w*) of 0.1 mm zyrcounium-sylica beads (300 μg beads added to 300 μL of buffer + the volume of the pellet). The bead beating cycle was performed for 1 min at 4000 rpm in order to avoid overheating. After this step, the samples were centrifuged for 5 min at 12,000× *g* at 4 °C to chill and disperse the foam. This cycle was repeated 6 times. After the 6th cycle, the samples were centrifuged for 20 min and the supernatant was saved in another tube and further processed for proteomics analysis.

2.3. Trypsin Digestion and Mass Spectrometry Analysis

Protein Digestion was performed according to the Filter-aided sample preparation (FASP) protocol described by Wiśniewski et al. [10] and optimized by Distler et al. [11] combining both the purification and digestion of the proteins.

Briefly, reduction (DTT 8 mM in urea buffer-8 M urea and 100 mM Tris), alkylation (IAA 50 mM in urea buffer 8 M urea and 100 mM Tris) and digestion by trypsin (final trypsin concentration of 0.01 μg/μL) were performed on filter tubes (Nanosep centrifugal device with Omega membrane-10 K MWCO).

Then, 0.25 μg of each digested samples were loaded in triplicate on a Symmetry C18 5 μm, 180 μm × 20 mm precolumn (Waters Corp., Milford, MA, USA) and subsequently separated by a 120 min reversed phase gradient at 300 nL/min (linear gradient, 2–40% ACN over 90 min) using a HSS T3 C18 1.8 μm, 75 μm × 150 mm nanoscale LC column (Waters Corp.) maintained at 40 °C.

Tryptic peptides were separated on an ACQUITY MClass System (Waters Corp.) and then separated using a High Definition Synapt G2-Si Mass spectrometer (Waters Corp) directly coupled to the chromatographic system.

The protein expression was evaluated by a high definition expression configuration mode (HDMS^E), a data-independent acquisition (DIA) protocol where ion mobility separation (IMS) was integrated into LC-MS^E workflow as described by Marini F. et al. [12].

The mass spectrometer parameters were set as: positive survey polarity of electrospray source (ES+), acquisition mode mass range 50–2000 *m/z*, capillary source voltage 3.2 kV, source T 80 °C, cone voltage 40 eV, TOF resolution power 20,000, precursor ion charge state 0.2–4, trap collision energy 4 eV, transfer collision energy 2 eV precursor MS scan time 0.5 s, and fragment MS/MS scan time 1.0 s. All spectra were acquired in IMS cycles with wave height at 40 V, wave velocity of 650 m/s, transfer wave height of 4 V, and transfer wave velocity of 175 m/s.

Data were post-acquisition lock mass corrected using the doubly charged monoisotopic ion of [Glu1]-Fibrinopeptide B (Waters), sampled every 30 s.

2.4. Bioinformatics and Metaproteomics

The LC-MS raw data from three replicate experiments for each sample/extraction were processed using the software ProteinLynx Global Server v. 3.0.3 (PLGS, Waters Corp.). The qualitative identification of proteins was obtained by searching two different databases: (i) *bacteria* (UniProt KB/Swiss-Prot Protein Knowledgebase restricted to *all Bacteria taxonomy*) and (ii) The Comprehensive AMR Database (CARD, https://card.mcmaster.ca/) as FASTA files [13,14].

Search parameters were set as: automatic tolerance for precursor ions and for product ions, minimum 1 fragment ions matched per peptide, minimum 3 fragment ions matched per protein, minimum 2 peptide matched per protein, 1 missed cleavage, carbamydomethylation of cysteines and oxidation of methionines as fixed and variable modifications, and a false discovery rate (FDR) of the identification algorithm under 1%.

The protein identifications were based on the detection of more than two fragment ions per peptide, more than two peptides measured per protein.

In addition, in order to validate the proteins of interest obtained by DIA analysis, a targeted label-free strategy was carried out using the freely available Skyline tool (MacCoss Lab Software, https://skyline.ms/project/home/software/Skyline/begin.view).

The qualitative and functional metaproteomics analysis was achieved using the peptides list obtained with PLGS. The obtained list was analyzed with UNIPEPT (https://unipept.ugent.be/) for each different extraction for the qualitative analysis [15].

The Venny 2.1.0 online tool (https://bioinfogp.cnb.csic.es/tools/venny/) was used for comparing lists with Venn Diagrams.

3. Results

3.1. Cow Milk Microbiome Analysis

As described in the methods section, the first experimental step was necessary to enrich the bacterial fraction. The sample preparation with bacterial enrichment was performed according to the scheme in Figure 1. Raw unpasteurized milk was vigorously agitated to detach the bacterial fraction from the lipids fraction. The samples were subsequently centrifuged to collect the bacterial pellet. This allowed a consistent enrichment of bacteria in a 30-min workflow.

The extraction procedure was performed separately on the first and second sample (extraction 1 and 2, respectively). Each extraction was then analyzed in triplicate via LC-MS/MS DIA integrated with ion mobility separation (IMS).

In order to identify the whole bacterial proteome, the obtained MS datasets were analyzed using different databases: UniProt KB/Swiss-Prot restricted to all reviewed *Bacteria* protein sequences (UniProt KB) and the Comprehensive Antibiotic Resistance Database (CARD) [14].

The technical replicates of the two different extractions were analyzed independently, and the results are shown in Figure 2. The composition of the microbiota showed a low degree of variability between the two extractions. This similarity was consistent up to the genre level (Firmicutes phylum, Lactobacillus genus). However, a higher degree of variability was found when the metaproteomics analysis was undertaken at the species level.

The peptide lists were obtained by searching the raw datasets against the whole bacterial database. The lists were then analyzed to determine the main molecular functions performed by the microbiome just before bacterial lysis. The 10 most probable functions (attributed by Unipept) executed by the whole milk microbiota are listed in Table 1.

Table 1. The 10 most represented Gene Ontology (GO) molecular functions of the analyzed microbiome.

Peptides	GO Term	Name
4001	GO:0005524	ATP binding
1985	GO:0003677	DNA binding
1595	GO:0046872	metal ion binding
1164	GO:0000287	magnesium ion binding
1061	GO:0008270	zinc ion binding
620	GO:0003899	DNA-directed 5′-3′ RNA polymerase activity
514	GO:0016787	hydrolase activity
511	GO:0000049	tRNA binding
481	GO:0005525	GTP binding
473	GO:0046933	proton-transporting ATP synthase activity, rotational mechanism

3.2. Resistome Proteins Analysis

The same raw MS dataset was then searched against the CARD 15 database. Figure 3 shows the Venn diagram of the proteins identified in the two extractions using the CARD 15 database. Based on the analytical parameters described in the methods, 35 proteins were identified combining both extractions. Specifically, 29 proteins were common to the two extractions corresponding to 82.9%

while 5.7% (2 proteins) and 11.4% (4 proteins) were found specifically in extraction 1 and in extraction 2, respectively.

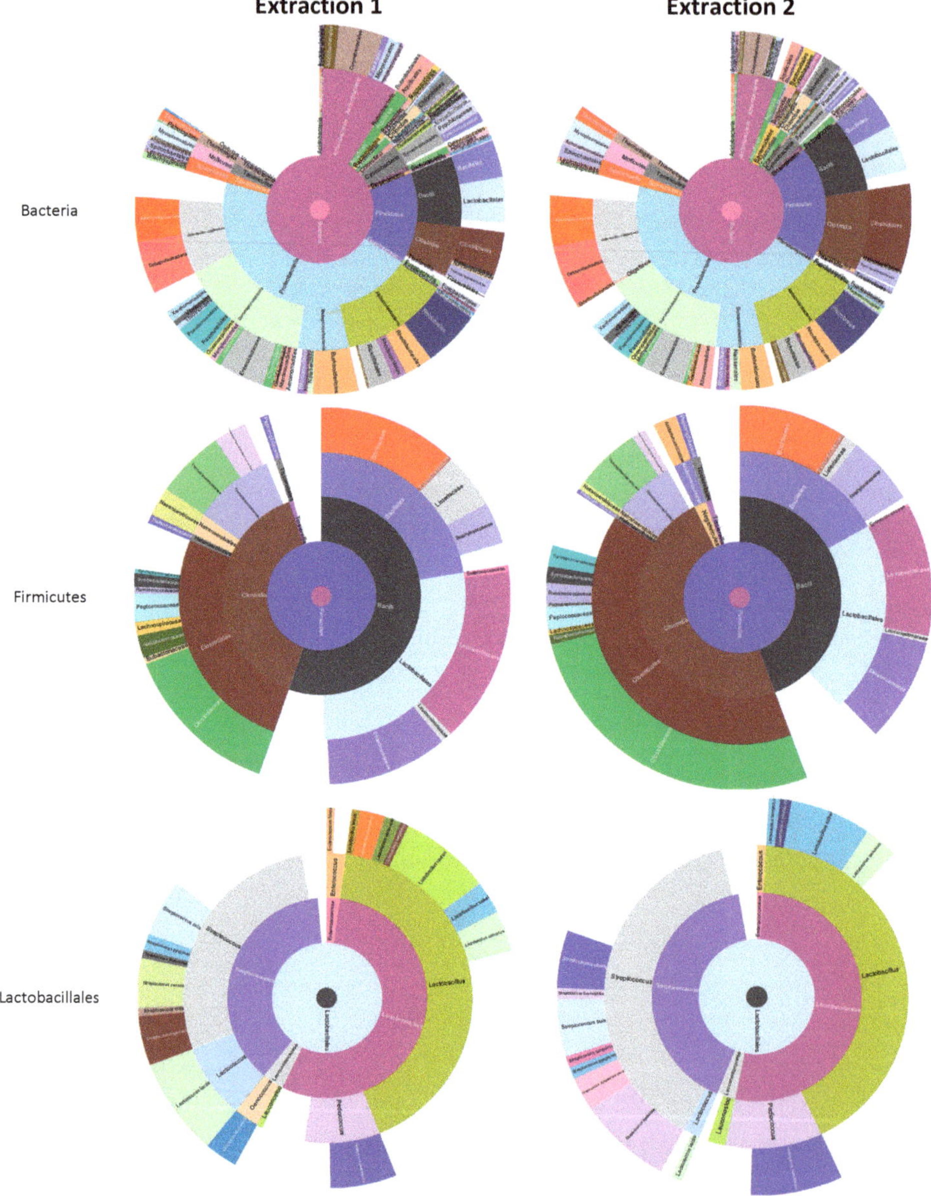

Figure 2. Metaproteomics (Unipept) analysis at the level of the Bacteria domain, Firmicutes phylum, and Lactobacillales order obtained using the peptides identified by searching against the UniProt database restricted to all reviewed bacterial entries.

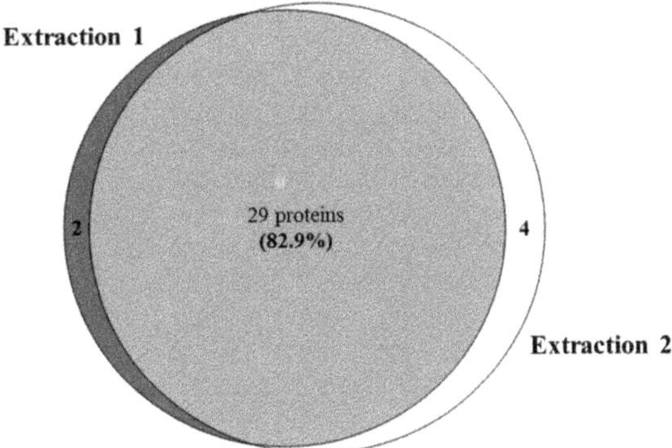

Figure 3. Venn diagram of the proteins identified by searching against the "The Comprehensive Antibiotic Resistance Database" CARD 15.

Table 2 shows the proteins commonly detected in both extractions. Those are mainly represented by orthologs of β-lactamases from several bacterial species, e.g., *Klebsiella pneumoniae* and *Escherichia coli*. Among other proteins with AMR potential identified using the CARD15 database there is an isoform of the Aminoglycoside N(6′)-acetyltransferase of *Enterococcus hirae*.

Table 2. List of the different β-lactamase isoforms detected using the CARD 15 resistome database.

29 Common Elements in "Extraction 1" and "Extraction 2":			
Protein.Entry	Protein.Accession	Protein.Description	Uniprot
ARO:3001066	AAA87176.1	SHV-7 [*Escherichia coli*]	Q46759
ARO:3001077	AAF34333.1	SHV-19 [*Klebsiella pneumoniae*]	Q9LAR9
ARO:3001078	AAF34334.1	SHV-20 [*Klebsiella pneumoniae*]	Q9LAR8
ARO:3001079	AAF34335.1	SHV-21 [*Klebsiella pneumoniae*]	Q9LAR7
ARO:3001076	AAF64386.1	SHV-18 [*Klebsiella pneumoniae*]	Q9LAJ9
ARO:3001073	AAG17550.1	SHV-14 [*Klebsiella pneumoniae*]	Q9F918
ARO:3001087	AAG49894.1	SHV-29 [*Klebsiella pneumoniae*]	Q9AHN9
ARO:3001092	AAK64187.1	SHV-34 [*Escherichia coli*]	Q93LM8
ARO:3001093	AAL68926.1	SHV-35 [*Klebsiella pneumoniae*]	Q8VP57
ARO:3001088	AAT75225.1	SHV-30 [*Enterobacter cloacae*]	Q6DLX7
ARO:3001146	ABN49111.1	SHV-94 [*Klebsiella pneumoniae*]	A3FFR3
ARO:3001148	ABN49112.1	SHV-96 [*Acinetobacter baumannii*]	A3FFR4
ARO:3001182	AEK80394.1	SHV-140 [*Klebsiella pneumoniae*]	G1EC65
ARO:3001183	AFC60795.1	SHV-141 [*Klebsiella pneumoniae*]	H9CTU8
ARO:3001188	AFQ23955.1	SHV-149 [*Klebsiella pneumoniae*]	J7I2U9
ARO:3001190	AFQ23957.1	SHV-151 [*Klebsiella pneumoniae*]	J7I6M7
ARO:3001193	AFQ23960.1	SHV-154 [*Klebsiella pneumoniae*]	J7I2V5
ARO:3001195	AFQ23962.1	SHV-156 [*Klebsiella pneumoniae*]	J7I6N3
ARO:3001197	AFQ23964.1	SHV-158 [*Klebsiella pneumoniae*]	J7I616
ARO:3001198	AFQ23965.1	SHV-159 [*Klebsiella pneumoniae*]	J7I2W1
ARO:3001200	AFQ23967.1	SHV-161 [*Klebsiella pneumoniae*]	J7I616
ARO:3001202	AFQ23969.1	SHV-163 [*Klebsiella pneumoniae*]	J7I622
ARO:3001357	AHA80959.1	SHV-173 [*Klebsiella pneumoniae*]	V5N2H6
ARO:3001364	AJO16042.1	SHV-182 [*Klebsiella pneumoniae*]	A0A0C5C1Y7
ARO:3003156	AJO16047.1	SHV-189 [*Klebsiella pneumoniae*]	A0A0C5C1Z0
ARO:3001204	BAM28879.1	SHV-167 [*Klebsiella pneumoniae*]	I7GSH3
ARO:3002589	CAE50925.1	AAC(6′)-Iid [*Enterococcus hirae*]	Q70E72
ARO:3001337	CAQ03504.1	SHV-99 [*Klebsiella pneumoniae*]	B7FDD8
ARO:3003155	CEA29751.1	SHV-188 [*Klebsiella pneumoniae*]	A0A0A1ISX2

All the β-lactamase isoforms that are present in the analyzed sample are shown in the phylogenetic tree in Figure 4.

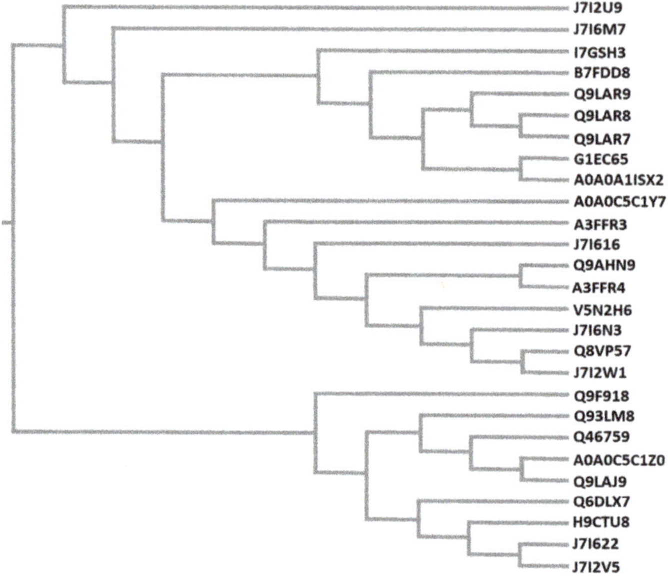

Figure 4. Phylogenetic tree displaying all β-lactamase isoforms detected in the analysed samples using the FastTree function of GenomeNet (https://www.bic.kyoto-u.ac.jp/).

In order to validate the DIA results, a targeted label-free strategy was applied to analyze the peptides related to the identified proteins (Table 3).

Table 3. J7I2U9 and J7I2V5 β-lactamase isoforms specific peptides. Peptides exclusively related to J7I2U9 and J7I2V5 β-lactamase isoforms are reported.

Protein.Entry	Protein Description	Uniprot Accession	Peptides Included Exclusively	RT Mean	%CV
ARO:3001188	SHV-149; [*Klebsiella pneumoniae*]	J7I2U9	LSESRLSGSVGMIEMDLASGR	63.23	1.28
			LSGSVGMIEMDLASGR	72.80	1.14
			LSGSVGMIEMDLASGRTLTAWR	73.01	1.07
			SVLPAGWFIADKTGAGER	65.34	1.16
			TGAGERGAR	79.06	0.95
ARO:3001193	SHV-154; [*Klebsiella pneumoniae*]	J7I2V5	LSESQLSGSVGMIEMDLASGR	64.68	1.15
			LSESQLSGSVGMIEMDLASGRTLTAWR	91.27	0.76

As shown in Figure 4, it has been possible to differentiate different isoforms of β-lactamase. The most divergent β-lactamase proteoforms showed a 1.4% variability. This produced a difference detectable in 7 tryptic peptides as in Figure 5 and Table 3.

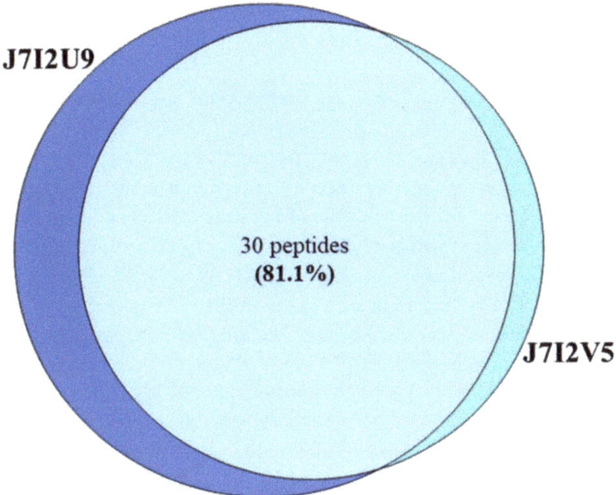

Figure 5. Venn diagram representing the distribution of the tryptic peptides shared between the J7IU9 and J7I2V5 isoforms of β-lactamase.

In Table 3, the peptides typical of J7I2U9 and J7I2V5 isoforms are listed with their respective retention times.

4. Discussion

Antibiotic resistant bacteria are naturally present in most of the microbial ecosystems. Their prevalence is greater in niches where antibiotics are used as in human beings, farm animals, pets, and closely related environments [16–19]. Cows are large animals that carry different biological environments rich in many diverse microbiomes. As consequence, they can be carriers of high numbers of antibiotic resistant bacteria and genes. Resistance genes and bacteria commonly do not represent a problem for dairy products because of the hygiene procedures adopted during food processing before their sale (pasteurization, heat treatments, microfiltration, and fermentations during cheese-making) [17]. However, inter- and intra-specific recombination may lead to the creation of single and multi-drug resistant bacteria that might be harmful for the environment and human health [20–22]. Once resistance genes are introduced inside an organism, it is difficult to track their flow because of their high rate of genetic recombination [18]. Therefore, it also becomes difficult to link such gene transfers to an eventual antimicrobial resistant infection that occurs in humans or animals.

The mammary gland is probably a sterile environment only before colostrogenesis and milk secretion [23,24]. Once these physiological processes are started and colostrum and milk accumulate into the mammary gland, it becomes an opened environment, and it is colonized by a bacterial microflora [24]. There is a high level of similarity between the milk and the intra mammary microbiome, therefore, the milk microbiome represents a good source of information about the intra-mammary environment. Such environment is altered during mastitis events and produces major changes in the milk microbiome and in the composition/integrity of the milk proteome [25,26].

The most frequently used approach for the study of the milk microbiome is 16 s rRNA sequencing which has also been applied to the study of mastitis [27–29]. In the past, it was widely accepted that mastitis is caused by one or a maximum of two bacterial pathogens. However, with the contribution of NGS technologies, it was possible to demonstrate that dysbiosis can be considered as a causative factor for both intra-mammary infections and mastitis itself [24].

Although NGS methods are highly sensitive and accurate in providing information about the composition of the microbiota, they mostly fail to provide information about the functionality of expressed genes. With this technology it is not possible to understand whether these genes are expressed up to the protein level or not. On the contrary, metaproteomic approaches detect protein expression and function.

Efficient bacterial enrichment represents the first step for a successful metaproteomic analysis. For this reason, we adopted a method to enrich the bacterial fraction according to Brewster and Paul [9]. Bacterial binding to the cream layer can be counterproductive for the analysis of the whole bacterial consortium because a relevant part of the microbiome could partition into the cream layer. The agitation step introduced at the beginning of the workflow allowed the collection of the bacterial pellet with a simple centrifugation step. As demonstrated by the aforementioned authors [9], this step facilitates the bacterial detachment from the cream layer resulting in 95% recovery of the viable form. The remaining 5% may still partition with the cream layer or loose viability. As specified in the Methods section, three subsequent centrifugation steps allowed the collection of the bacterial fraction of 160 mL of raw, non-pasteurized, and non-homogenized milk.

Bead beating for bacterial lysis, the FASP method [10,30] for the purification of the protein digestion and DIA IMS mass spectrometry analysis allowed enough dataset depth to study the composition of the bacterial consortia. This analysis was possible up to the genus level providing a qualitative picture of the raw milk's microbiome (Figure 2). The two different extractions yielded overlapping results highlighting proteobacteria and firmicutes as the two main fila present and bacilli and clostridia as the two dominant classes of firmicutes phylum. The genre lactobacillus, together with streptococcus, were the main genres with aerobic metabolism of firmicutes phylum. ATP binding, DNA binding, and metal ion binding were the three main represented functions that emerged as dominant in this analysis.

On the side of AMR, as can be seen in Table 1, computational analysis carried out in the experiment allowed the detection of at least two proteins involved in AMR. One of them is the β-lactamase that belongs from *Klebsiella pneumoniae*, *Escherichia coli*, or *Acinetobacter baumannii*. β-lactamase producing bacteria can be found in the environment as water sources [31], wastewater [32], and drinking water [33]. This poses the concern about the presence of AMR genes as environmental pollutants that could easily enter the animal and human feeding chain [34]. In this case, the detection of β-lactamase produced by the milk microbiome provides proof that these resistance genes are present among the libraries of this bacterial consortium. Surprisingly, these genes are being translated and expressed to protein level at a considerable amount that can be found with our culture-and induction-free proteomics experiments. This supports the hypothesis that a constant level of β-lactamic metabolizing activity might be often present in complex microbiomes. As Figures 4 and 5 demonstrate, β-lactamic activity was not only due to one isoform of β-lactamase, but to several isoforms. The most divergent isoforms of β-lactamase have 98.6% homology and are different by 4 amminoacids, including one arginine and one lysine substitution, which contribute to the different tryptic digestion profile.

Lactamase activity was not the only resistance mechanism that was detected. Even if in minor amounts, Aminoglycoside N(6′)-acetyltransferase presence was found in the metaproteome of milk. This protein catalyzes the acetylation of aminoglycosides conferring resistance to antibiotics containing the purpurosamine ring including amikacin, kanamycin and tobramycin [35].

5. Conclusions

The presented results demonstrated the presence of proteins clearly involved in bacterial resistance. All experiments were performed without any antibiotic induction except for the ones that might be already present in the given ecological niche. The separation of bacterial proteins using a modified precipitation and extraction method combined with bottom-up proteomics allowed the detection of different β-lactamase isoforms. The simultaneous metaproteomics study provided useful information about the taxonomy and the physiological functions of the microbiota.

This method could be easily applied to the study of AMR pattern, bacterial composition and functionality of complex microbiomes. In the field of animal production, it could present an important analytical tool for the study of bulk milk. This study is limited by its application to the characterization of the metaproteome and the resistome of bulk milk of the research facility of University of Milan. Thus, even if the environment is well controlled, does not take into account the possible temporal and geographical variability.

Author Contributions: Conceptualization, C.P., A.S., P.R.; Data curation, C.P., V.G., A.S., D.B., E.G.; Formal analysis, C.P., V.G., A.S., B.T.; Funding acquisition, C.P., P.R.; Investigation, C.P., V.G., A.S., B.T.; Methodology, C.P., V.G., A.S., A.U.; Project administration, C.P., P.R.; Resources, C.P., P.R.; Writing—original draft, C.P., V.G., A.S.; Writing—review & editing, C.P., V.G., E.G., A.S., B.T., L.B., A.Z., R.C., D.B., A.U., P.R. All authors have read and agreed to the published version of the manuscript.

Funding: This research was funded by the grant "Brains to South", Fondazione CON IL SUD, 2018-PDR-00912. (C.P., P.R.) and Italian Ministry of Research and University (MUR) PRIN project nr. 2017Mz5KWM_001- Safe milk: omic science for milk safety and quality (L.B., A.S.).

Conflicts of Interest: The authors declare no conflict of interest. The funders had no role in the design of the study; in the collection, analyses, or interpretation of data; in the writing of the manuscript, or in the decision to publish the results.

References

1. Cox, G.; Wright, G.D. Intrinsic antibiotic resistance: Mechanisms, origins, challenges and solutions. *Int. J. Med. Microbiol.* **2013**, *303*, 287–292. [CrossRef]
2. Olivares, J.; Bernardini, A.; Egarcia-Leon, G.; Corona, F.; Sanchez, M.B.; Martínez, J. The intrinsic resistome of bacterial pathogens. *Front. Microbiol.* **2013**, *4*, 103. [CrossRef]
3. Bhullar, K.; Waglechner, N.; Pawlowski, A.; Koteva, K.; Banks, E.D.; Johnston, M.D.; Balrton, H.A.; Wright, G.D. Antibiotic resistance is prevalent in an isolated cave microbiome. *PLoS ONE* **2012**, *7*, e34953. [CrossRef]
4. Zhang, X.; Li, L.; Butcher, J.; Stintzi, A.; Figeys, D. Advancing functional and translational microbiome research using meta-omics approaches. *Microbiome* **2019**, *7*, 154. [CrossRef]
5. Soggiu, A.; Piras, C.; Mortera, S.L.; Alloggio, I.; Urbani, A.; Bonizzi, L.; Roncada, P. Unravelling the effect of clostridia spores and lysozyme on microbiota dynamics in Grana Padano cheese: A metaproteomics approach. *J. Proteom.* **2016**, *147*, 21–27. [CrossRef]
6. Mortera, S.L.; Soggiu, A.; Vernocchi, P.; Del Chierico, F.; Piras, C.; Carsetti, R.; Marzano, V.; Britti, D.; Urbani, A.; Roncada, P.; et al. Metaproteomic investigation to assess gut microbiota shaping in newborn mice: A combined taxonomic, functional and quantitative approach. *J. Proteom.* **2019**, *203*, 103378. [CrossRef]
7. Soggiu, A.; Piras, C.; Gaiarsa, S.; Bendixen, E.; Panitz, F.; Bendixen, C.; Sassera, D.; Brasca, M.; Bonizzi, L.; Roncada, P. Draft genome sequence of Clostridium tyrobutyricum strain DIVETGP, isolated from cow's milk for Grana Padano production. *Genome Announc.* **2016**, *3*, 2164. [CrossRef]
8. Hogan, J.S.; National Mastitis Council. *Laboratory Handbook on Bovine Mastitis*; National Mastitis Council: Madison, WI, USA, 1999.
9. Brewster, J.D.; Paul, M. Short communication: Improved method for centrifugal recovery of bacteria from raw milk applied to sensitive real-time quantitative PCR detection of *Salmonella* spp. *J. Dairy Sci.* **2016**, *99*, 3375–3379. [CrossRef]
10. Wiśniewski, J.R.; Zougman, A.; Nagaraj, N.; Mann, M. Universal sample preparation method for proteome analysis. *Nat. Methods* **2009**, *6*, 359–362. [CrossRef]
11. Distler, U.; Kuharev, J.; Navarro, P.; Tenzer, S. Label-free quantification in ion mobility–enhanced data-independent acquisition proteomics. *Nat. Protoc.* **2016**, *11*, 795–812. [CrossRef]
12. Marini, F.; Carregari, V.C.; Greco, V.; Ronci, M.; Iavarone, F.; Persichilli, S.; Castagnola, M.; Urbani, A.; Pieroni, L. Exploring the HeLa Dark Mitochondrial Proteome. *Front. Cell Dev. Biol.* **2020**, *8*, 137. [CrossRef]
13. Jia, B.; Raphenya, A.R.; Alcock, B.; Waglechner, N.; Guo, P.; Tsang, K.K.; Lago, B.A.; Dave, B.M.; Pereira, S.; Sharma, A.N.; et al. CARD 2017: Expansion and model-centric curation of the comprehensive antibiotic resistance database. *Nucleic Acids Res.* **2017**, *45*, D566–D573. [CrossRef]
14. Alcock, B.P.; Raphenya, A.R.; Lau, T.T.Y.; Tsang, K.K.; Bouchard, M.; Edalatmand, A.; Huynh, W.; Nguyen, A.-L.V.; Cheng, A.A.; Liu, S.; et al. CARD 2020: Antibiotic resistome surveillance with the comprehensive antibiotic resistance database. *Nucleic Acids Res.* **2020**, *48*, D517–D525. [CrossRef]

15. Singh, R.G.; Tanca, A.; Palomba, A.; Van Der Jeugt, F.; Verschaffelt, P.; Uzzau, S.; Martens, L.; Dawyndt, P.; Mesuere, B. Unipept 4.0: Functional Analysis of Metaproteome Data. *J. Proteome Res.* **2019**, *18*, 606–615. [CrossRef]

16. World Health Organization. *Antimicrobial Resistance Global Report on Surveillance*; WHO Press, World Health Organization: Geneva, Switzerland, June 2014.

17. Noyes, N.R.; Yang, X.; Linke, L.M.; Magnuson, R.J.; Dettenwanger, A.; Cook, S.; Geornaras, I.; Woerner, E.D.; Gow, S.P.; McAllister, A.T.; et al. Resistome diversity in cattle and the environment decreases during beef production. *Elife* **2016**, *5*, e13195. [CrossRef]

18. Liu, J.; Taft, D.H.; Maldonado-Gomez, M.X.; Johnson, D.; Treiber, M.L.; Lemay, D.G.; Depeters, E.J.; Mills, D.A. The fecal resistome of dairy cattle is associated with diet during nursing. *Nat. Commun.* **2019**, *10*, 4406. [CrossRef]

19. Noyes, N.R.; Yang, X.; Linke, L.M.; Magnuson, R.J.; Cook, S.R.; Zaheer, R.; Yang, H.; Woerner, D.R.; Geornaras, I.; McArt, J.A.; et al. Characterization of the resistome in manure, soil and wastewater from dairy and beef production systems. *Sci. Rep.* **2016**, *6*, 1–12. [CrossRef]

20. Michael, G.B.; Freitag, C.; Wendlandt, S.; Eidam, C.; Feßler, A.T.; Lopes, G.V.; Kadlec, K.; Schwarz, S. Emerging issues in antimicrobial resistance of bacteria from food-producing animals. *Future Microbiol.* **2015**, *10*, 427–443. [CrossRef]

21. Rovira, P.; McAllister, T.; Lakin, S.M.; Cook, S.R.; Doster, E.; Noyes, N.R.; Weinroth, M.D.; Yang, X.; Parker, J.K.; Boucher, C.; et al. Characterization of the microbial resistome in conventional and "raised without antibiotics" beef and dairy production systems. *Front. Microbiol.* **2019**, *10*, 1980. [CrossRef]

22. Cuny, C.; Arnold, P.; Hermes, J.; Eckmanns, T.; Mehraj, J.; Schoenfelder, S.; Ziebuhr, W.; Zhao, Q.; Wang, Y.; Feßler, A.T.; et al. Occurrence of cfr-mediated multiresistance in staphylococci from veal calves and pigs, from humans at the corresponding farms, and from veterinarians and their family members. *Vet. Microbiol.* **2017**, *200*, 88–94. [CrossRef]

23. Tolle, A. The Microflora of the Udder. In: Factors Influencing the Bacteriological Quality of Raw Milk. *Int. Dairy J.* **1980**, *120*, 4.

24. Derakhshani, H.; Fehr, K.B.; Sepehri, S.; Francoz, D.; De Buck, J.; Barkema, H.W.; Plaizier, J.C.; Khafipour, E. Invited review: Microbiota of the bovine udder: Contributing factors and potential implications for udder health and mastitis susceptibility. *J. Dairy Sci.* **2018**, *101*, 10605–10625. [CrossRef]

25. Hale, O.J.; Morris, M.; Jones, B.; Reynolds, C.K.; Cramer, R. Liquid Atmospheric Pressure Matrix-Assisted Laser Desorption/Ionization Mass Spectrometry Adds Enhanced Functionalities to MALDI MS Profiling for Disease Diagnostics. *ACS Omega* **2019**, *4*, 12759–12765. [CrossRef]

26. Piras, C.; Ceniti, C.; Hartmane, E.; Costanzo, N.; Morittu, V.M.; Roncada, P.; Britti, D.; Cramer, R. Rapid liquid AP-MALDI MS profiling of lipids and proteins from goat and sheep milk for speciation and colostrum analysis. *Proteomes* **2020**, *8*, 20. [CrossRef]

27. Oikonomou, G.; Machado, V.S.; Santisteban, C.; Schukken, Y.H.; Bicalho, R.C. Microbial diversity of bovine mastitic milk as described by pyrosequencing of metagenomic 16s rDNA. *PLoS ONE* **2012**, *7*, e47671. [CrossRef]

28. Kuehn, J.S.; Gorden, P.J.; Munro, D.; Rong, R.; Dong, Q.; Plummer, P.J.; Wang, C.; Phillips, G.J. Bacterial community profiling of milk samples as a means to understand culture-negative bovine clinical mastitis. *PLoS ONE* **2013**, *8*, e61959. [CrossRef]

29. Oikonomou, G.; Bicalho, M.L.; Meira, E.; Rossi, R.E.; Foditsch, C.; Machado, V.S.; Teixeira, A.G.V.; Santisteban, C.; Schukken, Y.H.; Bicalho, R.C. Microbiota of cow's milk; distinguishing healthy, sub-clinically and clinically diseased quarters. *PLoS ONE* **2014**, *9*, e85904. [CrossRef]

30. Nagaraj, N.; Kulak, N.A.; Cox, J.; Neuhauser, N.; Mayr, K.; Hoerning, O.; Vorm, O.; Mann, M. System-wide perturbation analysis with nearly complete coverage of the yeast proteome by single-shot ultra HPLC runs on a bench top orbitrap. *Mol. Cell. Proteom.* **2012**, *11*, 1–284. [CrossRef]

31. Chavez, M.V.; Caicedo, L.D.; Castillo, J.E. Occurrence of β-Lactamase-Producing Gram-Negative Bacterial Isolates in Water Sources in Cali City, Colombia. *Int. J. Microbiol.* **2019**, *2019*, 1375060.

32. Adelowo, O.O.; Ikhimiukor, O.O.; Knecht, C.; Vollmers, J.; Bhatia, M.; Kaster, A.-K.; Müller, J.A. A survey of extended-spectrum beta-lactamase-producing Enterobacteriaceae in urban wetlands in southwestern Nigeria as a step towards generating prevalence maps of antimicrobial resistance. *PLoS ONE* **2020**, *15*, e0229451. [CrossRef]

33. Adesoji, A.T.; Ogunjobi, A.A. Detection of extended spectrum beta-lactamases resistance genes among bacteria isolated from selected drinking water distribution channels in southwestern Nigeria. *BioMed Res. Int.* **2016**, *2016*, 7149295. [CrossRef] [PubMed]

34. Mesa, R.J.; Blanc, V.; Blanch, A.R.; Cortés, P.; Gonzalez, J.J.; Lavilla, S.; Miró, E.; Muniesa, M.; Saco, M.; Tórtola, M.T.; et al. Extended-spectrum β-lactamase-producing Enterobacteriaceae in different environments (humans, food, animal farms and sewage). *J. Antimicrob. Chemother.* **2006**, *58*, 211–215. [CrossRef] [PubMed]

35. Tenover, F.C.; Filpula, D.; Phillips, K.L.; Plorde, J.J. Cloning and sequencing of a gene encoding an aminoglycoside 6′-N-acetyltransferase from an R factor of Citrobacter diversus. *J. Bacteriol.* **1988**, *170*, 471–473. [CrossRef]

Publisher's Note: MDPI stays neutral with regard to jurisdictional claims in published maps and institutional affiliations.

Article

Prevalence and Antibiotic Resistance Characteristics of Extraintestinal Pathogenic *Escherichia coli* among Healthy Chickens from Farms and Live Poultry Markets in China

Ming Zou [1], Ping-Ping Ma [1], Wen-Shuang Liu [1], Xiao Liang [1], Xu-Yong Li [2], You-Zhi Li [3] and Bao-Tao Liu [1,*]

[1] College of Veterinary Medicine, Qingdao Agricultural University, Qingdao 266109, China; zoudnet@163.com (M.Z.); maping20210219@163.com (P.-P.M.); liuwenshuang0719@163.com (W.-S.L.); liangxiao4000@163.com (X.L.)
[2] College of Agronomy, Liaocheng University, Liaocheng 252000, China; lixuyong@lcu.edu.cn
[3] Shandong Veterinary Drug Quality Inspection Institute, Jinan 250022, China; liyouzhi2009@126.com
* Correspondence: liubaotao@qau.edu.cn; Tel.: +86-532-58957734

Citation: Zou, M.; Ma, P.-P.; Liu, W.-S.; Liang, X.; Li, X.-Y.; Li, Y.-Z.; Liu, B.-T. Prevalence and Antibiotic Resistance Characteristics of Extraintestinal Pathogenic *Escherichia coli* among Healthy Chickens from Farms and Live Poultry Markets in China. *Animals* **2021**, *11*, 1112. https://doi.org/10.3390/ani11041112

Academic Editors: Paola Roncada and Bruno Tilocca

Received: 8 February 2021
Accepted: 7 April 2021
Published: 13 April 2021

Publisher's Note: MDPI stays neutral with regard to jurisdictional claims in published maps and institutional affiliations.

Simple Summary: Chicken meat has been proved to be a suspected source of extraintestinal pathogenic *Escherichia coli* (ExPEC), causing several diseases in humans, and bacteria in healthy chickens can contaminate chicken carcasses at the slaughter; however, reports about the prevalence and molecular characteristics of ExPEC in healthy chickens are still rare. In this study, among 926 *E. coli* isolates from healthy chickens in China, 22 (2.4%) were qualified as ExPEC and these ExPEC isolates were clonally unrelated. A total of six serogroups were identified in this study, with O78 being the most predominant type, and all the six serogroups had been frequently reported in human ExPEC isolates in many countries. All the 22 ExPEC isolates were multidrug-resistant and most isolates carried both $bla_{\text{CTX-M}}$ and *fosA3* resistance genes. Notably, plasmid-borne colistin resistance gene *mcr-1* was identified in six ExPEC isolates, among which two carried additional carbapenemase gene bla_{NDM}, compromising both the efficacies of the two critically important drugs for humans, carbapenems and colistin. These results highlight that healthy chickens can serve as a potential reservoir for multidrug resistant ExPEC isolates, including *mcr-1*-containing ExPEC.

Abstract: Chicken products and chickens with colibacillosis are often reported to be a suspected source of extraintestinal pathogenic *Escherichia coli* (ExPEC) causing several diseases in humans. Such pathogens in healthy chickens can also contaminate chicken carcasses at the slaughter and then are transmitted to humans via food supply; however, reports about the ExPEC in healthy chickens are still rare. In this study, we determined the prevalence and characteristics of ExPEC isolates in healthy chickens in China. A total of 926 *E. coli* isolates from seven layer farms (371 isolates), one white-feather broiler farm (78 isolates) and 17 live poultry markets (477 isolates from yellow-feather broilers) in 10 cities in China, were isolated and analyzed for antibiotic resistance phenotypes and genotypes. The molecular detection of ExPEC among these healthy chicken *E. coli* isolates was performed by PCRs, and the serogroups and antibiotic resistance characteristics of ExPEC were also analyzed. Pulsed-field gel electrophoresis (PFGE) and Multilocus sequence typing (MLST) were used to analyze the genetic relatedness of these ExPEC isolates. We found that the resistance rate for each of the 15 antimicrobials tested among *E. coli* from white-feather broilers was significantly higher than that from brown-egg layers and that from yellow-feather broilers in live poultry markets ($p < 0.05$). A total of 22 of the 926 *E. coli* isolates (2.4%) from healthy chickens were qualified as ExPEC, and the detection rate (7.7%, 6/78) of ExPEC among white-feather broilers was significantly higher than that (1.6%, 6/371) from brown-egg layers and that (2.1%, 10/477) from yellow-feather broilers ($p < 0.05$). PFGE and MLST analysis indicated that clonal dissemination of these ExPEC isolates was unlikely. Serogroup O78 was the most predominant type among the six serogroups identified in this study, and all the six serogroups had been frequently reported in human ExPEC isolates in many countries. All the 22 ExPEC isolates were multidrug-resistant (MDR) and the resistance rates to ampicillin (100%) and sulfamethoxazole-trimethoprim (100%) were the highest, followed by tetracycline (95.5%) and doxycycline (90.9%). $bla_{\text{CTX-M}}$ was found in 15 of the 22 ExPEC isolates including 10 harboring additional fosfomycin resistance gene *fosA3*. Notably, plasmid-borne colistin resistance gene *mcr-1* was identified in six ExPEC isolates in this study. Worryingly, two ExPEC isolates were found to carry both *mcr-1* and *bla*$_{NDM}$, compromising both the efficacies of carbapenems and colistin. The presence of ExPEC isolates in healthy chickens, especially those carrying *mcr-1* and/or *bla*$_{NDM}$, is alarming and will pose a threat to the health of consumers. To our knowledge, this is the first report of *mcr-1*-positive ExPEC isolates harboring *bla*$_{NDM}$ from healthy chickens.

Keywords: characteristics; extraintestinal pathogenic *E. coli*; healthy chickens; multidrug-resistant

1. Introduction

Escherichia coli is a commensal member of the intestinal tract of warm-blooded animals and most *E. coli* strains are harmless; however, a subgroup has possessed the ability to cause diseases, especially extraintestinal infections caused by the extraintestinal pathogenic *Escherichia coli* (ExPEC) [1]. ExPEC strains could colonize the human gastrointestinal tract, not causing disease; however, diverse infections occur when they enter a normally sterile body site [2]. For example, ExPEC strains have been the leading cause of urinary tract infections primarily affecting women [3] and been also the most common cause of bloodstream infections in humans [4]. Importantly, ExPEC infections would impose a large economic burden due to both medical costs and lost productivity, besides their association with morbidity and mortality [5].

A molecular definition of ExPEC is *E. coli* isolate harboring at least two of five virulence markers: *papA* and/or *papC*, *sfa/foc*, *afa/dra*, *kpsM* II and *iutA* [6], and this molecular criteria has been widely applied in epidemiological studies. ExPEC strains have been found in various water sources, including environmental water [7], wastewater [8] and drinking water [9]. Retail meats proved to be classic vehicles for several foodborne pathogens, are also commonly contaminated with ExPEC strains [10,11], posing a potential risk to consumers. In recent years, researchers found that human and animal-source ExPEC shared highly similar virulence genes and clonal backgrounds [12,13] and animal-source ExPEC were capable to adhere or invade human intestinal epithelial [14], suggesting that food-producing animals have been a potential source of human ExPEC. Investigations of the ExPEC within poultry were mainly focused on the avian pathogenic *Escherichia coli* (APEC), a subset of ExPEC, from poultry with colibacillosis [15,16] and APEC mainly caused respiratory and systemic disease in poultry; however, the molecular definition criteria of APEC in those studies was different from that of ExPEC [6,17]. Recently, ExPEC isolates within diseased chickens were also reported [18,19]. Notably, the feces of healthy chickens also carried ExPEC isolates [14,20], and the fecal ExPEC isolates could contaminate chicken carcasses at slaughter, including from rupture of the digestive system during processing, and then transmitted to humans by the food chain or direct human-animal contact [21]. However, studies focusing on ExPEC isolates in healthy chickens are still rare, especially in China, which has huge chicken production.

The presence of antibiotic resistance, one of the ten threats to global health for 2019 as determined by the World Health Organization, among ExPEC isolates has been another big concern. Antibiotic resistance genes, such as extended-spectrum β-lactamases (ESBLs)-encoding genes have been reported in ExPEC isolates [22] and ExPEC including those from poultry can also acquire different resistance genes [18], which would inevitably reduce the therapeutic options, increase morbidity and mortality of ExPEC infections, and eventually bring an increased risk to public health [23,24]. Although antibiotics do not select virulent strains such as ExPEC intrinsically [25], the heavy use of antibiotics in food-producing animals could facilitate the dissemination of ExPEC because such pathogens from environment and diseased animals have been often reported to be multidrug resistant (MDR) [7,14]. However, reports focusing on the antibiotic resistance among ExPEC isolates in healthy chickens remain rare, especially in China which has the largest consumption of antibiotics in the world [26].

Therefore, intense research efforts are warranted to fully understand the characteristics of ExPEC isolates from healthy animals to devise new strategies to prevent their dissemination. In this study, we investigated the prevalence of ExPEC isolates among healthy chickens from farms and live poultry markets in 10 cities in China, and the phenotypic and genotypic characteristics of antimicrobial resistance in these ExPEC isolates were also analyzed.

2. Materials and Methods

2.1. Sampling and Bacterial Isolation

From May 2015 to February 2017, a total of 926 fecal samples were collected from healthy chickens of seven layer farms (371 samples), one white-feather broiler farm (78 samples) and 17 live poultry markets (477 samples) in 10 cities of three provinces (Shandong, Anhui and Shanxi) in China, and 813 samples used in this study were from Shandong province because Shandong has the first largest broiler and layer production in China (Table S1). The white-feather broilers were five-weeks old when the 78 fecal samples were collected, and all the 371 layer fecal samples were from 70-weeks old brown-egg layers before being rejected. All the chickens in the eight farms would be slaughtered within one week after sampling. The 477 fecal samples collected from 17 live poultry markets were from yellow-feather broilers about 12 weeks old sold for consumption and all the yellow-feather broilers from each market were sampled. The chickens in these farms and live poultry markets had been fed with non-medicated feed for at least two weeks before we collected these samples.

A total of 2 g of fecal sample was suspended in 18 ml of trypticase soy broth (BectonDickinson Co., Cockeysville, MD, USA) and incubated aerobically overnight at 37 °C. The broth was then diluted in series of 1:10 and streaked onto MacConkey agar (Qingdao Haibo Microorganism Reagent Co., Ltd., Qingdao, China) followed by incubation for 18 h at 37 °C. Suspected colonies were streaked onto eosin methylene blue (EMB) agar (Qingdao Haibo Microorganism Reagent Co., Ltd., Qingdao, China), and one colony with typical *E. coli* morphology was selected from each sample. The *E. coli* isolates were identified by classical biochemical methods as previously described [27] and confirmed by API 20E system (bioMérieux, Marcy l'Étoile, France). Each farm or live poultry market was sampled only one time when we surveyed the prevalence of antibiotic resistance of *E. coli* and the sampling period covered the four seasons.

2.2. Antimicrobial Susceptibility Tests

The minimal inhibitory concentrations (MICs) of 17 antimicrobials—namely, meropenem, cefotaxime, ceftiofur, ampicillin, ciprofloxacin, enrofloxacin, levofloxacin, nalidixic acid, amikacin, gentamicin, kanamycin, streptomycin, tigecycline, doxycycline, tetracycline, florfenicol and fosfomycin—for these isolates were determined by the agar dilution method following the guidelines of the Clinical and Laboratory Standards Institute [28]. The MIC of colistin was determined according to the method of 2017 EUCAST (available at http://www.eucast.org/clinical_breakpoints/ (accessed on 29 June 2018)). The resistant breakpoints for colistin and tigecycline were recommended by the 2017 EUCAST (http://www.eucast.org/clinical_breakpoints/ (accessed on 28 July 2018)), while the breakpoints for the remaining antimicrobials were recommended by the CLSI [28,29]. *E. coli* ATCC 25,922 was used as the control strain. Ceftiofur, florfenicol and tigecycline were from Solarbio Life Sciences Co. (Beijing, China) and the remain antimicrobials used in this study were purchased from China National Institutes for Drug Control (Beijing, China).

2.3. Detection of Antibiotic Resistance Genes

Mobilized colistin resistance genes (*mcr-1* to *mcr-9*) among all *E. coli* isolates in this study were identified using multiplex PCRs as previously described [30,31]. Carbapenemase-encoding genes including bla_{IMP}, bla_{VIM}, bla_{NDM}, bla_{SPM}, bla_{AIM}, bla_{DIM}, bla_{GIM}, bla_{SIM}, bla_{KPC}, bla_{BIC} and bla_{OXA-48} were detected by PCRs [32]. Plasmid-mediated tigecycline resistance determinant *tet*(X4) was amplified as previously described [33]. All *E. coli* isolates were also screened for the presence of plasmid-mediated quinolone resistance (PMQR) genes (*qnrA*, *qnrB*, *qnrS*, *qnrC*, *qnrD*, *qepA*, and *oqxAB*) [34–38]. The presence of extended-spectrum β-lactamases (ESBLs) ($bla_{CTX-M-1G}$, $bla_{CTX-M-9G}$, $bla_{CTX-M-2G}$, and $bla_{CTX-M-25G}$) and plasmid-mediated AmpC β-lactamases (pAmpC) (bla_{CMY-2} and bla_{DHA-1}) both conferring resistance to cephalosporins was also analyzed as previously described [39–42]. Resistance genes *rmtB*, *fosA* and *fosA3* were also screened as previously reported [43,44].

2.4. Detection and Serotyping of ExPEC Isolates

All isolates were investigated for the following five key virulence markers: *papA* and/or *papC* (P fimbriae; counted as 1), *sfa* and/or *foc* (S and F1C fimbriae, respectively), *afa* and/or *dra* (Afimbrial and Dr-binding adhesion, respectively), *kpsM* II (group 2 capsule), and *iutA* (aerobactin system). The isolates carrying ≥ 2 of the above 5 ExPEC-defining markers were classified as ExPEC [6]. All PCR amplicons were sequenced to confirm these virulence genes.

After PCR identification of ExPEC, the 30 most prevalent serogroups including O1, O2, O4, O6, O7, O8, O9, O15, O18, O21, O22, O25, O26, O45, O55, O78, O83, O86, O101, O103, O111, O113, O117, O121, O138, O145, O149, O157, O158 and O165, were screened among these ExPEC isolates by PCRs as previously described [45].

2.5. Pulse-Field Gel Electrophoresis (PFGE) and Multilocus Sequence Typing (MLST)

To determine the genetic relationship between ExPEC isolates, PFGE was carried out as previously described [46]. Briefly, the ExPEC isolates were grown on Luria–Bertani agar (Qingdao Haibo Microorganism Reagent Co., Ltd., Qingdao, China) overnight at 37 °C and diluted to an optical density of 0.5. Subsequently, the bacteria dilutions were embedded in SeaKem Gold agarose (Lonza, Rockland, ME, United States) and culture plugs were lysed with 100 µg mL^{-1} protease K (Solarbio, Beijing, China) by incubation in a shaking water bath at 55 °C for 2 h. Then, the lysed plugs were washed using sterilized water and Tris–EDTA buffer, respectively. The plugs were then digested with *Xba*I (TaKaRa, Dalian, China) and subjected to PFGE analysis using Chef Mapper electrophoresis system (Bio-Rad Laboratories). The gels were run at 6.0 V cm^{-1} with an initial/final switch time of 2.16 s/54.17 s for 19 h. PFGE patterns were analyzed with BioNumerics software version 7.0 (Applied Maths, Kortrijk, Belgium) by using Dice coefficients and the unweighted-pair group method to achieve dendrograms with a 1.5% band position tolerance. *Salmonella enterica* serotype Braenderup H9812 standards served as size markers.

The ExPEC isolates were also subtyped by the multilocus sequence typing (MLST) method using seven house-keeping genes (*adk*, *fumC*, *gyrB*, *icd*, *mdh*, *recA* and *purA*) of *E. coli* as previously described [47]. All the PCR amplicons were sequenced and imported into the *E. coli* MLST database website (https://pubmlst.org/bigsdb?db=pubmlst_escherichia_seqdef&page=sequenceQuery (accessed on 29 July 2020)).

2.6. Statistical Analysis

Differences in proportions were compared using the χ^2 test implemented in SPSS software (Version 17.0; SPSS Inc., Chicago, IL, USA). All tests of significance were two-tailed, and a value of $p \leq 0.05$ was considered statistically significant.

3. Results

3.1. Antimicrobial Susceptibilities

A total of 926 *E. coli* isolates were obtained including 78 isolates from white-feather broilers, 371 isolates from brown-egg layers and 477 isolates from yellow-feather broilers in 17 live poultry markets (Table 1). As shown in Table 1, resistances to tetracyclines were observed most often among the total 926 *E. coli* isolates in this study, and 89.3% and 83.8% of the isolates were resistant to tetracycline and doxycycline, respectively, although none of the isolates were resistant to the newly tetracyclines drug, tigecycline, a last-resort treatment for infections caused by MDR Gram-negative bacteria in humans. For the β-lactam drugs, the rate of resistance to ampicillin was the highest (87.1%), followed by ceftiofur (44.7%), cefotaxime (41.8%), and meropenem (4.9%). Among aminoglycosides, resistance to streptomycin was the greatest (60.7%), followed by kanamycin (50.5%), gentamicin (31.9%), and amikacin (8.9%). For quinolones, the old quinolone drug nalidixic acid possessed the highest resistance rate (77.1%), and the rates of resistance to the three fluoroquinolones (enrofloxacin, ciprofloxacin, and levofloxacin) varied from 36.3% to 58.4%. Moreover, 20.6% and 69.1% of these isolates were resistant to fosfomycin and florfenicol, respectively. Worryingly, 17.0% of the isolates were resistant to colistin, a critically important antimicrobial for humans (Table 1).

Table 1. Comparison of the resistance rates of *E. coli* isolates from chickens of different origins.

Antimicrobials	Resistance Rates of Isolates (%) *			
	Total (*n* = 926)	Layer Farms (*n* = 371)	White-Feather Broiler Farms (*n* = 78)	Live Poultry Markets (*n* = 477)
β-lactams	-	-	-	-
AMP	87.1	83.0 [a]	100 [b]	88.3 [c]
CTF	44.7	39.4 [a]	98.7 [b]	40.0 [a]
CTX	41.8	26.1 [a]	100 [b]	41.5 [c]
MEM	4.9	1.9 [a]	48.7 [b]	0.0 [c]
Quinolones	-	-	-	-
NAL	77.1	77.6 [a]	96.2 [b]	73.6 [a]
ENR	58.4	59.3 [a]	89.7 [b]	52.6 [a]
CIP	48.4	45.8 [a]	83.3 [b]	44.7 [a]
LEV	36.3	40.2 [a]	51.3 [a]	30.8 [b]
Tetracyclines	-	-	-	-
TET	89.3	87.3 [a]	97.4 [b]	89.5 [a]
DOX	83.8	81.1 [a]	97.4 [b]	83.6 [a]
TIG	0	0	0	0
Aminoglycosides	-	-	-	-
STR	60.7	63.9 [a]	83.3 [b]	54.5 [c]
KAN	50.5	46.6 [a]	96.2 [b]	46.1 [a]
GEN	31.9	26.1 [a]	73.1 [b]	29.6 [a]
AMK	8.9	5.1 [a]	35.9 [b]	7.3 [a]
Polypeptides	-	-	-	-
COL	17.0	4.9 [a]	73.1 [b]	17.2 [c]
Others	-	-	-	-
FFC	69.1	68.7 [a]	94.9 [b]	65.2 [a]
FOS	20.6	10.5 [a]	78.2 [b]	19.1 [c]

* The different lowercase letters in the same line were considered significantly different ($p \leq 0.05$) between two groups using a χ2 test with SPSS software version 19.0. OLA, olaquindox; COL, colistin; FFC, florfenicol; DOX, doxycycline; AMP, ampicillin; CTX, cefotaxime; CTF, ceftiofur; CIP, ciprofloxacin; LEV, levofloxacin; FOS, fosfomycin; MEM, meropenem; NAL, nalidixic acid; GEN, gentamicin; ENR, enrofloxacin; KAN, kanamycin; STR, streptomycin; AMK, amikacin; TET, tetracycline. TIG, tigecycline.

For isolates from white-feather broilers, brown-egg layers and yellow-feather broilers (live poultry markets), respectively, the rates of resistance to ampicillin, tetracycline and doxycycline were all above 80.0% (Table 1). Notably, except levofloxacin and tigecycline, the resistance rate for each of the remaining 15 antimicrobials tested among *E. coli* from white-feather broilers, was significantly higher than that from brown-egg layers and that from yellow-feather broilers ($p < 0.05$) (Table 1). The rate of resistance to ampicillin, cefotaxime, colistin, and fosfomycin among *E. coli* from yellow-feather broilers was significantly higher than that from brown-egg layers, respectively ($p < 0.05$). For meropenem, levofloxacin and streptomycin, respectively; however, the *E. coli* isolates from brown-egg layers possessed significantly higher resistance rate than that from yellow-feather broilers ($p < 0.05$) (Table 1).

3.2. Detection of Resistance Genes

Among the 926 *E. coli* isolates, bla_{NDM} was found in 45 (4.9%) isolates, and no other carbapenemase-encoding genes was found in this study (Table 2). For the ESBLs-encoding genes, $bla_{CTX-M-9G}$ found in 222 (24.0%) of the total *E. coli* isolates was the most prevalent gene, consisting of 52 (14.0%) from 371 brown-egg layers, 54 (69.2%) from 78 white-feather broilers and 116 (24.3%) from 477 yellow-feather broilers of live poultry markets. There were 130 isolates (14.0%) carrying $bla_{CTX-M-1G}$, including 43, 32 and 55 isolates from layer farms, broiler farm and live poultry markets, respectively. A total of 22 isolates harbored both $bla_{CTX-M-1G}$ and $bla_{CTX-M-9G}$ and no other ESBLs-encoding genes was found in this study. pAmpC-encoding genes bla_{CMY-2} and bla_{DHA-1} were found in 53 (5.7%) and 3 (0.3%) of the 926 isolates. Among the PMQR determinants, $qnrS$ and $oqxAB$ found in 311 (33.6%) and 181 (19.5%) isolates, respectively, were the two most prevalent genes, followed by $qnrB$ (34, 3.7%) and $qnrD$ (21, 2.3%) (Table 2). There was no $qnrA$ and $qnrC$ found in this study. Of the 926 *E. coli* isolates, plasmid-borne fosfomycin resistance (PFR) genes were found in 191 isolates, and the number of isolates harboring $fosA3$ and $fosA$ was 189 and 2, respectively. Notably, 157 (17.0%) of the isolates were found to harbor $mcr-1$, consisting of 22 (5.9%) from 371 brown-egg layers, 53 (67.9%) from 78 white-feather broilers and 82 (17.2%) from 477 yellow-feather broilers, and no other mcr genes were found in this study. In addition, $rmtB$ was present in 35 isolates (2.8%). Luckily, none of the 926 isolates carried the plasmid-mediated tigecycline-resistance determinant $tet(X4)$.

The detection rate of bla_{NDM}, $bla_{CTX-M-9G}$, $bla_{CTX-M-1G}$, $mcr-1$, $qnrS$, $fosA3$ and $rmtB$ in *E. coli* from white-feather broilers, respectively, was significantly higher than that from layer farms and that from live poultry markets ($p < 0.05$) (Table 2). For $bla_{CTX-M-9G}$, bla_{CMY-2}, $mcr-1$, $oqxAB$, $qnrB$, $qnrD$, $fosA3$ and $rmtB$, respectively, the yellow-feather broilers from live poultry markets possessed significantly higher detection rate than that from brown-egg layers ($p < 0.05$) (Table 2).

3.3. Prevalence of ExPEC Isolates and Their Serogroups

In the present study, 22 (2.4%) of the 926 chicken isolates were qualified as ExPEC. As shown in Table 3, six ExPEC isolates were found in the white-feather broiler farm, and a total of six ExPEC isolates were also detected in four of the seven layer farms. The detection rates of ExPEC ranged from 0.7% to 6.3% among the four layer farms. Chickens harboring ExPEC in live poultry markets were found in three of the six cities we collected samples from and the detection rates of ExPEC among the yellow-feather broilers from poultry markets were 4.4% (3/68) in city Linyi, 3.0% (6/199) in city Qingdao and 1.8% (1/55) in city Yantai (Table 3). The six ExPEC isolates in Qingdao were from three live poultry markets (Table 3 and Figure 1). Notably, the detection rate (7.7%, 6/78) of ExPEC among white-feather broilers was significantly higher than that (1.6%, 6/371) from brown-egg layers and that (2.1%, 10/477) from yellow-feather broilers ($p < 0.05$). There was no significant difference between the detection rate (1.6%, 6/371) of ExPEC among brown-egg layers and that among yellow-feather broilers (2.1%, 10/477) ($p = 0.611$) (Table 3).

Table 2. Prevalence of resistance genes among the 926 *E. coli* isolates from chickens.

Resistance Genes	No. of Positive Isolates (%) *			
	Total (n = 926)	Layer Farms (n = 371)	White-Feather Broiler Farms 9 (n = 78)	Live Poultry Markets (n = 477)
Carbapenemases	-	-	-	-
bla_{NDM}	45 (4.9)	11 (3.0) [a]	34 (43.6) [b]	0 (0) [c]
ESBLs	-	-	-	-
$bla_{CTX-M-9G}$	222 (24.0)	52 (14.0) [a]	54 (69.2) [b]	116 (24.3) [c]
$bla_{CTX-M-1G}$	130 (14.0)	43 (11.6) [a]	32 (41.0) [b]	55 (11.5) [a]
pAmpC	-	-	-	-
bla_{CMY-2}	53 (5.7)	11 (3.0) [a]	3 (3.8) [ab]	39 (8.2) [b]
bla_{DHA-1}	3 (0.3)	1 (0.3) [a]	0 (0.0) [a]	2 (0.4) [a]
MCR	-	-	-	-
mcr-1	157 (17.0)	22 (5.9) [a]	53 (67.9) [b]	82 (17.2) [c]
PMQR	-	-	-	-
oqxAB	181 (19.5)	41 (11.1) [a]	22 (28.2) [b]	118 (24.7) [b]
qnrB *	34 (3.7)	7 (1.9) [a]	1 (1.3) [ab]	26 (5.5) [b]
qnrS	311 (33.6)	140 (37.7) [a]	12 (15.4) [b]	159 (33.3) [a]
qnrD *	21 (2.3)	3 (0.8) [a]	0 (0.0) [ab]	18 (3.8) [b]
PFR	-	-	-	-
fosA3	189 (20.4)	39 (10.5) [a]	61 (78.2) [b]	89 (18.7) [c]
fosA	2 (0.2)	0 (0.0) [a]	0 (0.0) [a]	2 (0.4) [a]
Others	-	-	-	-
rmtB	35 (3.8)	2 (0.5) [a]	19 (24.4) [b]	14 (2.9) [c]

* The different lowercase letters in the same line were considered significantly different ($p \leq 0.05$) between two groups using a χ2 test with SPSS software version 19.0.

Table 3. Prevalence and origins of the 22 ExPEC isolates in this study.

Location	Origins	No. of Farms/Markets	Year	ExPEC Isolates (%)/ Total Isolates
Weifang in Shandong	White-feather broiler farm	1	2015	6 (7.7%)/78
Hefei in Anhui	Layer farm	1	2015	1 (6.3%)/16
Liaocheng in Shandong	Layer farm	1	2015	2 (4.9%)/41
Binzhou in Shandong	Layer farms	2	2016	1 (0.7%)/135
Xi'an in Shanxi	Layer farm	1	2015	2 (2.1%)/97
Qingdao in Shandong	Layer farms	2	2017	0/82
Linyi in Shandong	Live poultry market	1	2015	3 (4.4%)/68
Qingdao in Shandong	Live poultry markets	11	2015	6 (3.0%)/199
Yantai in Shandong	Live poultry markets	2	2015	1 (1.8%)/55
Zaozhuang in Shandong	Live poultry market	1	2015	0/57
Zibo in Shandong	Live poultry market	1	2015	0/67
Weifang in Shandong	Live poultry market	1	2015	0/31
-	White-feather broiler farm	1	-	6 (7.7%)/78
-	Layer farms	7	-	6 (1.6%)/371
-	Live poultry markets	17	-	10 (2.1%)/477

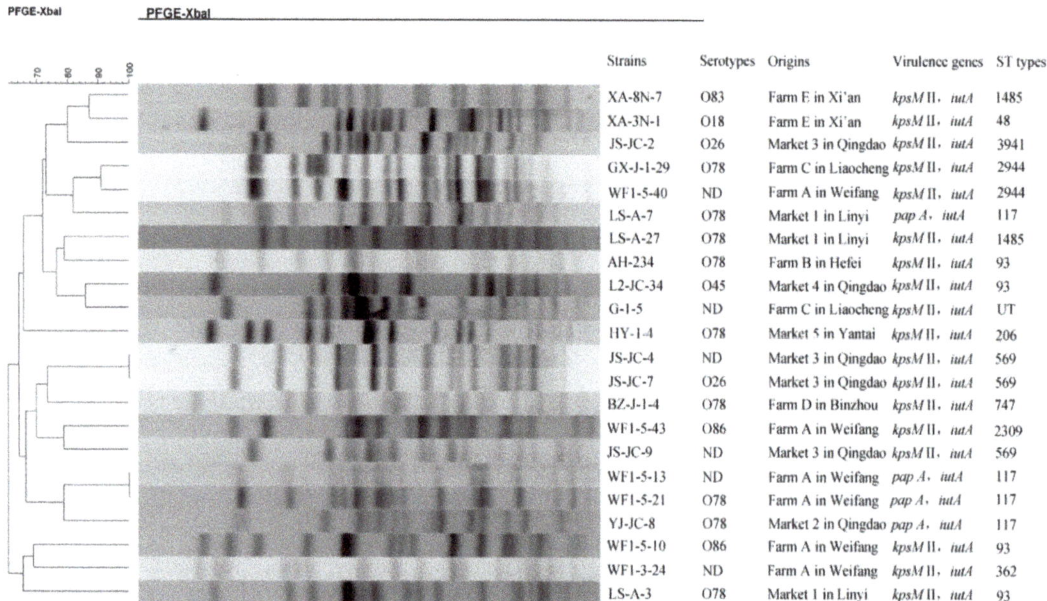

Strains	Serotypes	Origins	Virulence genes	ST types
XA-8N-7	O83	Farm E in Xi'an	*kpsM* II, *iutA*	1485
XA-3N-1	O18	Farm E in Xi'an	*kpsM* II, *iutA*	48
JS-JC-2	O26	Market 3 in Qingdao	*kpsM* II, *iutA*	3941
GX-J-1-29	O78	Farm C in Liaocheng	*kpsM* II, *iutA*	2944
WF1-5-40	ND	Farm A in Weifang	*kpsM* II, *iutA*	2944
LS-A-7	O78	Market 1 in Linyi	*papA*, *iutA*	117
LS-A-27	O78	Market 1 in Linyi	*kpsM* II, *iutA*	1485
AH-234	O78	Farm B in Hefei	*kpsM* II, *iutA*	93
L2-JC-34	O45	Market 4 in Qingdao	*kpsM* II, *iutA*	93
G-1-5	ND	Farm C in Liaocheng	*kpsM* II, *iutA*	UT
HY-1-4	O78	Market 5 in Yantai	*kpsM* II, *iutA*	206
JS-JC-4	ND	Market 3 in Qingdao	*kpsM* II, *iutA*	569
JS-JC-7	O26	Market 3 in Qingdao	*kpsM* II, *iutA*	569
BZ-J-1-4	O78	Farm D in Binzhou	*kpsM* II, *iutA*	747
WF1-5-43	O86	Farm A in Weifang	*kpsM* II, *iutA*	2309
JS-JC-9	ND	Market 3 in Qingdao	*kpsM* II, *iutA*	569
WF1-5-13	ND	Farm A in Weifang	*papA*, *iutA*	117
WF1-5-21	O78	Farm A in Weifang	*papA*, *iutA*	117
YJ-JC-8	O78	Market 2 in Qingdao	*papA*, *iutA*	117
WF1-5-10	O86	Farm A in Weifang	*kpsM* II, *iutA*	93
WF1-3-24	ND	Farm A in Weifang	*kpsM* II, *iutA*	362
LS-A-3	O78	Market 1 in Linyi	*kpsM* II, *iutA*	93

Figure 1. Characteristics and PFGE dendrogram patterns of the 22 ExPEC isolates from healthy chickens in this study.

Among the 22 ExPEC isolates, *iutA* was the most prevalent ExPEC-defining marker, followed by *kpsM* II (18 isolates) and *papA* (4 isolates) (Figure 1). All ExPEC isolates carried two of the five ExPEC-defining markers, and no other markers were found in our study. After the PCR-based serotyping method was applied to the 22 ExPEC isolates, the serogroups of 16 isolates were successfully identified and they belonged to six serogroups (O78, O26, O86, O18, O45 and O83). O78 detected in nine of the 22 ExPEC isolates (40.9%) was the most prevalent serogroup, followed by O26 (9.1%, 2/22) and O86 (9.1%, 2/22) (Figure 1).

3.4. Antimicrobial Resistance Phenotypes and Genotypes of the ExPEC Isolates

An antimicrobial susceptibility test showed that all 22 ExPEC isolates were resistant to ampicillin (100%) and sulfamethoxazole-trimethoprim (100%), followed by resistance to tetracycline (95.5%), and doxycycline (90.9%) (Table 4 and Figure 2). All the rates of resistance to florfenicol, streptomycin, kanamycin and nalidixic acid among these isolates were 81.8%. For the third-generation cephalosporins, the rates of resistance to cefotaxime and ceftiofur were both 72.7% (16/22), and the resistance rates to fluoroquinolones ranged from 45.5% to 59.1% (Figure 2). The number of isolates resistant to amikacin and fosfomycin were six (27.3%) and ten (45.5%), respectively. A total of six (27.3%) and two (9.1%) ExPEC isolates were resistant to the two critically important antibiotics colistin and meropenem, respectively. Notably, two ExPEC isolates WF1-5-13 and WF1-5-40 were resistant to both colistin and meropenem (Table 4). Luckily, no isolate was resistant to tigecycline. Detailed results of the antibiotic resistance profiles for the 22 ExPEC isolates were presented in Table 4. Interestingly, the rate of resistance to cefotaxime, ceftiofur, fosfomycin, amikacin, kanamycin and streptomycin in ExPEC isolates was significantly higher than that of non-ExPEC isolates in this study, respectively ($p < 0.05$) (Figure 2). Worryingly, all the ExPEC isolates in this study were MDR (resistance $\geq$ 3 three classes of antibiotics) in nature (Table 4).

Table 4. Resistance phenotypes and genotypes of the 22 ExPEC isolates in this study.

Strain (ST Types)	Source	City	Serotype #	ExPEC-Defining Markers	Resistance Phenotype	Resistance Genes
WF1-3-24	White-feather broiler farm A	Weifang	ND	*KpsM II, iutA*	*mcr-1, bla*CTX-M-1G, *fosA3, rmtB, oqxAB, floR*	COL, CTX, CTF, AMP, FOS, AMK, GEN, KAN, STR, NAL, CIP, ENR, LEV, FFL, DOX, TET, OLA, SXT
WF1-5-10	White-feather broiler farm A	Weifang	O86	*KpsM II, iutA*	*bla*CTX-M-9G, *fosA3, rmtB, floR*	CTX, CTF, AMP, AMK, GEN, KAN, STR, NAL, CIP, ENR, LEV, FFL, DOX, TET, SXT
WF1-5-13	White-feather broiler farm A	Weifang	ND	*papA, iutA*	*mcr-1, bla*NDM-5, *bla*CTX-M-9G, *bla*CTX-M-1G, *fosA3, floR*	COL, MEM, CTX, CTF, AMP, FOS, KAN, STR, NAL, CIP, ENR, LEV, FFL, DOX, TET, SXT
WF1-5-21	White-feather broiler farm A	Weifang	O78	*papA, iutA*	*mcr-1, bla*CTX-M-9G, *fosA3, rmtB, floR*	COL, CTX, CTF, AMP, FOS, AMK, GEN, KAN, STR, NAL, FFL, DOX, TET, SXT
WF1-5-40	White-feather broiler farm A	Weifang	ND	*KpsM II, iutA*	*mcr-1, bla*NDM-1, *bla*CTX-M-9G, *fosA3, rmtB, floR*	COL, MEM, CTX, CTF, AMP, FOS, AMK, GEN, KAN, STR, FFL, DOX, TET, SXT
WF1-5-43	White-feather broiler farm A	Weifang	O86	*KpsM II, iutA*	*bla*CTX-M-9G, *fosA3*	CTX, CTF, AMP, FOS, KAN, STR, NAL, CIP, ENR, LEV, DOX, TET, SXT
AH-234	Layer farm B	Hefei	O78	*KpsM II, iutA*	*bla*CTX-M-9G, *bla*CMY-2, *floR*	CTX, CTF, AMP, AMK, GEN, KAN, STR, NAL, CIP, ENR, LEV *, FFL, TET, SXT
G-1-5	Layer farm C	Liaocheng	ND	*KpsM II, iutA*	*mcr-1*	COL, AMP, KAN, STR, NAL, DOX, TET, SXT
GX-J-1-29	Layer farm C	Liaocheng	O78	*KpsM II, iutA*	*qnrS, floR*	AMP, FFL, DOX, TET, SXT
BZ-J-1-4	Layer farm D	Binzhou	O78	*KpsM II, iutA*	*bla*CTX-M-1G, *fosA3*	CTX, CTF, AMP, FOS, STR, NAL, FFL, DOX, TET, SXT
XA-8N-7	Layer farm E	Xi'an	O83	*KpsM II, iutA*	*bla*CTX-M-1G, *floR*	CTX, CTF, AMP, KAN, NAL, ENR, LEV, FFL, DOX, TET, SXT
XA-3N-1	Layer farm E	Xi'an	O18	*KpsM II, iutA*	*bla*CTX-M-1G, *floR*	CTX, CTF, AMP, KAN, STR, NAL, CIP, ENR, LEV, FFL, DOX, TET, SXT
LS-A-3	Live poultry market 1	Linyi	O78	*KpsM II, iutA*	*bla*CTX-M-9G, *fosA3, floR*	CTX, CTF, AMP, FOS, KAN, STR, NAL, CIP, ENR, LEV, FFL, DOX, TET, SXT
LS-A-7	Live poultry market 1	Linyi	O78	*papA, iutA*	*bla*CTX-M-1G, *fosA3, oqxAB, floR*	CTX, CTF, AMP, FOS, KAN, STR, NAL, CIP, ENR, LEV, FFL, DOX, TET, OLA, SXT
LS-A-27	Live poultry market 1	Linyi	O78	*KpsM II, iutA*	*bla*CTX-M-9G, *floR*	CTX, CTF, AMP, KAN, STR, NAL, ENR, FFL, DOX, TET, SXT
YJ-JC-8	Live poultry market 2	Qingdao	O78	*papA, iutA*	*oqxAB*	AMP, AMK, GEN, KAN, STR, NAL, CIP, ENR, LEV, FFL, DOX, TET, OLA, SXT
JS-JC-2	Live poultry market 3	Qingdao	O26	*KpsM II, iutA*	*floR*	CTX, CTF, AMP, FOS, GEN, KAN, STR, NAL, CIP, ENR, LEV, FFL, DOX, TET, SXT

Table 4. *Cont.*

Strain (ST Types)	Source	City	Serotype #	ExPEC-Defining Markers	Resistance Phenotype	Resistance Genes
JS-JC-4	Live poultry market 3	Qingdao	ND	*KpsM II, iutA*	*qnrS*	AMP, KAN, STR, NAL, DOX, TET, SXT
JS-JC-7	Live poultry market 3	Qingdao	O26	*KpsM II, iutA*	*qnrS*	AMP, STR, DOX, TET, SXT
JS-JC-9	Live poultry market 3	Qingdao	ND	*KpsM II, iutA*	$bla_{CTX-M-9G}$, *qnrS*	CTX, CTF, AMP, KAN, STR, FFL, DOX, TET, SXT
L2-JC-34	Live poultry market 4	Qingdao	O45	*KpsM II, iutA*	*mcr-1*, $bla_{CTX-M-9G}$, *fosA3*, *floR*	COL, CTX, CTF, AMP, FOS, GEN, KAN, NAL, FFL, SXT
HY-1-4	Live poultry market 5	Yantai	O78	*KpsM II, iutA*	-	AMP, NAL, ENR, FFL, DOX, TET, SXT

\# ND, not determined; * Intermediate resistance; AMP, ampicillin; MEM, meropenem; CTX, cefotaxime; CTF, ceftiofur; CAZ, Ceftazidime; ETP, ertapenem; IPM, imipenem; NAL, nalidixic acid; CIP, ciprofloxacin; ENR, enrofloxacin; LEV, levofloxacin; STR, streptomycin; KAN, kanamycin; GEN, gentamicin; AMK, amikacin; TET, tetracycline; DOX, doxycycline; FOS, fosfomycin; FFC, florfenicol; TIG, tigecycline; SXT, sulfamethoxazole-trimethoprim (SMZ-TMP).

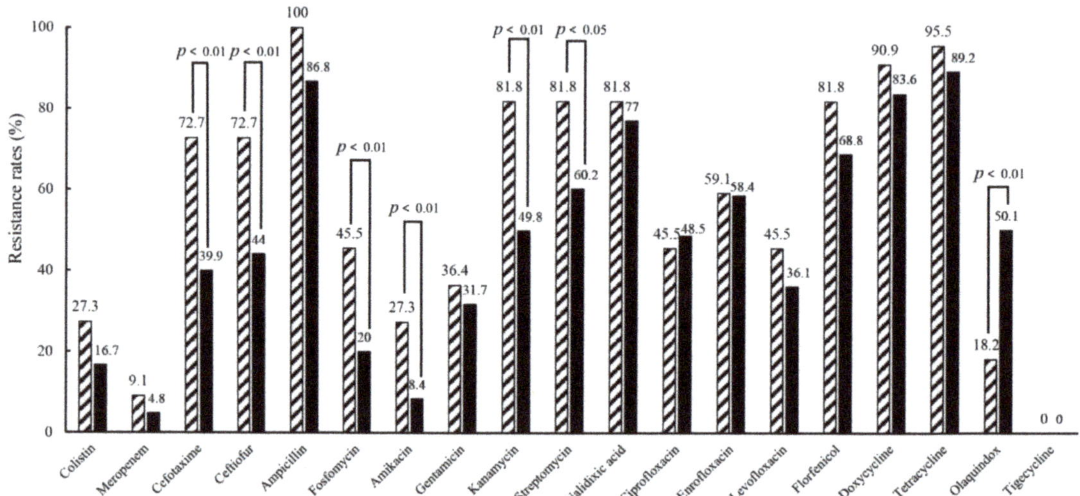

Figure 2. Comparison of the antibiotic resistance rates between the ExPEC and non-ExPEC isolates in this study.

Among the 16 ExPEC isolates resistant to cefotaxime, bla_{CTX-M} was found in 15 isolates (10 $bla_{CTX-M-9G}$ and 6 $bla_{CTX-M-1G}$), including one isolate WF1-5-13 carrying both $bla_{CTX-M-9G}$ and $bla_{CTX-M-1G}$ (Table 4). $bla_{CTX-M-9G}$ and bla_{CMY-2}-type pAmpC-encoding gene were present in isolate AH234 from layer farm B, accounting for the resistances to the third-generation cephalosporins. *fosA3* was found in ten ExPEC isolates resistant to fosfomycin and this gene was only distributed among bla_{CTX-M}-positive isolates. For the six colistin and two meropenem resistant ExPEC isolates, the presence of *mcr-1* and bla_{NDM} could account for their corresponding resistance, respectively. Notably, the two ExPEC isolates (WF1-5-13 and WF1-5-40) harboring both *mcr-1* and bla_{NDM}, carried additional bla_{CTX-M}, *fosA3* and *floR* genes, and isolate WF1-5-40 also possessed *rmtB* (Table 4).

3.5. Genetic Relationships of the ExPEC Isolates

All 22 ExPEC isolates could be successfully analyzed by PFGE and 20 different PFGE profiles were obtained indicating that clonal dissemination of these ExPEC isolates was unlikely (Figure 1). As shown in Figure 1, 12 types were identified by the MLST subtyping method along with one new ST (6-6-804-10-9-1-6 in isolate G-1-5) not previously registered in the *E. coli* MLST database. The most prevalent ST types were ST117 (four) and ST93 (four), followed by ST569 (three), ST1485 (two) and ST2944 (two). Most isolates sharing the same ST types had different PFGE profiles and were from different farms or markets of different cities (Figure 1). For example, the four isolates (AH234, L2-JC-34, WF1-5-10 and LS-A-3) belonging to ST93 were from four chicken markets/farms in four cities and had different PFGE patterns. Notably, the two ExPEC isolates WF1-5-13 and WF1-5-21 from the same chicken farm in city Weifang shared identical PFGE pattern and ST type; however, different resistance genotypes and phenotypes were found in these two isolates (Figure 1 and Table 4). Different resistance phenotypes were also found in isolates JS-JC-4 and JS-JC-7, which were from the same market and possessed identical PFGE pattern and ST type (Figure 1 and Table 4).

4. Discussion

The presence of ExPEC colonizing healthy chickens could be a huge threat to both animal and human health. For China, having the largest consumption of antibiotics in the world, the prevalence and antibiotic resistance of ExPEC among healthy chickens urgently need to be studied. In this study, we investigated the resistance of *E. coli* isolates from healthy chickens of seven layer farms, one white-feather broiler farm and 17 live poultry markets in China, and the ExPEC among these commensal isolates were also characterized.

Both the rates of resistance to tetracycline and doxycycline among isolates in this study were above 80.0%, consistent with that from chickens in China after 2012 [48]. The high resistance rates to these drugs could be attributed to the heavy usage of tetracyclines in poultry, because oxytetracycline, tetracycline, chlortetracycline and doxycycline have been heavily used for decades in animal production including poultry [49]. Tigecycline, a last-resort treatment for human infections caused by MDR Gram-negative bacteria, has never been used in animal husbandry. Luckily, isolate resistant to tigecycline was not found in our *E. coli* isolated during 2015–2017; however, the heavy usage of tetracyclines in animals could increase the prevalence of newly mobile tigecycline-resistance gene *tet*(X4) in *E. coli* and this should be paid more attention [33]. The resistance rates to meropenem (4.9%) and colistin (17.0%), two critically important antimicrobials in human medicine, among the 926 isolates in our study could be well accounted for the presence of bla_{NDM} (4.9%) and *mcr-1* (17.0%), respectively (Tables 1 and 2). *qnrS* was the most prevalent PMQR gene in this study, differing from that in humans [50] and animals [51], in which the most prevalent PMQR gene was *oqxAB*. In some countries, *qnrB* was the most prevalent type [52]. The prevalence of CTX-M-type ESBLs (35.6%) in healthy chickens in this study was similar to that (38.5%) in *E. coli* isolates from chickens in China [48], but lower than that in chicken production of India [53]. Notably, except levofloxacin and tigecycline, the resistance rate for each of the 15 antimicrobials tested among *E. coli* from white-feather broilers, was significantly higher than that from brown-egg layers and that from yellow-feather broilers ($p < 0.05$) (Table 1). Such phenomena could be attributed to that consumption of antibiotics in white-feather broilers is the largest among the three types of chickens. The rate of resistance to ampicillin, cefotaxime, colistin, and fosfomycin among *E. coli* from yellow-feather broilers was significantly higher than that from brown-egg layers, respectively ($p < 0.05$). This might because that β-lactams and colistin are often used in early feeding period of the yellow-feather broilers while almost all antimicrobials are forbidden in layer farms during the laying period.

Based on the molecular criteria of Johnson et al. [6], 2.4% (22/926) of the healthy chicken fecal *E. coli* isolates were qualified as ExPEC in this study. The ExPEC could asymptomatically colonize the gut of a fraction of healthy animal population and survive in extra-intestinal environments, causing diseases in animals and humans through the food chain [54]. The threat to human health posed by the healthy chicken ExPEC isolates in this study could been further proved by that healthy poultry ExPEC were capable to adhere or invade human intestinal epithelial [14]. In this study, virulence markers *iutA* and *KpsM* II were the two most prevalent genes among the ExPEC isolates from healthy chickens, consistent with the finding about MDR *E. coli* in healthy chickens in Brazil [14]. Besides isolation methods, geographic locations and management practices, different classification criteria for ExPEC has been the main factor contributing to differences in frequency of ExPEC between studies. We will focus the studies using the same PCR-based screening method for ExPEC as that in our study. The detection rate of ExPEC in our samples was 2.4%, similar to that (4.7%, 5/108) among chicken egg *E. coli* isolates ($p > 0.05$), but lower than that (21.5%, 130/606) in chicken meat isolates reported in the USA ($p < 0.05$) [55]. Notably, the prevalence of ExPEC isolates (2.4%, 22/926) in our study was also significantly lower than that (13.2%, 40/304) from farmed healthy chickens in Quebec, Canada [20]. This might be explained by that boiled DNA extracts from total cultures of samples were initially screened for all possible ExPEC strains in the previous study, contributing to a high recovery rate of ExPEC. In this study, the detection rate (7.7%, 6/78) of ExPEC among white-feather broilers was significantly higher than that (1.6%, 6/371) from brown-egg layers and that (2.1%, 10/477) from yellow-feather broilers in live poultry markets ($p < 0.05$). The phenomenon could be attributed to the selective pressure on the dissemination of ExPEC posed by antimicrobials frequently used in white-feather broilers. This was proved by that highly similar PFGE patterns were found in the two ExPEC isolates WF1-5-13 and WF1-5-21 from the same Farm (Figure 1).

In this study, the most two prevalent ST types of the 22 ExPEC isolates were ST117 and ST93 (Figure 1). Since ST117 and ST93 types of ExPEC had been found to be associated with meningitis of humans in Brazil in 1999 (http://enterobase.warwick.ac.uk/species/ecoli/search_strains?query=st_search (accessed on 7 October 2020)), both ST types of *E. coli* have caused sepsis among humans [56,57] and disseminated among humans around the world including the European countries and China (http://enterobase.warwick.ac.uk/species/ecoli/search_strains?query=st_search (accessed on 7 October 2020)). Besides ST2944, all other STs in this study have been also found in human isolates (http://enterobase.warwick.ac.uk/species/ecoli/search_strains?query=st_search (accessed on 7 October 2020)). Serogroups of the ExPEC isolates from animals were rarely studied, although the APEC isolates from diseased poultry were often serotyped. In this study, O78 was the most predominant serogroup among ExPEC isolates from healthy chickens, followed by O26 and O86. This was slightly different from that of a previous study about APEC from Korea in which O78 was the most prevalent serogroup followed by O2 and O53 [15], both of which were not identified in our study. Notably, O78 was also a common serogroup among human ExPEC isolates from neonatal meningitis in Europe [58]. The distribution rate of each serogroup in this study was also different from that of APEC obtained between 2005 and 2008 in Guangdong, China [59]. The presence of serogroup O86 in two ExPEC from healthy chickens in this study are of interest since this serogroup was only identified in human ExPEC strains in Brazil rather than in strains from poultry in a previous report [57]. In the present study, one ExPEC isolate from healthy chicken belonged to O18 serogroup, which was frequently found in APEC from diseased avian in the United States and also the main serogroup of ExPEC isolates causing newborn meningitis in the Europe [58,60,61]. O83 and O45, which were prevalent in neonates with *E. coli* meningitis from the Netherlands [61] and France [62], respectively, were also found in our ExPEC isolates. All these results show that the ExPEC isolates from healthy chickens in this study might transmit to humans and the prevalence of ExPEC isolates in healthy animals should be monitored in the future.

Notably, all the ExPEC isolates in this study were MDR isolates. Although the use of antibiotics in animals does not select ExPEC strains intrinsically [25], it will favor the dissemination of ExPEC with MDR phenotypes among healthy animals, contributing to the emergence of MDR ExPEC in human infections [22]. Almost all ExPEC isolates in this study were resistant to ampicillin (100%), sulfamethoxazole-trimethoprim (100%), tetracycline (95.5%), and doxycycline (90.9%), which were all used extensively in animal husbandry in China, further favoring the dissemination of ExPEC among animals and humans by co-selection. In recent years, ExPEC isolates producing ESBLs or AmpC in human infections have been increasing [63], and such pathogens have been also found in healthy poultry in Brazil recently [14]. In our study, CTX-M-type ESBLs were found in 15 of the 22 ExPEC isolates from healthy chickens. The increase in ESBLs or AmpC among ExPEC from poultry will inevitably reduce the therapeutic options of ExPEC infections in humans, because cephalosporins are important to human medicine. Fosfomycin has been widely recommended for treating uncomplicated urinary tract infection especially caused by ESBLs-producing or fluoroquinolone-resistant ExPEC isolates [64]. However, ten ExPEC isolates in this study harbored the fosfomycin resistance gene *fosA3* and all ten isolates also carried *bla*$_{\text{CTX-M}}$, among which six were resistant to fluoroquinolones (Table 4). Such pathogens in healthy chickens will pose a great threat to human health because they will compromise the efficacies of fosfomycin, fluoroquinolones and cephalosporins. For ExPEC from different markets/farms, three isolates (YJ-JC-8, WF1-5-21 and LS-A-7) carrying virulence genes *papA* and *iutA* belonged to serogroup O78 and they were all ST117 type, however, different PFGE and resistance profiles were present in the three isolates (Figure 1 and Table 4). This indicates that the ExPEC isolates have undergone a complex evolutionary process resulting in genetically diverse isolates although they share identical ST type and serogroup at the beginning. Even for ExPEC isolates with identical PFGE pattern and ST type from the same farm, such as WF1-5-13 and WF1-5-21, different resistance genotypes and phenotypes were also observed, further proving the complex evolutionary process within ExPEC.

Worryingly, besides *mcr-1* and *bla*$_{\text{NDM}}$, the two ExPEC isolates WF1-5-13 and WF1-5-40 carried additional *bla*$_{\text{CTX-M}}$, *fosA3* and *floR* genes, with isolate WF1-5-40 also harboring *rmtB* (Table 4). The presence of such ExPEC isolates co-harboring *bla*$_{\text{NDM}}$ and *mcr-1* in healthy chickens in this study will threaten the health of consumers because such pathogens will not only compromise the efficacies of cephalosporins, fosfomycin and aminoglycosides, but also threaten the usage of carbapenems and colistin, two critically important antimicrobials used for serious infections caused by MDR ExPEC [65]. *mcr* has been also found in two ExPEC isolates from diseased poultry in Brazil [18] and two ExPEC isolates from healthy ducks in China [66]. All four ExPEC isolates carrying *mcr* from animals in the two previous reports were susceptible to carbapenems, although NDM-producing ExPEC isolates susceptible to colistin have been reported in humans [67]. To the best of our knowledge, this is the first report about *mcr-1*-positive ExPEC isolates harboring *bla*$_{\text{NDM}}$ from healthy chickens.

5. Conclusions

In conclusion, we observed that the resistances in *E. coli* from white-feather broilers were more serious than those from layer farms and those from live-poultry markets in China, respectively. This study also reported that 2.4% of these *E. coli* isolates from healthy chickens were qualified as ExPEC using a molecular detection method. The most predominant serogroup of these ExPEC isolates was O78, followed by O26 and O86, and almost all serogroups identified in our study were frequently reported in human ExPEC isolates in many countries, suggesting that ExPEC isolates from healthy poultry could be a source of potentially virulent ExPEC causing multiple diseases in humans. Notably, all the ExPEC isolates in this study possessed MDR phenotypes and most showed resistances to cephalosporins and fosfomycin, which made co-selection of these ExPEC possible when corresponding drugs were used. More worryingly, six ExPEC isolates in this study carried *mcr-1*, including two harboring both *bla*$_{\text{NDM}}$ and *mcr-1*, which could compromise both the efficacies of carbapenems and colistin. The presence of MDR ExPEC isolates in healthy chickens, especially those carrying *mcr-1* and/or *bla*$_{\text{NDM}}$, is alarming and will pose a serious health threat to consumers. Interventions need to be taken to reduce these pathogens in the chicken intestine and prevent clinical ExPEC infections in humans by reducing transmission via poultry products. Further studies are required for monitoring the prevalence of MDR ExPEC in healthy chickens in China and other countries. To our knowledge, this is the first report of *mcr-1*-positive ExPEC isolates harboring bla_{NDM} from healthy chickens.

Supplementary Materials: The following are available online at https://www.mdpi.com/article/10.3390/ani11041112/s1, Table S1: Origins of the samples collected in this study.

Author Contributions: B.-T.L. designed the study, secured the funds, and supervised the whole work. M.Z. collected the samples and carried out the laboratory work. P.-P.M., W.-S.L. and X.L. also contributed in the laboratory work. B.-T.L., M.Z., P.-P.M. and X.-Y.L. wrote the original draft of the manuscript. B.-T.L., M.Z. and Y.-Z.L. did the data analysis and interpretation. B.-T.L. and M.Z. critically reviewed and edited the manuscript. All authors have read and agreed to the published version of the manuscript.

Funding: This study was supported by grants from the Scientific and Technological Projects of Qingdao (19-6-1-94-nsh and 21-1-4-ny-11-nsh), the Shandong Major Science and Technology Innovation Projects (2019JZZY010719), Major Agricultural Application Technology Innovation Project of Shandong Province (SD2019XM004) and the Scientific and Technological Projects of Qingdao (21-1-4-ny-10-nsh).

Institutional Review Board Statement: All animal experiments were carried out in accordance with guidelines issued by the Qingdao Agricultural University Animal Care and Use Committee (approval number, QDAUA-2015-033).

Data Availability Statement: No new data were created or analyzed in this study. Data sharing is not applicable to this article.

Conflicts of Interest: The authors declare no conflict of interest.

References

1. Mellata, M. Human and Avian Extraintestinal Pathogenic *Escherichia coli*: Infections, Zoonotic Risks, and Antibiotic Resistance Trends. *Foodborne Pathog. Dis.* **2013**, *10*, 916–932. [CrossRef] [PubMed]
2. Johnson, J.R.; Russo, T.A. Extraintestinal pathogenic *Escherichia coli*: "The other bad *E coli*". *J. Lab. Clin. Med.* **2002**, *139*, 155–162. [CrossRef] [PubMed]
3. Flores-Mireles, A.L.; Walker, J.N.; Caparon, M.; Hultgren, S.J. Urinary tract infections: Epidemiology, mechanisms of infection and treatment options. *Nat. Rev. Microbiol.* **2015**, *13*, 269–284. [CrossRef] [PubMed]
4. Laupland, K.B. Incidence of bloodstream infection: A review of population-based studies. *Clin. Microbiol. Infect.* **2013**, *19*, 492–500. [CrossRef] [PubMed]
5. Russo, T.A.; Johnson, J.R. Medical and economic impact of extraintestinal infections due to *Escherichia coli*: Focus on an increasingly important endemic problem. *Microbes Infect.* **2003**, *5*, 449–456. [CrossRef]
6. Johnson, J.R.; Murray, A.C.; Gajewski, A.; Sullivan, M.; Snippes, P.; Kuskowski, M.A.; Smith, K.E. Isolation and molecular characterization of nalidixic acid-resistant extraintestinal pathogenic *Escherichia coli* from retail chicken products. *Antimicrob. Agents Chemother.* **2003**, *47*, 2161–2168. [CrossRef]
7. Gomi, R.; Matsuda, T.; Matsumura, Y.; Yamamoto, M.; Tanaka, M.; Ichiyama, S.; Yoneda, M. Whole-Genome Analysis of Antimicrobial-Resistant and Extraintestinal Pathogenic *Escherichia coli* in River Water. *Appl. Environ. Microbiol.* **2016**, *83*, e02703-16. [CrossRef] [PubMed]
8. Zhi, S.; Stothard, P.; Banting, G.; Scott, C.; Huntley, K.; Ryu, K.; Otto, S.; Ashbolt, N.; Checkley, S.; Dong, T.; et al. Characterization of water treatment-resistant and multidrug-resistant urinary pathogenic *Escherichia coli* in treated wastewater. *Water Res.* **2020**, *182*, 115827. [CrossRef]
9. Rayasam, S.D.G.; Ray, I.; Smith, K.R.; Riley, L.W. Extraintestinal Pathogenic *Escherichia coli* and Antimicrobial Drug Resistance in a Maharashtrian Drinking Water System. *Am. J. Trop. Med. Hyg.* **2019**, *100*, 1101–1104. [CrossRef]
10. Johnson, J.R.; Porter, S.B.; Johnston, B.; Thuras, P.; Clock, S.; Crupain, M.; Rangan, U. Extraintestinal Pathogenic and Antimicrobial-Resistant *Escherichia coli*, Including Sequence Type 131 (ST131), from Retail Chicken Breasts in the United States in 2013. *Appl. Environ. Microbiol.* **2017**, *83*, e02956-16. [CrossRef]
11. Díaz-Jiménez, D.; García-Meniño, I.; Fernández, J.; García, V.; Mora, A. Chicken and turkey meat: Consumer exposure to multidrug-resistant Enterobacteriaceae including *mcr*-carriers, uropathogenic *E. coli* and high-risk lineages such as ST131. *Int. J. Food Microbiol.* **2020**, *331*, 108750. [CrossRef]

12. Jakobsen, L.; Spangholm, D.J.; Pedersen, K.; Jensen, L.B.; Emborg, H.-D.; Agersø, Y.; Aarestrup, F.M.; Hammerum, A.M.; Frimodt-Møller, N. Broiler chickens, broiler chicken meat, pigs and pork as sources of ExPEC related virulence genes and resistance in *Escherichia coli* isolates from community-dwelling humans and UTI patients. *Int. J. Food Microbiol.* **2010**, *142*, 264–272. [CrossRef] [PubMed]

13. Jakobsen, L.; Garneau, P.; Kurbasic, A.; Bruant, G.; Stegger, M.; Harel, J.; Jensen, K.S.; Brousseau, R.; Hammerum, A.M.; Frimodt-Møller, N. Microarray-based detection of extended virulence and antimicrobial resistance gene profiles in phylogroup B2 *Escherichia coli* of human, meat and animal origin. *J. Med Microbiol.* **2011**, *60*, 1502–1511. [CrossRef] [PubMed]

14. Ferreira, J.C.; Filho, R.A.C.P.; Kuaye, A.P.Y.; Andrade, L.N.; Chang, Y.-F.; Darini, A.L.C. Virulence potential of commensal multidrug resistant *Escherichia coli* isolated from poultry in Brazil. *Infect. Genet. Evol.* **2018**, *65*, 251–256. [CrossRef]

15. Bin Kim, Y.; Yoon, M.Y.; Ha, J.S.; Seo, K.W.; Noh, E.B.; Son, S.H.; Lee, Y.J. Molecular characterization of avian pathogenic *Escherichia coli* from broiler chickens with colibacillosis. *Poult. Sci.* **2020**, *99*, 1088–1095. [CrossRef]

16. Subedi, M.; Luitel, H.; Devkota, B.; Bhattarai, R.K.; Phuyal, S.; Panthi, P.; Shrestha, A.; Chaudhary, D.K. Antibiotic resistance pattern and virulence genes content in avian pathogenic *Escherichia coli* (APEC) from broiler chickens in Chitwan, Nepal. *BMC Veter. Res.* **2018**, *14*, 113. [CrossRef]

17. Johnson, T.J.; Wannemuehler, Y.; Doetkott, C.; Rosenberger, S.C.; Nolan, L.K. Identification of minimal predictors of avian pathogenic *Escherichia coli* virulence for use as a rapid diagnostic tool. *J. Clin. Microbiol.* **2008**, *46*, 3987–3996. [CrossRef] [PubMed]

18. Saidenberg, A.B.S.; Stegger, M.; Price, L.B.; Johannesen, T.B.; Aziz, M.; Cunha, M.P.; Moreno, A.M.; Knöbl, T. *mcr*-Positive *Escherichia coli* ST131-H22 from Poultry in Brazil. *Emerg. Infect. Dis.* **2020**, *26*, 1951–1954. [CrossRef]

19. Mohamed, M.A.; Shehata, M.A.; Rafeek, E. Virulence Genes Content and Antimicrobial Resistance in *Escherichia coli* from Broiler Chickens. *Veter. Med. Int.* **2014**, *2014*, 195189. [CrossRef]

20. Stromberg, Z.R.; Johnson, J.R.; Fairbrother, J.M.; Kilbourne, J.; Van Goor, A.; Curtiss, R.; Mellata, M. Evaluation of *Escherichia coli* isolates from healthy chickens to determine their potential risk to poultry and human health. *PLoS ONE* **2017**, *12*, e0180599. [CrossRef]

21. Bélanger, L.; Garenaux, A.; Harel, J.; Boulianne, M.; Nadeau, E.; Dozois, C.M. *Escherichia coli* from animal reservoirs as a potential source of human extraintestinal pathogenic *E. coli*. *FEMS Immunol. Med Microbiol.* **2011**, *62*, 1–10. [CrossRef]

22. Pitout, J.D. Extraintestinal Pathogenic *Escherichia coli*: A Combination of Virulence with Antibiotic Resistance. *Front. Microbiol.* **2012**, *3*, 9. [CrossRef] [PubMed]

23. Aslam, M.; Toufeer, M.; Narvaez Bravo, C.; Lai, V.; Rempel, H.; Manges, A.; Diarra, M.S. Characterization of Extraintestinal Pathogenic *Escherichia coli* isolated from retail poultry meats from Alberta, Canada. *Int. J. Food Microbiol.* **2014**, *177*, 49–56. [CrossRef] [PubMed]

24. Tumbarello, M.; Sanguinetti, M.; Montuori, E.; Trecarichi, E.M.; Posteraro, B.; Fiori, B.; Citton, R.; D'Inzeo, T.; Fadda, G.; Cauda, R.; et al. Predictors of mortality in patients with bloodstream infections caused by extended-spectrum-beta-lactamase-producing Enterobacteriaceae: Importance of inadequate initial antimicrobial treatment. *Antimicrob. Agents Chemother.* **2007**, *51*, 1987–1994. [CrossRef]

25. Johnson, J.R.; Kuskowski, M.A.; Gajewski, A.; Sahm, D.F.; Karlowsky, J.A. Virulence characteristics and phylogenetic background of multidrug-resistant and antimicrobial-susceptible clinical isolates of *Escherichia coli* from across the United States, 2000–2001. *J. Infect. Dis.* **2004**, *190*, 1739–1744. [CrossRef]

26. Collignon, P.; Voss, A. China, what antibiotics and what volumes are used in food production animals? *Antimicrob. Resist. Infect. Control.* **2015**, *4*, 16. [CrossRef]

27. Chen, X.; Gao, S.; Jiao, X.; Liu, X.F. Prevalence of serogroups and virulence factors of *Escherichia coli* strains isolated from pigs with postweaning diarrhoea in eastern China. *Vet. Microbiol.* **2004**, *103*, 13–20. [CrossRef]

28. CLSI. *Performance Standards for Antimicrobial Susceptibility Testing*; Twenty-Fifth Informational Supplement. CLSI Document M100-S25; Clinical and Laboratory Standards Institute: Wayne, PA, USA, 2015.

29. CLSI. *Performance Standards for Antimicrobial Disk and Dilution Susceptibility Tests for Bacterial Isolated from Animals*; Approved Standard-Fourth Edition and Supplement, VET01A4E and VET01S3E; Clinical and Laboratory Standards Institute: Wayne, PA, USA, 2015.

30. Rebelo, A.R.; Bortolaia, V.; Kjeldgaard, J.S.; Pedersen, S.K.; Leekitcharoenphon, P.; Hansen, I.M.; Guerra, B.; Malorny, B.; Borowiak, M.; Hammerl, J.A.; et al. Multiplex PCR for detection of plasmid-mediated colistin resistance determinants, *mcr-1*, *mcr-2*, *mcr-3*, *mcr-4* and *mcr-5* for surveillance purposes. *Eurosurveillance* **2018**, *23*, 17-00672. [CrossRef]

31. Borowiak, M.; Baumann, B.; Fischer, J.; Thomas, K.; Deneke, C.; Hammerl, J.A.; Szabo, I.; Malorny, B. Development of a Novel *mcr-6* to *mcr-9* Multiplex PCR and Assessment of *mcr-1* to *mcr-9* Occurrence in Colistin-Resistant *Salmonella enterica* Isolates From Environment, Feed, Animals and Food (2011–2018) in Germany. *Front. Microbiol.* **2020**, *11*, 80. [CrossRef]

32. Poirel, L.; Walsh, T.R.; Cuvillier, V.; Nordmann, P. Multiplex PCR for detection of acquired carbapenemase genes. *Diagn. Microbiol. Infect. Dis.* **2011**, *70*, 119–123. [CrossRef] [PubMed]

33. Sun, J.; Chen, C.; Cui, C.-Y.; Zhang, Y.; Liu, X.; Cui, Z.-H.; Ma, X.-Y.; Feng, Y.-J.; Fang, L.-X.; Lian, X.-L.; et al. Plasmid-encoded *tet*(X) genes that confer high-level tigecycline resistance in *Escherichia coli*. *Nat. Microbiol.* **2019**, *4*, 1457–1464. [CrossRef] [PubMed]

34. Cavaco, L.M.; Hasman, H.; Xia, S.; Aarestrup, F.M. *qnrD*, a novel gene conferring transferable quinolone resistance in *Salmonella enterica* serovar Kentucky and Bovismorbificans strains of human origin. *Antimicrob. Agents Chemother.* **2008**, *53*, 603–608. [CrossRef] [PubMed]

35. Wang, M.; Guo, Q.; Xu, X.; Wang, X.; Ye, X.; Wu, S.; Hooper, D.C.; Wang, M. New plasmid-mediated quinolone resistance gene, *qnrC*, found in a clinical isolate of *Proteus mirabilis*. *Antimicrob. Agents Chemother.* **2009**, *53*, 1892–1897. [CrossRef] [PubMed]

36. Yue, L.; Jiang, H.-X.; Liao, X.-P.; Liu, J.-H.; Li, S.-J.; Chen, X.-Y.; Chen, C.-X.; Lü, D.-H.; Liu, Y.-H. Prevalence of plasmid-mediated quinolone resistance *qnr* genes in poultry and swine clinical isolates of *Escherichia coli*. *Vet. Microbiol.* **2008**, *132*, 414–420. [CrossRef] [PubMed]

37. Yamane, K.; Wachino, J.-I.; Suzuki, S.; Arakawa, Y. Plasmid-mediated *qepA* gene among *Escherichia coli* clinical isolates from Japan. *Antimicrob. Agents Chemother.* **2008**, *52*, 1564–1566. [CrossRef] [PubMed]

38. Liu, B.-T.; Song, F.-J.; Zou, M. Characterization of Highly Prevalent Plasmids Coharboring *mcr-1*, *oqxAB*, and *bla*CTX-M and Plasmids Harboring *oqxAB* and *bla*CTX-M in *Escherichia coli* Isolates from Food-Producing Animals in China. *Microb. Drug Resist.* **2019**, *25*, 108–119. [CrossRef]

39. Brinas, L.; Moreno, M.A.; Zarazaga, M.; Porrero, C.; Saenz, Y.; García, M.; Dominguez, L.; Torres, C. Detection of CMY-2, CTX-M-14, and SHV-12 beta-lactamases in *Escherichia coli* fecal-sample isolates from healthy chickens. *Antimicrob. Agents Chemother.* **2003**, *47*, 2056–2058. [CrossRef]

40. Liu, J.H.; Wei, S.Y.; Ma, J.Y.; Zeng, Z.L.; Lu, D.H.; Yang, G.-X.; Chen, Z.-L. Detection and characterisation of CTX-M and CMY-2 beta-lactamases among *Escherichia coli* isolates from farm animals in Guangdong Province of China. *Int. J. Antimicrob. Agents* **2007**, *29*, 576–581. [CrossRef]

41. Pai, H.; Kang, C.-I.; Byeon, J.-H.; Lee, K.-D.; Park, W.B.; Kim, H.-B.; Kim, E.-C.; Oh, M.-D.; Choe, K.-W. Epidemiology and clinical features of bloodstream infections caused by AmpC-type-beta-lactamase-producing *Klebsiella pneumoniae*. *Antimicrob. Agents Chemother.* **2004**, *48*, 3720–3728. [CrossRef]

42. Kojima, A.; Ishii, Y.; Ishihara, K.; Esaki, H.; Asai, T.; Oda, C.; Tamura, Y.; Takahashi, T.; Yamaguchi, K. Extended-spectrum-beta-lactamase-producing *Escherichia coli* strains isolated from farm animals from 1999 to 2002: Report from the Japanese Veterinary Antimicrobial Resistance Monitoring Program. *Antimicrob. Agents Chemother.* **2005**, *49*, 3533–3537. [CrossRef]

43. Hou, J.; Huang, X.; Deng, Y.; He, L.; Yang, T.; Zeng, Z.; Chen, Z.; Liu, J.-H. Dissemination of the fosfomycin resistance gene *fosA3* with CTX-M beta-lactamase genes and *rmtB* carried on IncFII plasmids among *Escherichia coli* isolates from pets in China. *Antimicrob. Agents Chemother.* **2012**, *56*, 2135–2138. [CrossRef]

44. Chen, L.; Chen, Z.-L.; Liu, J.-H.; Zeng, Z.-L.; Ma, J.-Y.; Jiang, H.-X. Emergence of RmtB methylase-producing *Escherichia coli* and *Enterobacter cloacae* isolates from pigs in China. *J. Antimicrob. Chemother.* **2007**, *59*, 880–885. [CrossRef]

45. Debroy, C.; Roberts, E.; Fratamico, P.M. Detection of O antigens in *Escherichia coli*. *Anim. Heal. Res. Rev.* **2011**, *12*, 169–185. [CrossRef]

46. Gautom, R.K. Rapid pulsed-field gel electrophoresis protocol for typing of *Escherichia coli* O157:H7 and other gram-negative organisms in 1 day. *J. Clin. Microbiol.* **1997**, *35*, 2977–2980. [CrossRef] [PubMed]

47. Wirth, T.; Falush, D.; Lan, R.; Colles, F.; Mensa, P.; Wieler, L.H.; Karch, H.; Reeves, P.R.; Maiden, M.C.J.; Ochman, H.; et al. Sex and virulence in *Escherichia coli*: An evolutionary perspective. *Mol. Microbiol.* **2006**, *60*, 1136–1151. [CrossRef] [PubMed]

48. Wu, C.; Wang, Y.; Shi, X.; Wang, S.; Ren, H.; Shen, Z.; Wang, Y.; Lin, J.; Wang, S. Rapid rise of the ESBL and *mcr-1* genes in *Escherichia coli* of chicken origin in China, 2008–2014. *Emerg. Microbes Infect.* **2018**, *7*, 30. [CrossRef] [PubMed]

49. FDA. *Food and Drug Administration (2010) CVM Updates-CVM Reports on Antimicrobials Sold or Distributed for Food-Producing Animals*; Food Drug Admin: Silver Spring, MD, USA, 2010.

50. Ferjani, S.; Saidani, M.; Amine, F.S.; Boubaker, I.B.-B. Prevalence and characterization of plasmid-mediated quinolone resistance genes in extended-spectrum beta-lactamase-producing Enterobacteriaceae in a Tunisian hospital. *Microb. Drug Resist.* **2015**, *21*, 158–166. [CrossRef] [PubMed]

51. Liu, B.-T.; Yang, Q.-E.; Li, L.; Sun, J.; Liao, X.-P.; Fang, L.-X.; Yang, S.-S.; Deng, H.; Liu, Y.-H. Dissemination and characterization of plasmids carrying *oqxAB-bla*CTX-M genes in *Escherichia coli* isolates from food-producing animals. *PLoS ONE* **2013**, *8*, e73947. [CrossRef] [PubMed]

52. Chmielarczyk, A.; Pobiega, M.; De Champs, C.; Wojkowska-Mach, J.; Różańska, A.; Heczko, P.B.; Guillard, T.; Bulanda, M. The High Prevalence of Plasmid-Mediated Quinolone Resistance Among Very Low Birth-Weight Infants in Poland. *Microb. Drug Resist.* **2015**, *21*, 391–397. [CrossRef] [PubMed]

53. Brower, C.H.; Mandal, S.; Hayer, S.; Sran, M.; Zehra, A.; Patel, S.J.; Kaur, R.; Chatterjee, L.; Mishra, S.; Das, B.; et al. The Prevalence of Extended-Spectrum Beta-Lactamase-Producing Multidrug-Resistant *Escherichia Coli* in Poultry Chickens and Variation According to Farming Practices in Punjab, India. *Environ. Health Perspect.* **2017**, *125*, 077015. [CrossRef]

54. Clermont, O.; Olier, M.; Hoede, C.; Diancourt, L.; Brisse, S.; Keroudean, M.; Glodt, J.; Picard, B.; Oswald, E.; Denamur, E. Animal and human pathogenic *Escherichia coli* strains share common genetic backgrounds. *Infect. Genet. Evol.* **2011**, *11*, 654–662. [CrossRef] [PubMed]

55. Mitchell, N.M.; Johnson, J.R.; Johnston, B.; Curtiss, R., 3rd; Mellata, M. Zoonotic potential of *Escherichia coli* isolates from retail chicken meat products and eggs. *Appl. Environ. Microbiol.* **2015**, *81*, 1177–1187. [CrossRef]

56. Mora, A.; López, C.; Herrera, A.; Viso, S.; Mamani, R.; Dhabi, G.; Alonso, M.P.; Blanco, M.; Blanco, J.E.; Blanco, J. Emerging avian pathogenic *Escherichia coli* strains belonging to clonal groups O111:H4-D-ST2085 and O111:H4-D-ST117 with high virulence-gene content and zoonotic potential. *Vet. Microbiol.* **2012**, *156*, 347–352. [CrossRef]

57. Maluta, R.P.; Logue, C.M.; Casas, M.R.T.; Meng, T.; Guastalli, E.A.L.; Rojas, T.C.G.; Montelli, A.C.; Sadatsune, T.; de Carvalho Ramos, M.; Nolan, L.K.; et al. Overlapped sequence types (STs) and serogroups of avian pathogenic (APEC) and human extra-intestinal pathogenic (ExPEC) *Escherichia coli* isolated in Brazil. *PLoS ONE* **2014**, *9*, e105016. [CrossRef] [PubMed]
58. Moulin-Schouleur, M.; Répérant, M.; Laurent, S.; Brée, A.; Mignon-Grasteau, S.; Germon, P.; Rasschaert, D.; Schouler, C. Extraintestinal pathogenic *Escherichia coli* strains of avian and human origin: Link between phylogenetic relationships and common virulence patterns. *J. Clin. Microbiol.* **2007**, *45*, 3366–3376. [CrossRef] [PubMed]
59. Wang, X.-M.; Liao, X.-P.; Zhang, W.-J.; Jiang, H.-X.; Sun, J.; Zhang, M.-J.; He, X.-F.; Lao, D.-X.; Liu, Y.-H. Prevalence of serogroups, virulence genotypes, antimicrobial resistance, and phylogenetic background of avian pathogenic *Escherichia coli* in south of China. *Foodborne Pathog. Dis.* **2010**, *7*, 1099–1106. [CrossRef]
60. Johnson, T.J.; Wannemuehler, Y.; Johnson, S.J.; Stell, A.L.; Doetkott, C.; Johnson, J.R.; Kim, K.S.; Spanjaard, L.; Nolan, L.K. Comparison of extraintestinal pathogenic *Escherichia coli* strains from human and avian sources reveals a mixed subset representing potential zoonotic pathogens. *Appl. Environ. Microbiol.* **2008**, *74*, 7043–7050. [CrossRef]
61. Johnson, J.R.; Oswald, E.; O'Bryan, T.T.; Kuskowski, M.A.; Spanjaard, L. Phylogenetic distribution of virulence-associated genes among *Escherichia coli* isolates associated with neonatal bacterial meningitis in the Netherlands. *J. Infect. Dis.* **2002**, *185*, 774–784. [CrossRef]
62. Bonacorsi, S.; Clermont, O.; Houdouin, V.; Cordevant, C.; Brahimi, N.; Marecat, A.; Tinsley, C.; Nassif, X.; Lange, M.; Bingen, E. Molecular analysis and experimental virulence of French and North American *Escherichia coli* neonatal meningitis isolates: Identification of a new virulent clone. *J. Infect. Dis.* **2003**, *187*, 1895–1906. [CrossRef]
63. Denisuik, A.J.; Lagacé-Wiens, P.R.S.; Pitout, J.D.; Mulvey, M.R.; Simner, P.J.; Tailor, F.; Karlowsky, J.A.; Hoban, D.J.; Adam, H.J.; Zhanel, G.G.; et al. Molecular epidemiology of extended-spectrum beta-lactamase-, AmpC beta-lactamase- and carbapenemase-producing *Escherichia coli* and *Klebsiella pneumoniae* isolated from Canadian hospitals over a 5 year period: CANWARD 2007-11. *J. Antimicrob. Chemother.* **2013**, *68* (Suppl. 1), i57–i65. [CrossRef]
64. Karlowsky, J.A.; Denisuik, A.J.; Lagacé-Wiens, P.R.S.; Adam, H.J.; Baxter, M.R.; Hoban, D.J.; Zhanel, G.G. In Vitro activity of fosfomycin against *Escherichia coli* isolated from patients with urinary tract infections in Canada as part of the CANWARD surveillance study. *Antimicrob. Agents Chemother.* **2013**, *58*, 1252–1256. [CrossRef] [PubMed]
65. Rodríguez-Baño, J.; Gutiérrez-Gutiérrez, B.; Machuca, I.; Pascual, A. Treatment of Infections Caused by Extended-Spectrum-Beta-Lactamase-, AmpC-, and Carbapenemase-Producing Enterobacteriaceae. *Clin. Microbiol. Rev.* **2018**, *31*, e00079-17. [CrossRef] [PubMed]
66. Liu, F.; Zhang, R.; Yang, Y.; Li, H.; Wang, J.; Lan, J.; Li, P.; Zhu, Y.; Xie, Z.; Jiang, S. Occurrence and Molecular Characteristics of Mcr-1-Positive *Escherichia coli* from Healthy Meat Ducks in Shandong Province of China. *Animals* **2020**, *10*, 1299. [CrossRef]
67. Ranjan, A.; Shaik, S.; Mondal, A.; Nandanwar, N.; Hussain, A.; Semmler, T.; Kumar, N.; Tiwari, S.K.; Jadhav, S.; Wieler, L.H.; et al. Molecular Epidemiology and Genome Dynamics of New Delhi Metallo-beta-Lactamase-Producing Extraintestinal Pathogenic *Escherichia coli* Strains from India. *Antimicrob. Agents Chemother.* **2016**, *60*, 6795–6805. [CrossRef] [PubMed]

Article

S. aureus Biofilm Protein Expression Linked to Antimicrobial Resistance: A Proteomic Study

Cristian Piras [1,†], Pierluigi Aldo Di Ciccio [2,†], Alessio Soggiu [3], Viviana Greco [4,5], Bruno Tilocca [1], Nicola Costanzo [1], Carlotta Ceniti [1], Andrea Urbani [4,5], Luigi Bonizzi [3], Adriana Ianieri [6] and Paola Roncada [1,*]

[1] Department of Health Sciences, University "Magna Græcia" of Catanzaro, Campus Universitario "S. Venuta", Viale Europa, I-88100 Catanzaro, Italy; c.piras@unicz.it (C.P.); tilocca@unicz.it (B.T.); costanzo.nic@unicz.it (N.C.); ceniti@unicz.it (C.C.)

[2] Department of Veterinary Sciences, University of Torino, Largo Paolo Braccini 2, Grugliasco, 10095 Torino, Italy; pierluigialdo.diciccio@unito.it

[3] Surgical and Dental Sciences-One Health Unit, Department of Biomedical, University of Milano, Via Celoria 10, 20133 Milano, Italy; alessio.soggiu@unimi.it (A.S.); luigi.bonizzi@unimi.it (L.B.)

[4] Department of Basic Biotechnological Sciences, Intensivological and Perioperative Clinics, Università Cattolica del Sacro Cuore, Largo Francesco Vito 1, 00168 Roma, Italy; viviana.greco@unicatt.it (V.G.); andrea.urbani@unicatt.it (A.U.)

[5] Molecular and Genomic Diagnostics Unit, Fondazione Policlinico Universitario Agostino Gemelli IRCCS, Largo A. Gemelli 8, 00168 Roma, Italy

[6] Deparment of Food and Drug, University of Parma, Parco Area delle Scienze 27A, 43124 Parma, Italy; adriana.ianieri@unipr.it

* Correspondence: roncada@unicz.it

† Equal contribution.

Citation: Piras, C.; Di Ciccio, P.A.; Soggiu, A.; Greco, V.; Tilocca, B.; Costanzo, N.; Ceniti, C.; Urbani, A.; Bonizzi, L.; Ianieri, A.; et al. *S. aureus* Biofilm Protein Expression Linked to Antimicrobial Resistance: A Proteomic Study. *Animals* **2021**, *11*, 966. https://doi.org/10.3390/ani11040966

Academic Editor: Amit Vikram

Received: 3 March 2021
Accepted: 24 March 2021
Published: 31 March 2021

Publisher's Note: MDPI stays neutral with regard to jurisdictional claims in published maps and institutional affiliations.

Simple Summary: Biofilm formation represents one of the most effective forms of bacterial persistence in surfaces where nutrients are available or in the tissues of living hosts as humans or animals. Such persistence is due to the high rate of antimicrobial resistance of this shell conformation. It often represents a burden when the pathogen colonizes niches from where it is not removable such as food facilities, farm facilities or parts of living organisms. In this study, we investigated biofilm formation mechanisms and enhanced antimicrobial resistance of 6 different *S. aureus* strains. The detected mechanisms were primarily related to the control of catabolites, the production of proteins with moonlighting activities and the detoxification of compounds with antimicrobial activities (i.e., alcohol). Glycolysis and aerobic metabolisms were found to be less active in the biofilm conformation. Consequently, less H_2O_2 production from aerobic metabolism was translated into a measurable under-representation of catalase protein.

Abstract: Antimicrobial resistance (AMR) represents one of the most critical challenges that humanity will face in the following years. In this context, a "One Health" approach with an integrated multidisciplinary effort involving humans, animals and their surrounding environment is needed to tackle the spread of AMR. One of the most common ways for bacteria to live is to adhere to surfaces and form biofilms. *Staphylococcus aureus* (*S. aureus*) can form biofilm on most surfaces and in a wide heterogeneity of environmental conditions. The biofilm guarantees the survival of the *S. aureus* in harsh environmental conditions and represents an issue for the food industry and animal production. The identification and characterization of biofilm-related proteins may provide interesting insights into biofilm formation mechanisms in *S. aureus*. In this regard, the aims of this study were: (i) to use proteomics to compare proteomes of *S. aureus* growing in planktonic and biofilm forms in order to investigate the common features of biofilm formation properties of different strains; (ii) to identify specific biofilm mechanisms that may be involved in AMR. The proteomic analysis showed 14 differentially expressed proteins among biofilm and planktonic forms of *S. aureus*. Moreover, three proteins, such as alcohol dehydrogenase, ATP-dependent 6-phosphofructokinase, and fructose-bisphosphate aldolase, were only differentially expressed in strains classified as high biofilm producers. Differentially regulated catabolites metabolisms and the switch to lower oxygen-related metabolisms were related to the sessile conformation analyzed.

Keywords: *Staphylococcus aureus*; planktonic cells; biofilm; proteomics; food safety; antimicrobial resistance

1. Introduction

Humanity is already facing a challenge related to antimicrobial resistance (AMR). Such a burden will become worse due to the massive use of antimicrobials such as alcohol-based products for hands and workplaces sanitization necessary to mitigate the transmission of coronavirus disease 2019 (COVID-19). These key precautions may create an ecological pressure on microorganisms and contribute to the emergence of AMR in microbial populations that can colonize human body and the environment.

The use of biocides in the environment (such as farms and food industries) has already created a phenomenon known as AMR cross-resistance [1–3]. Biofilm formation contributes to enhance AMR resistance by physical and biochemical means [4]. A biofilm is defined as "a microbially derived sessile community characterized by cells that are irreversibly attached to a substrate or interface or to each other, are embedded in an auto-produced matrix of extracellular polymeric substances (which is composed of protein, DNA and polysaccharide) and exhibit an altered phenotype with respect to growth rate and gene transcription" [5]. It is well known that bacteria growing as biofilms might be physiologically distinct from the same bacteria growing as free-swimming planktonic cells [6,7].

Briefly, biofilms allow bacteria to better resist harsh environmental conditions [8]. Such a conformation can be found everywhere where there is a source of nutrients such as in the food-processing environment or zootechnical industry (food-processing equipment, milk collection and storage facilities) [9]. Biofilms-enhanced resistance to disinfectants/antimicrobials/antibiotics represents a threat for food industries and farms [10]. The biofilm, in fact, protects the bacteria from detaching by cleaning agents and from being killed by disinfectants [11]. However, biofilm protection mechanisms appear to be different from those responsible for resistance to conventional antibiotics [12]. First, the extracellular polymeric substances (EPS) matrix delays or prevents antimicrobial action, either by limiting disinfectants diffusion or by chemical interaction/inactivation with proteins and extracellular polysaccharides [13]. Other factors can play a role in this feature, such as the bacterial growth rate, the heterogeneity within the biofilm, the general stress response, quorum sensing mechanisms, the induction of a certain biofilm phenotype and the over-expression of efflux pumps [14]. In addition, biofilm activities include the upregulation of virulence factors and secretion of extracellular polymers [15]. Horizontal gene transfer plays an important role in AMR. The small intra-cellular distance typical of biofilms facilitates the spread of resistance genes and generates the presence of extracellular DNA in the biofilm matrix [16].

Among bacteria, *Staphylococcus aureus* (*S. aureus*) is able to form biofilm on most surfaces and under almost all the environmental conditions found in food industries [17]. It is a commensal and opportunistic pathogen and under certain conditions, may cause a wide range of infectious diseases such as skin infections, bacteremia, endocarditis, pneumonia and food poisoning. *S. aureus* biofilm mode of growth is regulated by complex genetic factors and can produce at least two different types of biofilm: ica operon-dependent (i.e., promoted by the ica operon) and ica operon-independent [17]. A study carried out by Resch et al. (2005) identified more than 160 genes that were significantly over-expressed during biofilm growth conditions. Those genes encoded for binding factors, polysaccharide intracellular adhesion (PIA) and peptidoglycan modeling factors [7]. Additionally, many proteins have been implicated as important components in cellular adhesion and biofilm matrix development [18]. These include surface-associated proteins (protein A), fibrinogen-binding proteins (FnBPA and FnBPB), biofilm-associated protein (Bap) and clumping factor B (ClfB).

Considering the concerns for food safety associated with *S. aureus* biofilms and the high cost of managing this issue in the food industry, a better knowledge of the mechanisms involved in *S. aureus* biofilm growth mode is essential. To date, several studies have focused on pathogenicity and only a few have addressed differences in protein expression of *S. aureus* due to biofilm formation [19,20]. The identification and characterization of proteins linked with biofilm could provide interesting insights on the mechanism and/or process of biofilm formation in *S. aureus*.

According to this premise, the aims of this study were: (i) to compare proteomes of *S. aureus* growing in planktonic and biofilm forms, in order to investigate the common features of biofilm formation properties of six different strains; and (ii) to identify possible biofilm mechanisms that may be involved in AMR. The employment of 6 different strains will help with the comprehension of biofilm formation mechanisms more representative of the *S. aureus* species rather than be focused on mechanisms typical of a single strain.

2. Materials and Methods

2.1. Bacterial Strains

A total of six biofilm-forming *S. aureus* strains were analyzed in this study. In details, three *S. aureus* reference strains (ATCC 35556, ATCC 12600, ATCC 29213) and three food-related isolates (wild-types) were used in the experiment. The food related-strains were isolated from food (n.1) and food handlers (n.2), respectively.

Stock cultures were stored at $-80\ °C$. All strains were incubated for 24 h at 37 °C in tryptone soy broth (TSB, Oxoid S.p.A., Milan, Italy) before each experiment. All these strains have been grown both in the planktonic and in the sessile form (biofilm cultures) and analyzed through 2D electrophoresis coupled with matrix-assisted laser desorption/ionization time-of-flight mass spectrometry (MALDI-TOF MS).

The sessile (biofilm cultures) and planktonic cells were prepared as follows:

- Biofilm cultures

The ability of *S. aureus* isolates to produce biofilms was determined according to the protocol described by Di Ciccio et al., 2015 [21]. In all cases, all experiments were repeated in triplicate. Briefly, polystyrene tissue culture plates (6 wells—961 mm^2) were used as substratum for biofilm formation at 37 °C. Cultures of *S. aureus* were prepared, from overnight tryptone soy agar (TSA, Oxoid S.p.A., Milan, Italy) growth, in TSB by incubating at selected temperature: 37 °C. Cultures were then washed three times with sterile phosphate-buffered saline (PBS; pH 7.3) (Sigma-Aldrich S.r.l., Milan, Italy) and diluted with fresh TSB to reach a concentration of about 10^8 colony-forming units (CFU) mL^{-1} by reading the optical density (OD) level at 550 nm (UV Mini-1240—Shimadzu, Long Beach, CA, USA). Three milliliters (ml) of the standardized inoculum were then added to polystyrene tissue culture plates (well—35 mm diameter). Samples were then incubated at 37 °C. After 24 h incubation, non-adherent cells were removed by washing each well three times with sterile PBS. After adding sterile PBS (3 mL), biofilm in wells was dislodged mechanically by scraping vigorously using a sterile cell-scraper. Finally, the cells were harvested by centrifugation (4000 rpm, 10 min., 4 °C, Beckman, J2-MC, centrifuge). The resulting pellets, washed and resuspended in sterile PBS, were centrifuged again (4000 rpm, 10 min., 4 °C). The cells were washed several times and pelleted by five centrifugations. Finally, the supernatant was removed and the pellet from the biofilm cultures grown was stored at $-80\ °C$ until use for proteomic studies (the pellets from the biofilm cultures had a weight of 50 mg).

- Planktonic cells

S. aureus reference strains (ATCC 35556, ATCC 12600, ATCC 29213) and food-related isolates *S. aureus* (281, 402, 184) were used. An overnight culture was created by inoculating a colony of *S. aureus* into 5 mL of TSB for 24 h at 37 °C. After incubation, the *S. aureus* culture was centrifuged for 10 min at 4000 rpm, 4 °C. The supernatant was then replaced with sterile PBS, and pellet was resuspended by thoroughly mixing with pipette. The cells were washed several times and pelleted by five centrifugations (4000 rpm, 10 min, 4 °C). Finally, the supernatant was removed and the pellet from the overnight cultures grown was stored at −80 °C until use for proteomic studies (the pellets from the planktonic cultures had weights: 50 mg).

2.2. Proteomic Analysis

- Protein Extraction and 2-Dimensional Electrophoresis (2-DE)

We diluted 50 milligrams of cellular pellet of the different *S. aureus* strains in 700 µL of 2DE buffer containing 7 M urea, 2 M thiourea, 4% CHAPS, 1% DTT, and protease inhibitors (GE-Healthcare) according to manufacturer instructions.

To ensure the complete disruption of the collected bacterial cells, the samples were processed with 6 cycles of 1-min bead beating interspersed by a cycle of centrifuge. For this purpose, into the sample was added the same amount (1:1 *v/w*) of 0.1 mm zyrcounium-silica beads (300 µg beads added to 300 µL of buffer + the volume of the pellet). The bead beating cycle was conducted at 4000 rpm for 1 min with the purpose to avoid overheating. Then, the samples were centrifuged at $12,000 \times g$ for 5 min at 4 °C in order to chill and disperse the foam. This operation was repeated 6 times. After the 6th cycle, samples were centrifuged for 20 min and the supernatant was stored in another tube for subsequent proteomics analysis.

Two-dimensional (2D) electrophoresis was run in all samples: 100 micrograms of protein were loaded on a 7 cm strip through active rehydration performed overnight at 50 V in a buffer containing 7 M urea, 2 M thiourea, 2% CHAPS, 0.5% ampholytes 3–10 Amersham, and 26 mM DTT. For isoelectric focusing (IEF), the following protocol was applied: 100 V/1 h linear, 250 V/2 h linear, 4000 V/5 h linear, 4000 V step/50,000 total volt-hours (VhT), using a protean IEF platform.

Once the final amount of VhT was reached, immobilized pH gradient (IPG) strips were frozen up to the next step or directly equilibrated in two steps of 15 min under gentle stirring. The first step of equilibration was performed in buffer (6 M UREA, 2% SDS, 0.05 M Tris-HCl pH 8.8, 20% glycerol) supplemented with 1% DTT *w/v* and the second step was performed in a buffer with the addition of 2.5% *w/v* iodoacetamide. The IPG strips were put in a 12% home-made acrylamide gel and IEF run under constant amperage of 15 mA per gel, until the bromophenol blue (BFB) reached the front. The gels were then eliminated from the plates, washed three times with double-distilled water and spotted overnight (ON) with Coomassie Brilliant Blue.

Using an Imagescanner III (GE Healthcare) the gels were digitalized. The image analysis was performed using SameSpots software (Version 4.5, Nonlinear Dynamics U.K.). All imported images were checked for quality (saturation, ending) and spots, with a *p*-value lower than 0.05, were manually excised for subsequent mass spectrometry (MS) analysis and protein identification. If the MALDI MS/MS identification was obtained with a MASCOT score higher than 40, the protein was analyzed via GO for the comprehension of its function/role.

- Protein Identification by Matrix-Assisted Laser Desorption/Ionization Time-of-Flight Mass Spectrometry (MALDI-TOF/TOF MS) Analysis

Protein identification was performed according to previous studies [22,23].

Briefly, after different steps of dehydration, reduction and alkylation, the excided single spots were digested with a solution of 0.01 µg/µL of porcine trypsin (Promega, Madison, WI, USA) at 37 °C o.n., and peptides were concentrated using C18 ZipTip (Millipore, Bedford, MA, USA). they were then co-crystallized with a solution of αciano-4-hydroxycinnamic acid and spotted on a Ground Steel plate (Bruker-Daltonics, Bremen, Germany).

The MS analysis was performed on a Ultraflex III MALDI-TOF/TOF spectrometer (Bruker-Daltonics) in positive reflectron mode.

External calibration was performed using the standard peptide mixture calibration (m/z: 1046.5418, 1296.6848, 1347.7354, 1619.8223, 2093.0862, 2465.1983, 3147.4710; Bruker-Daltonics).

FlexAnalysis 3.3 software (Bruker-Daltonics) was used for the selection of the monoisotopic peptide masses of each mass spectra. After an internal calibration on autolysis peaks of porcine trypsin (m/z: 842.509 and 2211.104) and exclusion of contaminant ions (known matrix and human keratin peaks), the created peak lists were analyzed by MASCOT v.2.4.1 algorithm (www.matrixscience.com, accessed on 23 March 2021) searching against SwissProt 2021_database restricted to Firmicutes and *Staphylococcus aureus* (11,196 sequences) taxonomy.

The parameters used for database search are the following: carbamidomethylation of cysteines and oxidation on methionine as fixed and variable modifications respectively; one missed cleavage site allowed for trypsin; 70 ppm as maximal tolerance.

Mascot protein scores greater than 50 were considered significant ($p < 0.05$) for protein identification assignment.

To confirm the identifications, MS/MS spectra were also acquired by switching the instrument in LIFT mode with $4–8 \times 10^3$ laser shots. For the fragmentation, precursor ions were manually selected, and the precursor mass window was automatically set. Spectra baseline subtraction, smoothing (Savitsky–Golay) and centroiding were operated using Flex-Analysis 3.3 software.

The parameters used for the database search are the following: carbamidomethylation of cysteines and oxidation on methionine as fixed and variable modifications respectively; one missed cleavage; 50 ppm and 0. 5 Da as mass tolerance for precursor ions and for fragments respectively. The taxonomy was restricted to *Staphylococcus aureus* (10,227 sequences).

The confidence interval for protein identification was set to 95% ($p < 0.05$) and only peptides with an individual ion score above the identity threshold were considered correctly identified.

3. Results

The proteomic analysis was performed in order to discover the mechanisms of biofilm formation common to all analyzed *S. aureus* strains. Six different strains with different biofilm formation indexes were analyzed in the planktonic form and the biofilm form. For each strain, biofilm formation, expressed as BPI, was calculated as follows: "BPI = [OD$_{\text{mean biofilm}}$ surface $(\text{mm}^2)^{-1}] \times 1000$". All isolates were defined in different categories (weak, moderate or strong producers) on the basis of their BPIs values (Table 1).

Table 1. Biofilm formation index (BPI) of *S. aureus* strains on polystyrene at 37° included in this study.

Strains	Source	BPI
A—*S. aureus* ATCC n.35556	ATCC n.35556	0.758
B—*S. aureus* ATCC n.12600	ATCC n.12600	0.405
C—*S. aureus* n.281	Food	1.019
D—*S. aureus* n.402	Food-handler	0.311
E—*S. aureus* n.184	Food-handler	0.290
F—*S. aureus* ATCC 29213	ATCC n.29213	0.260

The analyzed strains included: *S. aureus* ATCC 35556, already described as a strong biofilm producer [24,25]; *S. aureus* ATCC 12600, classified as moderate biofilm producer [21]; three food isolates of *S. aureus* classified as strong (281), moderate (402) and weak biofilm producer (184); *S. aureus* ATCC 29213 measured as weak biofilm producer. BPI on polystyrene at 37 °C was used as the measure for all the experimental procedures in this work. All the strains with BPI below 0.300 were considered weak biofilm producers. In these cases, the biofilm layer was phenotypically barely visible and not stable in its structure. Such a phenotype was confirmed by the extremely low BPI below 300. For this reason, four strains (A, B, C and D) were considered as part of the moderate/high biofilm-producing group, while the remaining two (E and F) showed a phenotype closer to the low/non-forming biofilm group.

Proteomics analysis was carried out to compare the sessile versus the planktonic phenotype; however, a separated analysis was performed, including only the moderated to strong biofilm producers. The differentially represented proteins were chosen according to the Progenesis SameSpots provided analysis of variance (ANOVA) *p*-value. The topmost significant ones were chosen to be analyzed via MALDI-TOF MS/MS peptide mass fingerprinting (PMF) and peptide fragment fingerprinting (PFF) if necessary. Of the chosen spots, only the ones successfully identified with a MASCOT score higher than 40 were considered for subsequent Gene Ontology (GO), metabolism and pathway analysis.

As reported in Table 2, 14 proteins were differentially expressed among *S. aureus* planktonic and sessile groups. Of these, 11 were differentially expressed when considering all the strains together with a *p*-value ≤ 0.05 (column: regulation in planktonic vs. sessile). Alcohol dehydrogenase, ATP-dependent 6-phosphofructokinase and Fructose-bisphosphate aldolase differential expression were significant for the medium/high biofilm-forming sub-group (high biofilm producers' column). This classification was done according to the observation of the datasets that clearly showed how the representation trend of some of the differentially expressed proteins was clearly not following the same path in the weak biofilm forming strains. As previously mentioned, this was the case for alcohol dehydrogenase, ATP-dependent 6-phosphofructokinase and fructose-bisphosphate aldolase.

If considering the entirety of the differentially regulated proteins, five were found to be over-represented in the sessile versus planktonic group, and 9 proteins were found to be under-represented. This low number of detected proteins might be due to the high heterogeneity of the different strain analyzed. Three of the five over-represented proteins were involved in carbon metabolism or in stress response. Interestingly, alcohol dehydrogenase and 30 s ribosomal proteins are involved in antimicrobials resistance mechanisms, i.e., detoxification.

On the other hand, under-represented proteins such as 2,3-bisphosphoglycerate-dependent phosphoglycerate mutase, alkyl hydroperoxide reductase subunit C, ATP-dependent 6-phosphofructokinase, catalase etc. were mostly involved in energy and oxygen-related metabolism.

All data are shown in Table 2 and the image of the differentially represented proteins is shown in Figure 1a. For each protein, it is represented the related figure from the image analysis software. Table 2 indicates the *p*-values obtained from the built-in ANOVA test of the Progenesis SameSpots software. For each protein it is provided with the UNIPROT name and accession number (first two columns of the table); the SameSpots coding number, which represents the code provided by the image analysis software; the Mascot score identification obtained by the combined MALDI peptide mass fingerprint together with the peptide fragment fingerprint for the MS/MS identification; the number of matched peptides and the mascot score; and the ANOVA *p*-value obtained by comparing the planktonic and sessile form of all strains and just moderate/high biofilm producers (last column, the values of normalized volume for each spot are provided in Supplementary Materials, Table S1).

Table 2. List of differentially represented proteins among the six different strains analyzed under planktonic and biofilm conditions. As in the last two columns, the analysis was performed, including all the analyzed strains and, subsequently, excluding the low biofilm producers (last column). OS= organism name. Every significant *p*-value (lower than 0.05) is printed in bold.

Uniprot Name/Accession Number	Uniprot Name	SameSpots Coding Number	Protein Name	Mascot Score	Sequence Coverage (%)	N of Matched Peptides	Regulation in Sessile vs. Planktonic	High Biofilm Producers (Sessile vs. Planktonic)
Q2FJ31	ADH_STAA3	820	alcohol dehydrogenase	66	35	9/59	↑0.070	↑**0.018**
A7X656	GPMA_STAA1	824	2,3-bisphosphoglycerate-dependent phosphoglycerate mutase OS = *Staphylococcus aureus*	116	51	11/52	↓**0.001**	↓**0.016**
P0A0H0	RS12_STAA8	836	30S ribosomal protein S12	68	37	4/45	↑**0.026**	0.135
Q2FJN4	AHPC_STAA3	838	Alkyl hydroperoxide reductase subunit C OS = *Staphylococcus aureus*	67	39	6/31	↓**0.028**	0.137
A6U2G5	PFKA_STAA2	840	ATP-dependent 6-phosphofructokinase OS = *Staphylococcus aureus*	121	35	11/31	↓0.084	↓**0.007**
A7WZR9	CLPP_STAA1	842	ATP-dependent Clp protease proteolytic subunit OS = *Staphylococcus aureus*	76	30	9/23	↓**0.008**	0.084
Q5HF38	CCPA_STAAC	845	Catabolite control protein A OS = *Staphylococcus aureus*	76	26	9/35	↑**0.001**	↑**0.004**
Q9L4S1	CATA_STAAU	846	Catalase OS = *Staphylococcus aureus* GN = katA PE = 3 SV = 1	40	16		↓**0.041**	↓**0.041**
Q9L4S1	CATA_STAAU	848	Catalase OS = *Staphylococcus aureus* GN = katA PE = 3 SV = 1	68	30	6/40	↓**0.021**	↓**0.027**
Q9L4S1	CATA_STAAU	851	Catalase OS = *Staphylococcus aureus* GN = katA PE = 3 SV = 1	98	47	11/40	↓**0.012**	↓**0.004**
Q2FDQ4	ALF1_STAA3	853	Fructose-bisphosphate aldolase class 1 OS = *Staphylococcus aureus*	112	40	17/67	↑0.096	↑**0.0002**
Q2YSZ4	GCSPB_STAAB	855	Probable glycine dehydrogenase (decarboxylating) subunit 2 OS = *Staphylococcus aureus*	68	26	9/50	↓**0.037**	↓**0.013**
A7X395	ENGB_STAA1	856	Probable GTP-binding protein EngB OS = *Staphylococcus aureus* _12042016_	68	16	4/16	↓**0.007**	0.060
Q7A551	Y1532_STAAN	858	putative universal stress protein SA153 (782) OS = *Staphylococcus aureus*	86	36	6/35	↑**0.010**	0.094

Figure 1a provides a graphic representation of the Coomassie Brilliant Blue stained entire proteins as detected by the image analysis software. The top four rows show high and moderate biofilm producers' spots, while the two rows at the bottom indicate the low biofilm producers.

Figure 1b shows the graphic representation of the most relevant differentially regulated proteins and metabolisms among the two analyzed *S. aureus* phenotypes. Biological functions were manually checked after each GO search and subsequently reported in the scheme in Figure 1b.

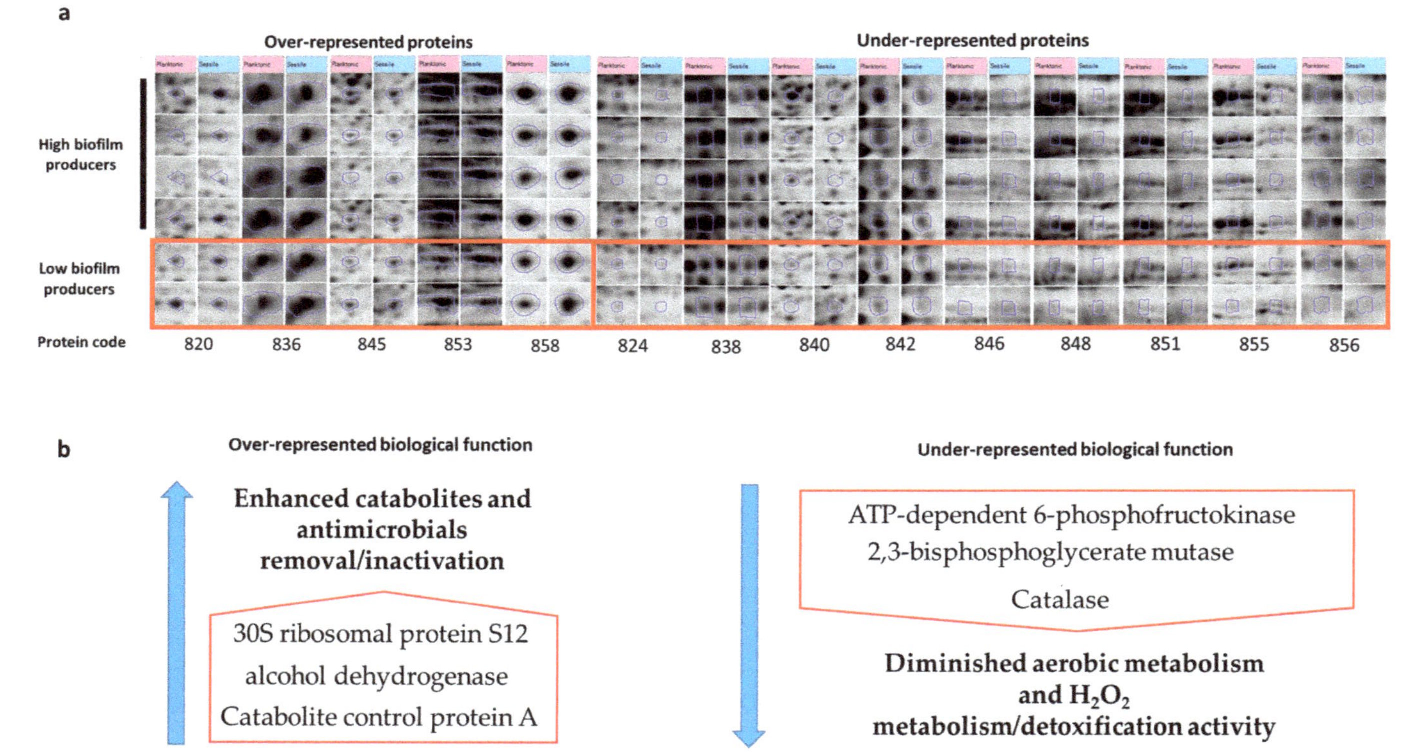

Figure 1. (**a**) Graphic representation of the differentially expressed proteins mostly relevant to the regulation of the described mechanisms/pathways. (**b**) Representation of the differentially regulated proteins and the related modulated mechanisms.

4. Discussion

Biofilms growth is the preferred strategy for the expansion and survival of many clinically and environmentally relevant microorganisms [5]. *S. aureus* is one well-known biofilm-forming pathogen capable of colonizing medical devices [26], food contact surfaces [21] and farm industry facilities [27]. In the biofilm form, *S. aureus* can successfully cope against strong stress conditions [28] and persist on the surfaces of food-processing plants [4], leading to recurrent contamination of both fresh and processed foods worldwide [29–32]. From this perspective, biofilm formation represents a severe threat because of its difficult removal linked to the extremely high tolerance to antimicrobials and antibiotics. Improving knowledge about its formation mechanisms and pathways is mandatory to better design possible and practical intervention strategies. Studies performed on single strains (strain-specific mechanisms) documented the over-representation of fibrinogen-binding protein and the accumulation-associated protein (Aap) in *S. aureus* cells growing embedded in the biofilm matrix in comparison to those growing in the planktonic form [33,34]. Also, increased production of staphylococcal accessory regulator A (SarA) was shown in biofilm formation [20].

All these and many other studies extensively describe the physiology of *S. aureus* biofilm formation that is specific to the strain analyzed. However, it is not considered that diverse *S. aureus* strains may have different mechanisms and pathways of biofilm formation.

In the current study, we employed a comparative proteomic approach to understand better the process of biofilm formation and the possible mechanisms involved in the enhancement of antimicrobial resistance. To achieve this result, we performed a differential proteomics analysis of planktonic versus sessile *S. aureus* isolates and ATCC strains. Six different strains with a wide range of biofilm formation indexes were employed in order to maximize the possibility to detect general mechanisms more representative of the *S. aureus* specie.

The whole comparison allowed the discovery of 14 proteins differentially regulated between the planktonic and sessile group and, three of those (alcohol dehydrogenase, ATP-dependent 6-phosphofructokinase and fructose-bisphosphate aldolase) were specific to the high biofilm-producing strains.

Ribosomal proteins are involved in biofilm regulation/formation and enhanced antimicrobial resistance [35,36]. Interestingly, changes at the ribosomal protein isoforms can shape the response to antibiotics by modifying the affinity of tetracyclines, chloramphenicol, macrolides (e.g., erythromycin) and aminoglycosides (e.g., streptomycin) for the transcription machinery. Hence, a switch in the composition of ribosomal subunits could be involved in biofilm formation and the different susceptibility to antimicrobial molecules [37].

Fructose bisphosphate aldolase and catabolite control protein A (ccpA) are over-represented in the biofilm conformation versus the sessile condition. The first is an essential enzyme of the glycolytic pathway with virulence functions shaped according to its cellular localization (i.e., moonlighting properties) [38]. As a moonlight protein, it is often expressed in the bacterial surface [39] where it has been linked to virulence in several bacterial pathogens, such as *Francisella* [40], by directly affecting cell migration through its interference with the actin polymerization process.

Similarly, fructose bisphosphate aldolase expression is induced in oxygen depletion conditions, and it has also been associated with transcriptional regulator functions [39]. Catabolite control protein A (ccpA) was found to be massively over-represented in both high and low biofilm producers growing in the sessile conditions. This might be explained by the requirements of the typical multi-layered packed structure of the biofilm, which needs a tight control of nutrients availability, catabolites and secondary metabolites (e.g., ethanol, reactive oxygen species (ROS) etc.). Indeed, nutrients depletion or catabolites accumulation would exert toxic/detrimental effects on the bacterial community itself. In Gram+ bacteria, ccpA expression regulates the synthesis of capsular polysaccharides, toxigenic exoproteins and promotes biofilm formation [25]. Similarly, *S. epidermidis* biofilm formation is positively regulated by ccpA and causes tricarboxylic acid (TCA) cycle repression [41]. This demonstrates that the management of carbon and energy flow by regulating the enzymes involved in glycolytic/fermentative metabolism [42] represents an essential element for the proper formation of biofilm. Accordingly, previous evidence reported that environmental acidification or other phenomena associated with rapid metabolism of carbohydrates occurring in bacteria growing in sessile conditions are regulated by ccpA throughout the modulation of pfka and gpma expression [42,43]. Moreover, the structural organization of the biofilm is likely to result in the accumulation of toxic secondary metabolites such as ethanol from fermentation processes. This may explain the detected increased expression of alcohol dehydrogenase (adh) in the sessile growing strains. The oxygen depletion in the biofilm's inner layers may cause a metabolic shift towards the mixed alcoholic fermentation with increased ethanol concentration that needs to be promptly detoxified by the induction of adh [44–46]. The hypothesis of the metabolic shift towards fermentation and ethanol production is also supported by the under-expression of PfkA and gpmA, which are active in pyruvate production. By limiting the production of pyruvate, sessile cells control the pathways towards any possible fermentative process. Thus, the reduced abundance of PfkA and gpmA in the sessile bacteria might represent the effect of a negative feedback modulation of the fermentative process to protect the bacteria from the toxicity of their secondary metabolites. Analogously, the reduced abundance of catalase, the enzyme active in ROS detoxification, may be a consequence of the reduced oxygen availability in the bacterial samples growing in biofilm form [47]. Such a reduction in the hydroperoxide detoxification power is confirmed by the down-regulation of 3 different catalase isoforms and of alkyl hydroperoxide reductase subunit C (Q2FJN4). This may help to explain the high power of low doses of hydrogen peroxide to dissolve the biofilm conformation [48].

5. Conclusions

The comparative top-down proteomics (2D-electrophoresis–MALDI TOF) approach used here identified some possible biofilm formation mechanisms of *S. aureus* strains with a wide range of biofilm formation indexes. Biofilm is one of the essential strategies for bacterial virulence and persistence over a wide variety of surfaces and unfavourable conditions and it facilitates survival and resistance in the presence of antimicrobial compounds [49]. Comparison of high- and low-biofilm forming strains in sessile and planktonic form highlighted common mechanisms as the catabolite control and the modulation of the detoxification machinery aimed at avoiding self-inhibition/toxicity (i.e., ethanol detoxification). Glycolysis and aerobic metabolisms seem to be down-regulated in favor of possible fermentation pathways that might be responsible for ethanol production and, possibly, for the induction of alcohol dehydrogenase production.

This study is characterized by using a top-down proteomics approach that led the differential quantification of intact proteofoms. On the other hand, this approach limits the detection of differentially represented, less-abundant proteins. Complementing these data with shotgun proteomics and metabolomics is desired to support the observed evidence and to discover potential biomolecular targets to contrast and/or attenuate this phenomenon.

Supplementary Materials: The following are available online at https://www.mdpi.com/article/10.3390/ani11040966/s1, Table S1: Raw normalized volume.

Author Contributions: Conceptualization, C.P., P.A.D.C., A.S., P.R.; Data curation, C.P., P.A.D.C., A.S., V.G.; Formal analysis, C.P., P.A.D.C., A.S., V.G.; Funding acquisition, A.I., L.B., A.U., P.R.; Investigation, C.P., P.A.D.C., A.S., V.G., B.T.; Methodology, C.P., P.A.D.C., A.S., V.G., P.R., Project administration, A.I., P.R.; Resources, A.I., P.R., L.B., A.U.; Writing—original draft, C.P., P.A.D.C., A.S., P.R.; Writing—review and editing, C.P., P.A.D.C., A.S., V.G., B.T., N.C., C.C., A.I., P.R. All authors have read and agreed to the published version of the manuscript.

Funding: This research received no external funding.

Institutional Review Board Statement: Not applicable.

Data Availability Statement: Not applicable.

Conflicts of Interest: The authors declare no conflict of interest.

References

1. Oniciuc, E.-A.; Likotrafiti, E.; Alvarez-Molina, A.; Prieto, M.; López, M.; Alvarez-Ordóñez, A. Food processing as a risk factor for antimicrobial resistance spread along the food chain. *Curr. Opin. Food Sci.* **2019**, *30*, 21–26. [CrossRef]
2. Guardabassi, L.; Larsen, J.; Weese, J.; Butaye, P.; Battisti, A.; Kluytmans, J.; Lloyd, D.; Skov, R. Public health impact and antimicrobial selection of meticillin-resistant staphylococci in animals. *J. Glob. Antimicrob. Resist.* **2013**, *1*, 55–62. [CrossRef]
3. Weese, J.S.; van Duijkeren, E. Methicillin-resistant *Staphylococcus aureus* and *Staphylococcus pseudintermedius* in veterinary medicine. *Vet. Microbiol.* **2010**, *140*, 418–429. [CrossRef]
4. Herrera, J.J.R.; Cabo, M.L.; González, A.; Pazos, I.; Pastoriza, L. Adhesion and detachment kinetics of several strains of *Staphylococcus aureus* subsp. aureus under three different experimental conditions. *Food Microbiol.* **2007**, *24*, 585–591. [CrossRef]
5. Donlan, R.M.; Costerton, J.W. Biofilms: Survival mechanisms of clinically relevant microorganisms. *Clin. Microbiol. Rev.* **2002**, *15*, 167–193. [CrossRef]
6. Sauer, K.; Camper, A.K.; Ehrlich, G.D.; Costerton, J.W.; Davies, D.G. *Pseudomonas aeruginosa* displays multiple phenotypes during development as a biofilm. *J. Bacteriol.* **2002**, *184*, 1140–1154. [CrossRef]
7. Resch, A.; Rosenstein, R.; Nerz, C.; Götz, F. Differential gene expression profiling of *Staphylococcus aureus* cultivated under biofilm and planktonic conditions. *Appl. Environ. Microbiol.* **2005**, *71*, 2663–2676. [CrossRef]
8. Costerton, J.W.; Stewart, P.S.; Greenberg, E.P. Bacterial biofilms: A common cause of persistent infections. *Science* **1999**, *284*, 1318–1322. [CrossRef]
9. Brooks, J.D.; Flint, S.H. Biofilms in the food industry: Problems and potential solutions. *Int. J. Food Sci. Technol.* **2008**, *43*, 2163–2176. [CrossRef]
10. Srey, S.; Jahid, I.K.; Ha, S.-D. Biofilm formation in food industries: A food safety concern. *Food Control* **2013**, *31*, 572–585. [CrossRef]
11. Wassenaar, T.; Ussery, D.; Nielsen, L.; Ingmer, H. Review and phylogenetic analysis of qac genes that reduce susceptibility to quaternary ammonium compounds in *Staphylococcus species*. *Eur. J. Microbiol. Immunol.* **2015**, *5*, 44–61. [CrossRef]
12. Glinel, K.; Thebault, P.; Humblot, V.; Pradier, C.-M.; Jouenne, T. Antibacterial surfaces developed from bio-inspired approaches. *Acta Biomater.* **2012**, *8*, 1670–1684. [CrossRef]
13. Simoes, M.; Simões, L.C.; Vieira, M.J. A review of current and emergent biofilm control strategies. *LWT Food Sci. Technol.* **2010**, *43*, 573–583. [CrossRef]
14. Mah, T.-F.C.; O'Toole, G.A. Mechanisms of biofilm resistance to antimicrobial agents. *Trends Microbiol.* **2001**, *9*, 34–39. [CrossRef]
15. Branda, S.S.; Vik, Å.; Friedman, L.; Kolter, R. Biofilms: The matrix revisited. *Trends Microbiol.* **2005**, *13*, 20–26. [CrossRef]
16. Flemming, H.-C.; Wingender, H.-C.F.J.; Szewzyk, U.; Steinberg, P.; Rice, S.A.; Kjelleberg, S.A.R.S. Biofilms: An emergent form of bacterial life. *Nat. Rev. Microbiol.* **2016**, *14*, 563. [CrossRef] [PubMed]
17. Das, S.; Vishakha, K.; Banerjee, S.; Mondal, S.; Ganguli, A. Sodium alginate-based edible coating containing nanoemulsion of Citrus sinensis essential oil eradicates planktonic and sessile cells of food-borne pathogens and increased quality attributes of tomatoes. *Int. J. Biol. Macromol.* **2020**, *162*, 1770–1779. [CrossRef] [PubMed]
18. Atshan, S.S.; Shamsudin, M.N.; Esekawi, Z.; Lung, L.T.T.; Talab, F.B.; Liew, Y.K.; Alreshidi, M.A.; Abduljaleel, S.A.; Hamat, R.A. Comparative proteomic analysis of extracellular proteins expressed by various clonal types of *Staphylococcus aureus* and during planktonic growth and biofilm development. *Front. Microbiol.* **2015**, *6*, 524. [CrossRef]
19. Bénard, L.; Litzler, P.Y.; Cosette, P.; Lemeland, J.F.; Jouenne, T.; Junter, G.A. Proteomic analysis of *Staphylococcus aureus* biofilms grown in vitro on mechanical heart valve leaflets. *J. Biomed. Mater. Res. Part A An Off. J. Soc. Biomater. Japanese Soc. Biomater. Aust. Soc. Biomater. Korean Soc. Biomater.* **2009**, *88*, 1069–1078.
20. Resch, A.; Leicht, S.; Saric, M.; Pásztor, L.; Jakob, A.; Götz, F.; Nordheim, A. Comparative proteome analysis of *Staphylococcus aureus* biofilm and planktonic cells and correlation with transcriptome profiling. *Proteomics* **2006**, *6*, 1867–1877. [CrossRef]
21. Di Ciccio, P.; Vergara, A.; Festino, A.; Paludi, D.; Zanardi, E.; Ghidini, S.; Ianieri, A. Biofilm formation by *Staphylococcus aureus* on food contact surfaces: Relationship with temperature and cell surface hydrophobicity. *Food Control* **2015**, *50*, 930–936. [CrossRef]
22. Soggiu, A.; Piras, C.; Greco, V.; Devoto, P.; Urbani, A.; Calzetta, L.; Bortolato, M.; Roncada, P. Exploring the neural mechanisms of finasteride: A proteomic analysis in the nucleus accumbens. *Psychoneuroendocrinology* **2016**, *74*, 387–396. [CrossRef] [PubMed]
23. Piras, C.; Soggiu, A.; Bonizzi, L.; Greco, V.; Ricchi, M.; Arrigoni, N.; Bassols, A.; Urbani, A.; Roncada, P. Identification of immunoreactive proteins of Mycobacterium avium subsp. paratuberculosis. *Proteomics* **2015**, *15*, 813–823. [CrossRef]
24. Cramton, S.E.; Gerke, C.; Schnell, N.F.; Nichols, W.W.; Götz, F. The intercellular adhesion (ica) locus is present in *Staphylococcus aureus* and is required for biofilm formation. *Infect. Immun.* **1999**, *67*, 5427–5433. [CrossRef]
25. Seidl, K.; Goerke, C.; Wolz, C.; Mack, D.; Berger-Bächi, B.; Bischoff, M. Staphylococcus aureus CcpA affects biofilm formation. *Infect. Immun.* **2008**, *76*, 2044–2050. [CrossRef] [PubMed]
26. Ziebuhr, W. Staphylococcus aureus and *Staphylococcus epidermidis*: Emerging pathogens in nosocomial infections. *Contrib. Microbiol.* **2001**, *8*, 102–107.

27. Weber, M.; Liedtke, J.; Plattes, S.; Lipski, A. Bacterial community composition of biofilms in milking machines of two dairy farms assessed by a combination of culture-dependent and–independent methods. *PLoS ONE* **2019**, *14*, e0222238. [CrossRef]

28. Enright, M.C. The evolution of a resistant pathogen–the case of MRSA. *Curr. Opin. Pharmacol.* **2003**, *3*, 474–479. [CrossRef]

29. Nostro, A.; Blanco, A.R.; Cannatelli, M.A.; Enea, V.; Flamini, G.; Morelli, I.; Roccaro, A.S.; Alonzo, V. Susceptibility of methicillin-resistant staphylococci to oregano essential oil, carvacrol and thymol. *FEMS Microbiol. Lett.* **2004**, *230*, 191–195. [CrossRef]

30. Braga, L.C.; Shupp, J.; Cummings, C.; Jett, M.; Takahashi, J.; Carmo, L.; Chartone-Souza, E.; Nascimento, A. Pomegranate extract inhibits *Staphylococcus aureus* growth and subsequent enterotoxin production. *J. Ethnopharmacol.* **2005**, *96*, 335–339. [CrossRef]

31. Marques, S.C.; Rezende, J.D.G.O.S.; Alves, L.A.D.F.; Silva, B.C.; Alves, E.; De Abreu, L.R.; Píccoli, R.H. Formation of biofilms by *Staphylococcus aureus* on stainless steel and glass surfaces and its resistance to some selected chemical sanitizers. *Braz. J. Microbiol.* **2007**, *38*, 538–543. [CrossRef]

32. Oulahal, N.; Brice, W.; Martial, A.; Degraeve, P. Quantitative analysis of survival of *Staphylococcus aureus* or *Listeria innocua* on two types of surfaces: Polypropylene and stainless steel in contact with three different dairy products. *Food Control* **2008**, *19*, 178–185. [CrossRef]

33. Wolf, C.; Hochgräfe, F.; Kusch, H.; Albrecht, D.; Hecker, M.; Engelmann, S. Proteomic analysis of antioxidant strategies of *Staphylococcus aureus*: Diverse responses to different oxidants. *Proteomics* **2008**, *8*, 3139–3153. [CrossRef]

34. Engelmann, S.; Hecker, M. Proteomic analysis to investigate regulatory networks in *Staphylococcus aureus*. In *Bacterial Pathogenesis*; Springer: Berlin/Heidelberg, Germany, 2008; pp. 25–45.

35. Sałamaszyńska-Guz, A.; Rose, S.; Lykkebo, C.A.; Taciak, B.; Bącal, P.; Uśpieński, T.; Douthwaite, S. Biofilm formation and motility are promoted by Cj0588-directed methylation of rRNA in *Campylobacter jejuni*. *Front. Cell. Infect. Microbiol.* **2018**, *7*, 533. [CrossRef] [PubMed]

36. Zheng, J.; Wu, Y.; Lin, Z.; Wang, G.; Jiang, S.; Sun, X.; Tu, H.; Yu, Z.; Qu, D. ClpP participates in stress tolerance, biofilm formation, antimicrobial tolerance, and virulence of *Enterococcus faecalis*. *BMC Microbiol.* **2020**, *20*, 30. [CrossRef]

37. Lilleorg, S.; Reier, K.; Volõnkin, P.; Remme, J.; Liiv, A. Phenotypic effects of paralogous ribosomal proteins bL31A and bL31B in *E. coli*. *Sci. Rep.* **2020**, *10*, 1–10. [CrossRef]

38. Shams, F.; Oldfield, N.J.; Wooldridge, K.G.; Turner, D.P.J. Fructose-1, 6-bisphosphate aldolase (FBA)—A conserved glycolytic enzyme with virulence functions in bacteria: 'Ill met by moonlight'. *Biochem. Soc. Trans.* **2014**, *42*, 1792–1795. [CrossRef] [PubMed]

39. Oldfield, N.J.; Shams, F.; Wooldridge, K.G.; Turner, D.P.J. Moonlighting Functions of Bacterial Fructose 1, 6-Bisphosphate Aldolases. In *Moonlighting Proteins: Novel Virulence Factors in Bacterial Infections*; John Wiley & Sons, Inc.: Hoboken, NJ, USA, 2017; pp. 321–331.

40. Ziveri, J.; Tros, F.; Guerrera, I.C.; Chhuon, C.; Audry, M.; Dupuis, M.; Barel, M.; Korniotis, S.; Fillatreau, S.; Gales, L.; et al. The metabolic enzyme fructose-1, 6-bisphosphate aldolase acts as a transcriptional regulator in pathogenic Francisella. *Nat. Commun.* **2017**, *8*, 1–15. [CrossRef]

41. Sadykov, M.R.; Hartmann, T.; Mattes, T.A.; Hiatt, M.; Jann, N.J.; Zhu, Y.; Ledala, N.; Landmann, R.; Herrmann, M.; Rohde, H.; et al. CcpA coordinates central metabolism and biofilm formation in *Staphylococcus epidermidis*. *Microbiology* **2011**, *157*, 3458. [CrossRef]

42. Rudra, P.; Boyd, J.M. Metabolic control of virulence factor production in *Staphylococcus aureus*. *Curr. Opin. Microbiol.* **2020**, *55*, 81–87. [CrossRef]

43. Kim, J.N.; Burne, R.A. CcpA and CodY coordinate acetate metabolism in *Streptococcus mutans*. *Appl. Environ. Microbiol.* **2017**, *83*. [CrossRef] [PubMed]

44. Fuchs, S.; Pané-Farré, J.; Kohler, C.; Hecker, M.; Engelmann, S. Anaerobic gene expression in *Staphylococcus aureus*. *J. Bacteriol.* **2007**, *189*, 4275–4289. [CrossRef] [PubMed]

45. Becker, P.; Hufnagle, W.; Peters, G.; Herrmann, M. Detection of differential gene expression in biofilm-forming versus planktonic populations of *Staphylococcus aureus* using micro-representational-difference analysis. *Appl. Environ. Microbiol.* **2001**, *67*, 2958–2965. [CrossRef] [PubMed]

46. Zhang, K.; Yang, X.; Yang, J.; Qiao, X.; Li, F.; Liu, X.; Wei, J.; Wang, L. Alcohol dehydrogenase modulates quorum sensing in biofilm formations of *Acinetobacter baumannii*. *Microb. Pathog.* **2020**, *148*, 104451. [CrossRef]

47. Shen, F.; Ge, C.; Yuan, P. Metabolomics Study Reveals Inhibition and Metabolic Dysregulation in *Staphylococcus aureus* Planktonic Cells and Biofilms Induced by Carnosol. *Front. Microbiol.* **2020**, *11*, 2315. [CrossRef]

48. Christensen, B.E.; Trønnes, H.N.; Vollan, K.; Smidsrød, O.; Bakke, R. Biofilm removal by low concentrations of hydrogen peroxide. *Biofouling* **1990**, *2*, 165–175. [CrossRef]

49. Vázquez-Sánchez, D.; Rodríguez-López, P. Biofilm formation of *Staphylococcus aureus*. In *Staphylococcus aureus*; Elsevier: Amsterdam, The Netherlands, 2018; pp. 87–103.

Article

Characterization of Extended-Spectrum Cephalosporin (ESC) Resistance in *Salmonella* Isolated from Chicken and Identification of High Frequency Transfer of bla_{CMY-2} Gene Harboring Plasmid In Vitro and In Vivo

Bo-Ram Kwon [1,†], Bai Wei [1,†], Se-Yeoun Cha [1], Ke Shang [1], Jun-Feng Zhang [1], Hyung-Kwan Jang [1,2,*] and Min Kang [1,2,*]

1　Department of Veterinary Infectious Diseases and Avian Diseases, College of Veterinary Medicine and Center for Poultry Diseases Control, Jeonbuk National University, Iksan 54596, Korea; aniscikwon@gmail.com (B.-R.K.); weibai116@hotmail.com (B.W.); kshmnk@hanmail.net (S.-Y.C.); shangke0624@gmail.com (K.S.); jfzhang018@gmail.com (J.-F.Z.)
2　Bio Disease Control (BIOD) Co., Ltd., Iksan 54596, Korea
*　Correspondence: hkjang@jbnu.ac.kr (H.-K.J.); minkang@jbnu.ac.kr (M.K.); Tel.: +82-63-850-0945 (H.-K.J.); Tel.: +82-63-850-0690 (M.K.); Fax: +82-63-858-9155 (H.-K.J.); Fax: +82-63-858-0686 (M.K.)
†　These authors contributed equally to this study.

Citation: Kwon, B.-R.; Wei, B.; Cha, S.-Y.; Shang, K.; Zhang, J.-F.; Jang, H.-K.; Kang, M. Characterization of Extended-Spectrum Cephalosporin (ESC) Resistance in *Salmonella* Isolated from Chicken and Identification of High Frequency Transfer of bla_{CMY-2} Gene Harboring Plasmid In Vitro and In Vivo. *Animals* **2021**, *11*, 1778. https://doi.org/10.3390/ani11061778

Academic Editors: Bruno Tilocca and Paola Roncada

Received: 1 May 2021
Accepted: 10 June 2021
Published: 14 June 2021

Publisher's Note: MDPI stays neutral with regard to jurisdictional claims in published maps and institutional affiliations.

Simple Summary: The prevalence of extended-spectrum cephalosporin (ESC)-resistant *Salmonella* is of great concern, as these strains with the same β-lactamase (*bla*) genes were found in human and poultry. The objective is to characterize ESC-resistant *Salmonella* isolated from chicken and to determine the transferability of β-lactamase gene-harboring plasmid in vitro and in vivo. ESC resistance genes in *Salmonella* isolated from chickens and presented a comprehensive analysis of the highly frequent transfer of the bla_{CMY-2} gene in vitro and in vivo. In addition, this study has demonstrated the ease with which a bla_{CMY-2} gene-harboring plasmid can be rapidly transferred between *Salmonella* and pathogenic *E. coli* within the intestinal tracts of mice, even without antimicrobial selective pressure. Given the potential risk of the frequent transfer of the bla_{CMY-2} gene via the food chain to the human digestive tract, the molecular mechanism involved in the dissemination and maintenance of ESC resistance genes should be studied as further research in greater detail, and enhanced surveillance should be implemented to prevent the widespread of ESC resistant strains.

Abstract: A total of 136 *Salmonella* isolates from chicken feces and meat samples of the top 12 integrated chicken production companies throughout Korea were collected. Among the 17 ESC-resistant *Salmonella*; $bla_{CTX-M-15}$ was the most prevalent gene and two strains carried $bla_{TEM-1}/bla_{CTX-M-15}$ and bla_{CMY-2}, respectively. The transferable $bla_{CTX-M-15}$ gene was carried by IncFII plasmid in three isolates and the bla_{CMY-2} gene carried by IncI1 plasmid in one isolate. bla_{CMY-2} gene-harboring strain was selected as the donor based on the high frequency of bla_{CMY-2} gene transfer in vitro and its transfer frequencies were determined at 10^{-3} transconjugants per recipient. The transfer of bla_{CMY-2} gene-harboring plasmid derived from chicken isolate into a human pathogen; enteroinvasive *Escherichia coli* (EIEC), presented in mouse intestine with about 10^{-1} transfer frequency without selective pressure. From the competition experiment; bla_{CMY-2} gene-harboring transconjugant showed variable fitness burden depends on the parent strains. Our study demonstrated direct evidence that the bla_{CMY-2} gene harboring *Salmonella* from chicken could frequently transfer its ESC-resistant gene to *E. coli* in a mouse intestine without antimicrobial pressure; resulting in the emergence of multidrug resistance in potentially virulent EIEC isolates of significance to human health; which can increase the risk of therapeutic inadequacy or failures.

Keywords: *Salmonella*; chicken; extended-spectrum cephalosporin; bla_{CMY-2}; mouse; frequent transfer

1. Introduction

Recently, an increasing occurrence of extended-spectrum cephalosporins (ESC)-resistant strains has been recognized as a serious threat to human health [1]. Resistance to β-lactam antimicrobials is mainly caused by the production of antimicrobial inactivation enzymes called β-lactamases [2]. Extended-spectrum β-lactamases (ESBL) and AmpC β-lactamase (AmpC) are the major β-lactamases detected in ESC-resistant strains worldwide [3]. These enzymes are frequently encoded by genes that are located on a plasmid, which is a mobile genetic element that can transfer horizontally within and between different bacterial species [4]. Various studies have suggested that food-producing animals as a reservoir for ESBL/AmpC-producing strains that could promote the transmission of resistance determinants to humans [2]. Similar or identical ESC-resistant isolates or ESBL/AmpC plasmids were found in chicken meat and patients, suggesting poultry and poultry products play a pivotal role in the spread of ESC resistance genes to humans [2].

The fact that the same plasmid is observed in several bacterial strains isolated from poultry and humans confirms that antimicrobial resistance genes can be transferred from poultry to humans [2]. A previous study observed the possibility using in vitro human gut simulation model that there is a transfer from food-borne ESC resistant isolates to other commensal and pathogenic bacteria [5]. However, there is a lack of actual evidence that ESC resistance genes and particularly the bla_{CMY-2} gene transfer from poultry to human-origin pathogenic isolates in vivo could cause considerable risks, such as the high possibility of inadequate treatment or therapeutic failures. Antimicrobial resistance, by the acquisition of a mobile genetic element or by mutation, is generally thought to induce a competitive fitness disadvantage on host bacteria in the absence of selective pressure for resistance phenotypes [6]. However, few studies have examined the fitness advantage of their host bacteria after acquired resistance plasmids [7].

This study aimed to clarify the characteristic of ESC-resistant *Salmonella* isolated from chicken and to determine the transferability of ESC resistance-determining plasmid in vitro and in vivo. We also examined the ability to donate ESC resistance genes and how frequently they are transferred from chicken isolates to human pathogens in the mammalian mouse intestine. As antimicrobial resistance is a widely acknowledged factor affecting plasmid persistence in the absence of selective pressure [8], we attempted to identify the contribution of ESC-resistant plasmid in in vitro fitness by competition between susceptible and resistant isolates. Our goals were to assess the interspecific horizontal gene transfer (HGT) of ESC resistance from animal-derived *Salmonella* to human-derived bacteria in vitro and in vivo, also to evaluate the impact of ESC resistance genes acquisition on bacteria fitness.

2. Materials and Methods

2.1. Bacterial Isolates

A total of 136 *Salmonella* isolates isolated from chicken feces and chicken meat samples from 2017 to 2018 were collected from the top 12 integrated chicken production companies throughout Korea. The isolation and serotyping of *Salmonella* were conducted as described previously [9]. Among 136 *Salmonella* isolates, those showing either ESBL or AmpC phenotype were used in this study. *Salmonella* strains that are resistant to ceftiofur are considered ESBL/AmpC-producing strains. To select a recipient for the in vivo transfer experiment, we obtained a total of 10 strains (Table S1), which were isolated from human patient's stool samples and categorized as pathogenic *Escherichia coli*, from the National Culture Collection for Pathogens (NCCP) South Korea.

2.2. Antimicrobial Susceptibility Test

The antimicrobial susceptibility of all isolates was evaluated by the minimum inhibitory concentrations (MICs) of the test antimicrobial agents amoxicillin/clavulanic acid (AMC), cefoxitin (FOX), cefepime (FEP), ceftazidime (TAZ), ceftiofur (XNL), trimethoprim/sulfamethoxazole (SXT), sulfisoxazole (FIS), chloramphenicol (CHL), ampicillin

(AMP), ciprofloxacin (CIP), nalidixic acid (NAL), streptomycin (STR), gentamicin (GEN), tetracycline (TET), meropenem (MERO), and colistin (COL) using the KRNV5F (TREK Diagnostic Systems, Korea). *Escherichia coli* ATCC 25922 was used as the reference strain for quality control. The susceptibility breakpoints of most antimicrobials were interpreted according to the CLSI guidelines [10]. Since CLSI breakpoints were not available for colistin, ceftiofur, and streptomycin, MICs were determined according to the European Committee on Antimicrobial Susceptibility Testing (EUCAST) guidelines [11] for colistin and to Centers for Disease Control and Prevention [12] for the ceftiofur and streptomycin.

2.3. Identification of β-Lactamases

After phenotypic screening, PCR were implemented regarding ceftiofur resistant isolates for detecting the presence of β-lactamase genes encoding CTX-M, TEM, and CMY-type following a previous protocol [13,14]. Genomic DNA templates for PCR were prepared using fresh *Salmonella* colonies on MacConkey agar (Difco laboratories, Sparks, Maryland, USA) plates by adding 100 µL of sterile distilled water and boiling in a heater block at 100 °C for 15 min. The sequencing reactions were performed by an external company (Solgent, Daejeon, Korea). The obtained amino acid sequences were compared with those in the GenBank nucleotide database using the BLAST online service, provided by the National Center for Biotechnology Information (www.ncbi.nlm.nih.gov/BLAST, accessed on 21 March 2020), to determine the specific types of β-lactamase genes.

2.4. Plasmid Replicon Typing

Plasmid DNA was extracted using HiYield™ Plasmid Mini Kit (RBC Bioscience, Taipei, Taiwan) according to the manufacturer's instructions. Plasmid incompatibility groups were determined by the PCR-based replicon typing (PBRT) method [4]. For plasmids such as IncF, IncI1, IncHI2, and IncHI plasmids were subtyped by plasmid MLST (pMLST) (http://pubmlst.org/plasmid/, accessed on 10 April 2020).

2.5. In Vitro Conjugation Experiment

A conjugation experiment was performed according to previously reported methods with some modification [15]. In vitro mating was performed in liquid media, and cephalosporin-susceptible *Escherichia coli* J53 (sodium azide-resistance) and selected *E. coli* NCCP isolate (certain antimicrobial resistance) were used as recipients. Briefly, overnight cultures of donor and recipient strains were re-cultured in logarithmic phase (OD 600 nm of 0.5) at 37 °C in fresh tryptic soy broth (Difco Laboratories, Detroit, MI, USA) medium for 4 h. Next, 1 mL of the donor and 4 mL of the recipient were mixed and incubated without shaking for 1 h at 37 °C. The culture was spread on MacConkey agar plates containing sodium azide (200 µg/mL) and ceftiofur (8 µg/mL) for detecting *E. coli* J53-derived transconjugants. MacConkey agar plates containing certain antimicrobial and ceftiofur (8 µg/mL) were used for detecting *E. coli* NCCP isolate-derived transconjugants. The experiment was repeated three times, and three putative transconjugant colonies were randomly selected from each experiment. For verifying the transconjugant, the transconjugant was evaluated by the MICs with the method described above, and the presence of a marker gene of an ESC-resistant plasmid was confirmed by PCR, as previously described [14]. Conjugation frequency was calculated as the ratio of the number of transconjugants per recipient (Tc/R). Recipient isolate counts were calculated by subtracting transconjugant colony counts from the number of colonies obtained on agar plates, which included both recipients and transconjugants.

2.6. In Vivo Transfer Experiment

When selecting the recipient for the transfer experiment in vivo, one recipient for the in vivo transfer experiment was selected based on the results of the conjugation frequency test. Enteroinvasive *E. coli* (EIEC) NCCP 13719 carried the virulence gene of *ipa*H [16], which was ceftiofur-susceptible but tetracycline-resistant and showed the highest frequency of transfer, was selected as the recipient for the in vivo transfer experiment.

The animal experiment was conducted in accordance with the requirements of the Animal Care and Ethics Committees of Jeonbuk National University and were approved by the National Association of Laboratory Animal Care (JBNU 2021-06). Female SPF 6-week-old BALB/c mice (Samtako, Osan, Korea) were randomly housed in four groups of five animals each, and each group was kept in a separated isolator (Three-Shine, Daejeon, Korea). Before the inoculation of donor and recipient, fecal samples from all mice in each group were pooled, and the absence of resistant strains was confirmed by spreading onto a plate that we used in this study. In addition, fecal samples were also checked to be free of the *bla*CMY-2 gene and *ipa*H gene. The experimental groups were as follows: streptomycin-treated control group (G1), streptomycin-treated and then donor-inoculated group (G2) for monitoring donor strain colonization, streptomycin-treated and then recipient-inoculated group (G3) for monitoring recipient strain colonization, and streptomycin-treated and then donor and recipient simultaneously inoculated group (G4). Before inoculation, streptomycin at a dose of 20 mg per mouse was pretreated to eliminate the circumstance of microbial competition and induce the colonization of inoculating isolates (Figure 1) [17]. Food and water were discontinued 4 h before oral administration of streptomycin. Then, food and water were made available to be consumed ad libitum. At 20 h after streptomycin administration, food and water were ceased again for 4 h before the mice were inoculated orally by gavage with 0.2 mL of 10^8 CFU/mL of donor and recipient. As for G4, the recipient was inoculated 30 min after inoculating the donor. Then, water was offered immediately and food was made available 2 h after infection ad libitum. On 1, 2, 4, and 7 days after infection, fresh fecal samples were collected from each mouse. The samples were weighed and diluted five-fold and then finally homogenized by vortexing in phosphate buffer saline (PBS). For colony counting, 10-fold serially diluted samples were inoculated onto appropriate agar plates, including antimicrobials for each group. For verification of transconjugant isolates, putative colonies were sub-cultured onto antimicrobial selective agar plates, and genomic DNA was extracted according to the boiling method as described above for using PCR analysis to test for the possession of marker gene (β-lactamase gene from donor and virulence gene from the selected recipient). As for G4, transfer frequency was calculated as the ratio of the number of transconjugants per recipient (Tc/R).

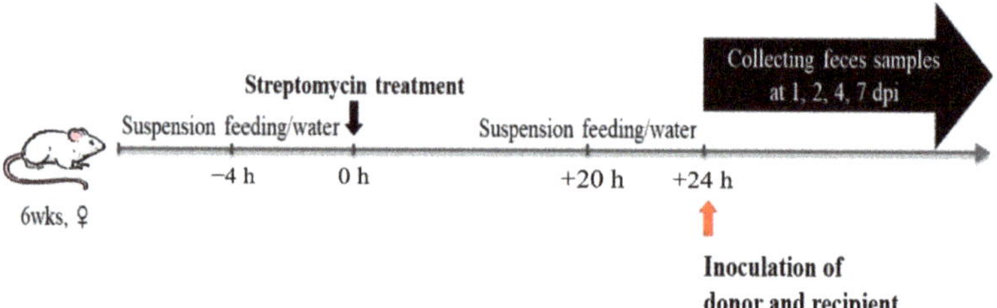

Figure 1. Schematic representation of in vivo transfer experiment set up.

2.7. Competition Experiment In Vitro

To assess the fitness effect of resistance plasmid carriage, competition assays between resistance plasmid-harboring transconjugant and its parental isolates, *E. coli* J53 and *E. coli* NCCP 13719, were conducted. The competition experiment was carried out as previously described [18] and repeated five times. Briefly, parental isolates were incubated overnight at 37 °C with shaking at $200\times g$ rpm in 10 mL of lysogeny broth (LB). Transconjugants were cultured in LB with the addition of 8 µg/mL ceftiofur to ensure the expression of ESC-resistant genes. Overnight cultures of two pairwise strains (*Escherichia coli* J53/transconjugant of *E. coli* J53, *E. coli* NCCP 13719/transconjugant of *E. coli* NCCP 13719) were adjusted to OD 600 nm of 0.5, were diluted 10^{-4}, and were then mixed 1:1 in

LB broth at 0 h. After 24 h of incubation at 37 °C, the mixed isolates were again 10^{-4} diluted into a fresh LB medium. This procedure was repeated every 24 h until the competition experiment had lasted for 72 h. The total number of isolates were determined by dropping 10 µL of properly diluted samples onto antimicrobial-free and antimicrobial-supplemented selective MacConkey agar plates in triplicate at 24 h, 48 h, and 72 h. The number of CFUs growing on the MacConkey agar plate including ceftiofur (8 µg/mL) was subtracted from the number of CFUs growing on the antimicrobial-free plate to determine the number of ESC-resistant gene-free isolates in the mixed population. To assess the relative fitness of transconjugants compared with its parental strains, an in vitro competitive ratio was calculated using a previously described method [17]. The competitive ratio was defined as the ratio of the number of CFU of the transconjugants vs. the parental strain at 24, 48, and 72 h.

3. Results

3.1. Characterization of Bacterial Strains

Based on the results of antimicrobial susceptibility assessment of 136 isolates, 17 out of 136 (12.5%) *Salmonella* spp. were consistent with an ESBL/AmpC phenotype and genotype (Table 1). The frequency of ESBL β-lactamase production with CTX-M gene was 11.8% (16/136) in *Salmonella* isolates, which was significantly ($p < 0.05$) higher than AmpC β-lactamase production (0.7%, 1/136) with CMY gene in *Salmonella* isolates. The four serovars isolated were *Salmonella* Enteritidis (52.9%), *Salmonella* Virchow (35.5%), *Salmonella* Albany (5.9%), and *Salmonella* spp. (5.9%). CTX-M (94.1%) was the most commonly detected β-lactamase family, and *S. enteritidis* had CTX-M and TEM gene combination while *S.* Albany was CMY-positive. All strains showed multidrug resistance. For all 17 strains, a conjugation experiment was implemented regarding the transfer of the ESC-resistant gene, and seven (41.2%) out of 17 strains were successfully conjugated in a wide range of frequencies from $<10^{-7}$ to $\geq 10^{-3}$ (transconjugant/recipient). Among the transconjugants, one strain, which the harbored bla_{CMY-2} gene, showed high transfer frequency ($\geq 10^{-3}$) [19,20]. On analysis using PCR-based replicon typing for conjugative plasmids, IncFIIS plasmids harboring CTX-M-15 were found in three *S.* Enteritidis isolates, and IncI1 plasmid harboring CMY-2 was found in one *S.* Albany isolate. IncI1 plasmids were further submitted to pMLST and assigned to a sequence type of ST12. The remaining three isolates were indicated as non-typeable plasmids.

3.2. In Vitro Transfer

Using the liquid mating method, transconjugant with its parental strains, *E. coli* J53 (Tc.J53), and *E. coli* NCCP 13719 (Tc.13719) displayed ESC resistance profile corresponding to the acquisition of the bla_{CMY-2} gene (Table 2). Tc.J53 and Tc.13719 expressing of *the* bla_{CMY-2} gene were resistant to ampicillin, cefoxitin, ceftiofur, ceftazidime, and amoxicillin/clavulanic acid (AMC), however, remind to susceptive to fourth-generation cephalosporins of cefepime, which is consistent with an AmpC phenotype. The bla_{CMY-2} carrying transconjugants of Tc.J53 did not show any resistance to non–β-lactam antibiotics, suggesting that no other resistance genes were located on this IncI1 plasmid.

The conjugation frequency of bla_{CMY-2} IncI1 plasmid from *Salmonella* to *E. coli* J53 and *E. coli* NCCP 13719 was determined to be $1 \times 10^{-3} \pm 1 \times 10^{-7}$ and $2 \times 10^{-3} \pm 1 \times 10^{-3}$ on an agar plate with 200 µg/mL sodium azide and 8 µg/mL ceftiofur, 100 µg/mL tetracycline and 8 µg/mL ceftiofur, respectively.

Table 1. Information on 17 cephalosporin-resistant *Salmonella* spp. isolated from chicken-related sources.

No.	Strain	Serovar	Source	Antimicrobial Resistance	Phenotype	ESC Resistance Gene	Conjugation	Plasmid Type
1	A17-KCI-MNK-002-2S	S. Albany	Feces	AMC/AMP/FOX/TAZ/XNL/SXT/FIS/CHL/NAL/TET	AmpC	CMY-2	a	I1
2	A17-KCI-OP-004-4S	S. spp	Feces	AMP/TAZ/XNL/FEP/FIS/NAL/STR/TET	ESBL	CTX-M-15	NA	ND
3	A17-KCI-CRBR-001-5	S. Enteritidis	Carcass	AMP/TAZ/XNL/FEP/FIS/NAL/STR/GEN/TET	ESBL	CTX-M-15	b	FIIS
4	A17-KCI-HMR-001-4	S. Enteritidis	Carcass	AMP/XNL/SXT/FIS/CHL/CIP/NAL/STR/GEN/TET/COL	ESBL	TEM-1, CTX-M-15	NA	ND
5	A17-KCI-HMR-002-1	S. Enteritidis	Carcass	AMP/TAZ/XNL/FEP/NAL/GEN/TET	ESBL	CTX-M-15	NA	ND
6	A17-KCI-HMR-002-2	S. Enteritidis	Carcass	AMP/TAZ/XNL/FEP/NAL/GEN/TET	ESBL	CTX-M-15	NA	ND
7	A17-KCI-HMR-002-3	S. Enteritidis	Carcass	AMP/TAZ/XNL/FEP/NAL/GEN/TET	ESBL	CTX-M-15	b	FIIS
8	A17-KCI-HMR-002-4	S. Enteritidis	Carcass	AMP/TAZ/XNL/FEP/NAL/GEN/TET	ESBL	CTX-M-15	b	FIIS
9	A17-KCI-HMR-002-5	S. Enteritidis	Carcass	AMP/TAZ/XNL/FEP/NAL/GEN/TET	ESBL	CTX-M-15	NA	ND
10	A18-KCI-HMR-001-1S	S. Virchow	Feces	AMP/TAZ/XNL/FEP/FIS/NAL/STR/TET	ESBL	CTX-M-15	b	Nontypeable
11	A18-KCI-HMR-001-3S	S. Virchow	Feces	AMP/TAZ/XNL/FEP/SXT/FIS/CHL/NAL/STR/TET	ESBL	CTX-M-15	NA	ND
12	A18-KCI-HMR-001-4S	S. Virchow	Feces	AMP/TAZ/XNL/FEP/FIS/NAL/STR/TET	ESBL	CTX-M-15	b	Nontypeable
13	A18-KCI-HMR-002-2S	S. Virchow	Feces	AMP/TAZ/XNL/FEP/FIS/NAL/STR/TET	ESBL	CTX-M-15	b	Nontypeable
14	A18-KCI-HMR-002-3S	S. Virchow	Feces	AMP/TAZ/XNL/FEP/SXT/FIS/CHL/NAL/STR/TET	ESBL	CTX-M-15	NA	ND
15	A18-KCI-HMR-002-5S	S. Virchow	Feces	AMP/TAZ/XNL/FEP/SXT/FIS/CHL/NAL/STR/TET	ESBL	CTX-M-15	NA	ND
16	A18-KCI-OP-003-2S	S. Enteritidis	Feces	AMP/TAZ/XNL/FEP/NAL/TET	ESBL	CTX-M-15	NA	ND
17	A18-KCI-OP-003-3S	S. Enteritidis	Feces	AMP/TAZ/XNL/FEP/NAL/TET	ESBL	CTX-M-15	NA	ND

AMC, Amoxicillin/clavulanic acid; AMP, Ampicillin; FOX, cefoxitin; TAZ, Ceftazidime; XNL, Ceftiofur; FEP, cefepime; SXT, Trimethoprim/sulfamethoxazole; FIS, sulfisoxazole; CHL, chloramphenicol; CIP, ciprofloxacin; NAL, nalidixic acid; STR, streptomycin; GEN, gentamicin; TET, tetracycline; COL, colistin. a, Conjugation frequency (Transconjugant/recipient) with $\geq 10^{-3}$; b, Conjugation frequency with $<10^{-7}$. NA, not available; ND, not done.

Table 2. Characteristic changes of transconjugant after conjugation and transfer frequency of the resistance gene.

Strain [a]		Species	MIC (µg/mL)							Phenotype	ESC Resistance Gene	Conjugation Frequency [b]
			FOX	XNL	TAZ	FEP	AMP	AMC	TET			
			≥ 32	≥ 8	≥ 16	≥ 16	≥ 32	$\geq 32/16$	≥ 16			
MNK	Donor	S. Albany	>32	>8	>16	1	>64	>32/16	32	AmpC	CMY-2	
J53	Recipient	E. coli	8	<0.5	<1	<0.25	<2	4/2	<2			
Tc.J53	Transconjugant		>32	>8	>16	<0.25	>64	>32/16	<2	AmpC	CMY-2	$1 \times 10^{-3} \pm 1 \times 10^{-7}$
13719 (EIEC)	Recipient	E. coli	8	<0.5	<1	<0.25	8	4/2	>128			
Tc.13719	Transconjugant		>32	>8	>16	<0.25	>64	>32/16	>128	AmpC	CMY-2	$2 \times 10^{-3} \pm 1 \times 10^{-3}$

FOX, cefoxitin; XNL, ceftiofur; TAZ, ceftazidime; FEP, cefepime; AMP, ampicillin; AMC, amoxicillin/clavulanic acid; TET, tetracycline. [a] MNK, *Salmonella* Albany A17-KCI-MNK-002-2S; J53, *Escherichia coli* J53; Tc.J53, transconjugant of *E. coli* J53; 13719, *E. coli* NCCP 13719; Tc.13719, transconjugant of *E. coli* NCCP 13719; EIEC, enteroinvasive *Escherichia coli*. [b] Transconjugant per recipient (Tc/R).

3.3. In Vivo Transfer

To get the information on the efficiency of bacterial intergenic plasmid transfer in the mammalian intestine and to better mimic the in vivo situation, streptomycin-treated mice were used in this study. One-day after inoculation with donor and recipient (G4), the concentration of donor from each mouse ranged from 10^7 CFU/g to 10^9 CFU/g and that of recipients ranged from 10^2 CFU/g to 10^4 CFU/g (Figures 2 and S1). The frequency of plasmid transfer at 1-day post-infection (dpi) from G4 was estimated at an average of $2 \times 10^{-1} \pm 4 \times 10^{-1}$ with the ratio of transconjugants per recipient. This showed that the bla_{CMY-2} IncI1 plasmid was indeed efficiently transferred from the *Salmonella* isolate to the EIEC in the gut of streptomycin-treated mice. The number of transconjugants did not reach detectable levels at 7 dpi in four out of five mice.

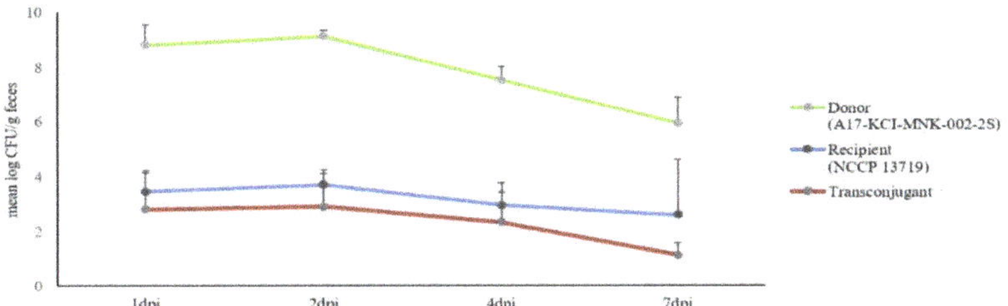

Figure 2. Bacterial counts of the donor, recipient, and transconjugant from mouse fecal samples in group 4 (G4), expressed as the log number of CFU per gram of feces.

3.4. Fitness Cost Assessment by Competition Experiment

The impact of harboring the bla_{CMY-2} gene on host fitness was evaluated by a pairwise competition experiment (Figure 3). Our results showed that the ceftiofur-sensitive strains were out-competing resistant strains in the absence of selective pressure with the value of transconjugant per parent strain at below 0, indicating that the bla_{CMY-2} gene-harboring plasmid-free strain dominated. In *E. coli* J53 with the bla_{CMY-2} gene-harboring plasmid, a slight fitness decrease was observed, and the fitness was stable following continuous passage until 72 h. A greater reduction in fitness was observed in *E. coli* NCCP 13719 compared with *E. coli* J53 in the serial passage, with a reduction of more than 3 log units for lasting 72 h. Regarding the value of the log ratio of resistant versus susceptible strains, the bla_{CMY-2} gene-harboring plasmid imposed a slight fitness cost to *E. coli* J53 from about log -0.89 at 24 h to -0.97 at 72 h. In contrast, the transconjugant from *E. coli* NCCP 13719 showed quite a high fitness cost from about log -2.02 at 24 h to -5.58 at 72 h.

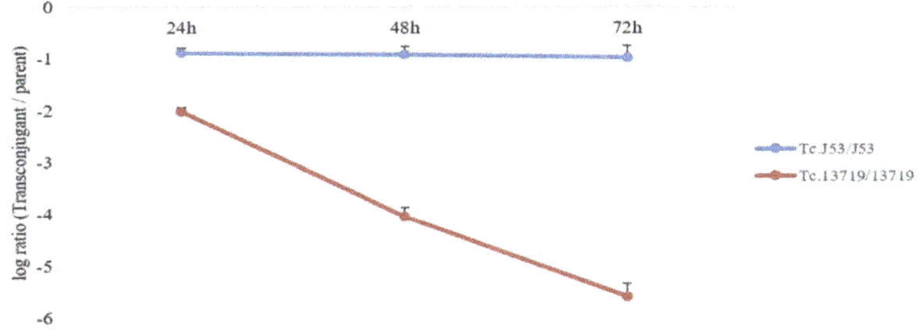

Figure 3. Competitive growth kinetics. Dynamics of replicate competition experiments for parent strains, *E. coli* J53 and *E. coli* NCCP 13719, and their transconjugant containing the bla_{CMY-2} gene.

4. Discussion

Since the first finding of the CTX-M-type gene from Korea in 2001, the prevalence of the $bla_{CTX-M-15}$-producing *Salmonella* in humans and chickens has rapidly increased over the years in Korea [3]. In the present study, the $bla_{CTX-M-15}$ gene was the most frequently detected but showed low frequencies at $<10^{-7}$ transconjugants per recipient, which is consistent with a previous study [21]. The $bla_{CTX-M-15}$ genes belonging to IncFII plasmid are known to be highly prevalent and involved in the concurrent transfer of antimicrobial resistance and virulence genes, which increases co-selection and probably leads to the emergence or outbreaks of virulent and multidrug-resistant (MDR) clones [22].

Conversely, the bla_{CMY-2} gene was observed from one strain in this study. The first report of the bla_{CMY-2} gene was described in the 1990s [23], and now, it is one of the most common and widely disseminated genes by plasmid-mediated AmpC β-lactamase from humans and chickens [2]. Regarding the transfer frequency of the bla_{CMY-2} gene between bacteria, the results of this study explain how frequently their resistance gene gets transferred to other bacterial species. Frequent transfer of the IncI1 plasmid carrying the bla_{CMY-2} gene was measured with the ratio of over 10^{-3} transconjugants per recipient in this study. This result is higher than previous findings wherein the transfer frequencies of bla_{CMY-2} and bla_{TEM-1} genes from the *Salmonella* isolated from poultry meat were in the ratio of 6.0×10^{-8} to 2.4×10^{-4} transconjugants per recipient [24]. The high frequency of transfer and their possibility to exchange genes within and between species might have resulted in the increasing prevalence of the bla_{CMY-2} gene in animals and humans, and its rapid dissemination may constitute a significant risk to public health. To our knowledge, this is the first description of the transfer of a bla_{CMY-2} gene-harboring plasmid from chicken-origin *Salmonella enterica* to pathogenic *E. coli* isolated from a human patient in a mammalian model. Identifying the transfer of antimicrobial-resistant plasmids and their frequency in a mouse model, which is an adequate way to predict the risk of the dissemination of antimicrobial resistance genes with a perspective of food safety. From this point of view, we used a streptomycin-pretreated mouse model, which provides more realistic results than any in vitro or gnotobiotic study because the normal microflora barrier and the present immune system give the tested animal model advantages in mimicking the human gut [25]. In this study, *E. coli* transconjugant appeared in all mice fecal samples 1 dpi in G4, and the high transfer frequency observed with the mean ratio of transconjugants per recipient was about 2×10^{-1} and per donor was 4×10^{-6}, which support statements on the rapid transfer of the bla_{CMY-2} gene. Although there is a lack of in vivo studies focused on the bla_{CMY-2} gene, several studies for conjugal transfer of ESBL genes have been reported. bla_{TEM} gene from *Salmonella* was transferred to *E. coli* recipient with the ratio of transconjugant per donor being 6.5×10^{-5} in mice without selective pressure [26], and $bla_{CTX-M-9}$ gene derived from chicken-origin *Salmonella* to *E. coli* at a frequency range of about 5.4×10^{-5} in gnotobiotic rats [27]. It is important to emphasize that it demonstrated not only the capability of transfer of bla_{CMY-2} gene with high frequency but also showed that the ratio of transconjugants per recipient in vivo was 2 log units higher than in vitro. Similar findings were reported wherein the rate of plasmid transfer between *Enterococcus faecium* strains was up to 8 log units higher in the germ-free mice model than in vitro [28]. The high frequencies of plasmid transfer in vivo may be due to the constant mixing of bacteria by the peristaltic movements in the gastrointestinal tract, stimulating a donor with more access to recipients than during in vitro mating, wherein the bacterial movement is much lesser [29]. These results emphasized the necessity of in vivo test for transferability and transfer frequency to figure out the potential risk of the presence of resistant strains in the digestive tract to humans.

Our result could be a direct evidence that the ESC resistant *Salmonella* from chicken-related products can transfer their resistance gene to other pathogens, thus leading to the possibility of inappropriate antimicrobial selection and limited treatment options resulting in therapeutic failure [5]. A case of treatment failure due to the emergence of resistance to ceftriaxone, a 3rd generation of cephalosporin, has been reported. The originally susceptible pathogen developed ceftriaxone resistance via the acquisition of a plasmid containing the ceftriaxone resistance gene during the 3rd ESC treatment, which finally caused therapeutic failure in the patient [30]. In addition, even if resistant bacteria transiently colonize, it may quickly transfer resistance plasmid into the human gastrointestinal tract; normal microbiota and the nutrient-rich environment make the gastrointestinal tract offer an ideal condition for gene exchange [28]. In this study, two days after inoculation, about $>10^4$ CFU/g of ceftiofur resistant *E. coli* isolates, regarded as normal flora-derived strain, were observed from one mouse in G2, which was inoculated only with *Salmonella* (donor). Likewise, the intestinal microbiota can act as a massive reservoir of antimicrobial resistance genes, thus prolonging the spread of MDR bacteria and resulting in therapeutic failure. Consequently, secondary infections would occur more often, indicating a serious threat to human health [31].

In vitro direct competition studies of the bla_{CMY-2} gene-harboring plasmid and two recipient *E. coli* showed that a variable fitness cost depends on the parent strains, and we observed that susceptible strains can outcompete resistant strains consistent with a previous study [17]. Normally, the acquisition of a plasmid often imposes a fitness burden on a bacterial cell [6]. Since *E. coli* NCCP 13719 in this study has the virulent gene *ipa*H, which may be encoded by a large plasmid, carrying another plasmid may present an adverse situation for the bacteria [6]. Conversely, the stable inheritance of bacterial plasmids without any selective pressure was also observed from transconjugants from *E. coli* J53 during the 72 h of experiment time. This phenomenon suggests that strains with low fitness costs even with the acquisition of plasmids from other strains may exist. For further studies, the mechanism of sustaining resistance plasmid with low fitness cost is expected to be a key research topic for suggesting the way to control the dissemination of antimicrobial resistance genes.

There are several limitations that we examined the characteristic of bla_{CMY-2} gene-harboring bacteria with a single strain; however, it may serve as fundamental data that defined the characteristics, and further studies with a greater number of resistant bacteria harboring the bla_{CMY-2} gene are required due to their increasing trend of emergence recently. In addition, mice are often naturally resistant to non-mice-origin *E. coli* colonization [32] as seen in our results, and thus, decreasing the number of bacteria is an inevitable phenomenon; however, it can be presented as a model that is sufficiently able to establish the transferability and frequency of antimicrobial resistance genes and emphasize that colonizing bacteria may transfer resistance plasmids readily in the intestinal tract [28]. To confirm the persistence of resistant genes through the colonization of antimicrobial-resistant bacteria in vivo and subsequent transfer of the gene to normal flora, changes in test strain or replacement of the in vivo model are required.

5. Conclusions

This study showed the prevalence of ESC resistance genes in *Salmonella* isolated from chickens and presented a comprehensive analysis of the highly frequent transfer of the bla_{CMY-2} gene in vitro and in vivo. In addition, this study has demonstrated the ease with which a bla_{CMY-2} gene-harboring plasmid can be rapidly transferred between *Salmonella* and pathogenic *E. coli* within the intestinal tracts of mice, even without antimicrobial selective pressure. Notably, we observed that once bla_{CMY-2} gene-harboring strains enter the mammalian intestinal tract, their dissemination could be more rapid and frequent than it would be in vitro, and even they could be transferred to the indigenous intestinal microbiota, threatening future treatments of infections. Since the use of cephalosporin in the poultry industry has increased over the last decade in Korea [33], the increasing emergence of ESBL/AmpC producing ESC resistant *Salmonella* spp. isolated from poultry is of concern. There is a risk for consumers related to exposure to ESBL/AmpC genes by contaminated food, so the application of guidelines for prudent antimicrobial usage in the poultry industry is urgently needed. Given the potential risk of the frequent transfer of the bla_{CMY-2} gene via the food chain to the human digestive tract, the molecular mechanism involved in the dissemination and maintenance of ESC resistance genes should be studied as further research in greater detail, and enhanced surveillance should be implemented to prevent the widespread of ESC resistant strains.

Supplementary Materials: The following are available online at: https://www.mdpi.com/article/10.3390/ani11061778/s1. Table S1. The list of pathogenic *Escherichia coli* isolates was obtained from the National Culture Collection for Pathogen (NCCP), based in Korea, Figure S1. Fecal excretion of the donor in group 2 (G2) and the recipient in group 3 (G3) (a), the donor, recipient, and transconjugant in group 4 (G4) (inoculation of donor and recipient, simultaneously) (b), expressed as the log number of CFU per gram of feces.

Author Contributions: B.W., M.K., and H.-K.J. contributed to the conception and design of experiments. S.-Y.C. and B.-R.K. contributed to the acquisition, analysis, and interpretation of data. B.-R.K., B.W., S.-Y.C., K.S., J.-F.Z., M.K. and H.-K.J. drafted and/or revised the article. All authors have read and agreed to the published version of the manuscript.

Funding: This work was supported by the Korea Institute of Planning and Evaluation for Technology in Food, Agriculture and Forestry (IPET) through Agriculture, Food and Rural Affairs Convergence Technologies Program for Educating Creative Global Leader (716002-7, 320005-4) funded by Ministry of Agriculture, Food and Rural Affairs (MAFRA). And this work was supported by the National Research Foundation of Korea (NRF) grant funded by the Korean government (MSIT) (No. 2020R1F1A1065136).

Institutional Review Board Statement: The study was conducted according to the guidelines of the Animal Care and Ethics Committees of Jeonbuk National University, and were approved by the National Association of Laboratory Animal Care (JBNU 2021-06).

Data Availability Statement: The data presented in this study are available from the corresponding author on reasonable request.

Conflicts of Interest: The authors declare no conflict of interest.

References

1. Mo, S.S.; Sunde, M.; Ilag, H.K.; Langsrud, S.; Heir, E. Transfer Potential of Plasmids Conferring Extended-Spectrum-Cephalosporin Resistance in *Escherichia coli* from Poultry. *Appl. Environ. Microbiol.* **2017**, *83*, 83. [CrossRef] [PubMed]
2. Doublet, B.; Praud, K.; Nguyen-Ho-Bao, T.; Argudín, M.A.; Bertrand, S.; Butaye, P.; Cloeckaert, A. Extended-spectrum -lactamase- and AmpC -lactamase-producing D-tartrate-positive *Salmonella enterica* serovar Paratyphi B from broilers and human patients in Belgium, 2008–2010. *J. Antimicrob. Chemother.* **2013**, *69*, 1257–1264. [CrossRef] [PubMed]
3. Na, S.H.; Moon, D.C.; Kang, H.Y.; Song, H.-J.; Kim, S.-J.; Choi, J.-H.; Yoon, J.W.; Yoon, S.-S.; Lim, S.-K. Molecular characteristics of extended-spectrum β-lactamase/AmpC-producing *Salmonella enterica* serovar Virchow isolated from food-producing animals during 2010–2017 in South Korea. *Int. J. Food Microbiol.* **2020**, *322*, 108572. [CrossRef]
4. Carattoli, A.; Bertini, A.; Villa, L.; Falbo, V.; Hopkins, K.; Threlfall, E.J. Identification of plasmids by PCR-based replicon typing. *J. Microbiol. Methods* **2005**, *63*, 219–228. [CrossRef]
5. Depoorter, P.; Persoons, D.; Uyttendaele, M.; Butaye, P.; de Zutter, L.; Dierick, K.; Herman, L.; Imberechts, H.; van Huffel, X.; Dewulf, J. Assessment of human exposure to 3rd generation cephalosporin resistant *E. coli* (CREC) through consumption of broiler meat in Belgium. *Int. J. Food Microbiol.* **2012**, *159*, 30–38. [CrossRef]
6. Gama, J.A.; Zilhão, R.; Dionisio, F. Conjugation efficiency depends on intra and intercellular interactions between distinct plasmids: Plasmids promote the immigration of other plasmids but repress co-colonizing plasmids. *Plasmid* **2017**, *93*, 6–16. [CrossRef]
7. Dionisio, F.; Conceição, I.; Marques, A.C.; Fernandes, L.; Gordo, I. The evolution of a conjugative plasmid and its ability to increase bacterial fitness. *Biol. Lett.* **2005**, *1*, 250–252. [CrossRef]
8. Sommer, M.O.A.; Dantas, G.; Church, G.M. Functional Characterization of the Antibiotic Resistance Reservoir in the Human Microflora. *Science* **2009**, *325*, 1128–1131. [CrossRef]
9. Shang, K.; Wei, B.; Kang, M. Distribution and dissemination of antimicrobial-resistant *Salmonella* in broiler farms with or without enrofloxacin use. *BMC Vet. Res.* **2018**, *14*, 1–14. [CrossRef]
10. Clinical Laboratory Standards Institute (CLSI). *Performance Standards for Antimicrobial Susceptibility Testing*, 27th ed.; CLSI Supplement M100; CLSI: Wayne, PA, USA, 2017.
11. European Committee on Antimicrobial Susceptibility Testing (EUCAST). *Breakpoint Tables for Interpretation of MICs and Zone Diameters*; European Committee on Antimicrobial Susceptibility Testing: Växjö, Sweden, 2019.

12. National Antimicrobial Resistance Monitoring System (NARMS). *National Antimicrobial Resistance Monitoring System for Enteric Bacteria: Human Isolates Surveillance Report for 2015 (Final Report)*; Department of Health and Human Services, CDC: Atlanta, GA, USA, 2018.

13. Dallenne, C.; da Costa, A.; Decré, D.; Favier, C.; Arlet, G. Development of a set of multiplex PCR assays for the detection of genes encoding important β-lactamases in *Enterobacteriaceae*. *J. Antimicrob. Chemother.* **2010**, *65*, 490–495. [CrossRef] [PubMed]

14. Pérez-Pérez, F.J.; Hanson, N.D. Detection of Plasmid-Mediated AmpC -Lactamase Genes in Clinical Isolates by Using Multiplex PCR. *J. Clin. Microbiol.* **2002**, *40*, 2153–2162. [CrossRef] [PubMed]

15. Mata, C.; Navarro, F.; Miró, E.; Walsh, T.; Mirelis, B.; Toleman, M. Prevalence of SXT/R391-like integrative and conjugative elements carrying blaCMY-2 in *Proteus mirabilis*. *J. Antimicrob. Chemother.* **2011**, *66*, 2266–2270. [CrossRef] [PubMed]

16. Barletta, F.; Mercado, E.H.; Lluque, A.; Ruiz, J.; Cleary, T.G.; Ochoa, T.J. Multiplex Real-Time PCR for Detection of *Campylobacter*, *Salmonella*, and *Shigella*. *J. Clin. Microbiol.* **2013**, *51*, 2822–2829. [CrossRef]

17. Barthel, M.; Hapfelmeier, S.; Quintanilla-Martínez, L.; Kremer, M.; Rohde, M.; Hogardt, M.; Pfeffer, K.; Rüssmann, H.; Hardt, W.-D. Pretreatment of Mice with Streptomycin Provides a *Salmonella enterica* Serovar Typhimurium Colitis Model That Allows Analysis of Both Pathogen and Host. *Infect. Immun.* **2003**, *71*, 2839–2858. [CrossRef]

18. Göttig, S.; Riedel-Christ, S.; Saleh, A.; Kempf, V.A.; Hamprecht, A. Impact of blaNDM-1 on fitness and pathogenicity of *Escherichia coli* and *Klebsiella pneumoniae*. *Int. J. Antimicrob. Agents* **2016**, *47*, 430–435. [CrossRef]

19. Bates, S.; Cashmore, A.M.; Wilkins, B.M. IncP Plasmids Are Unusually Effective in Mediating Conjugation of *Escherichia coli* and *Saccharomyces cerevisiae*: Involvement of the Tra2 Mating System. *J. Bacteriol.* **1998**, *180*, 6538–6543. [CrossRef] [PubMed]

20. Lampkowska, J.; Feld, L.; Monaghan, A.; Toomey, N.; Schjørring, S.; Jacobsen, B.; van der Voet, H.; Andersen, S.R.; Bolton, D.; Aarts, H.; et al. A standardized conjugation protocol to asses antibiotic resistance transfer between lactococcal species. *Int. J. Food Microbiol.* **2008**, *127*, 172–175. [CrossRef] [PubMed]

21. Thingholm, K.R.; Hertz, F.B.; Løbner-Olesen, A.; Frimodt-Møller, N.; Nielsen, K.L. *Escherichia coli* belonging to ST131 rarely transfers blactx-m-15 to fecal *Escherichia coli*. *Infect. Drug Resist.* **2019**, *12*, 2429–2435. [CrossRef] [PubMed]

22. Carattoli, A. Plasmids in Gram negatives: Molecular typing of resistance plasmids. *Int. J. Med. Microbiol.* **2011**, *301*, 654–658. [CrossRef]

23. Bauernfeind, A.; Stemplinger, I.; Jungwirth, R.; Giamarellou, H. Characterization of the plasmidic beta-lactamase CMY-2, which is responsible for cephamycin resistance. *Antimicrob. Agents Chemother.* **1996**, *40*, 221–224. [CrossRef]

24. Chen, S.; Zhao, S.; White, D.G.; Schroeder, C.M.; Lu, R.; Yang, H.; McDermott, P.F.; Ayers, S.; Meng, J. Characterization of Multiple-Antimicrobial-Resistant *Salmonella* Serovars Isolated from Retail Meats. *Appl. Environ. Microbiol.* **2004**, *70*, 1–7. [CrossRef] [PubMed]

25. Schjørring, S.; Struve, C.; Krogfelt, K.A. Transfer of antimicrobial resistance plasmids from *Klebsiella pneumoniae* to *Escherichia coli* in the mouse intestine. *J. Antimicrob. Chemother.* **2008**, *62*, 1086–1093. [CrossRef] [PubMed]

26. Lalruatdiki, A.; Dutta, T.K.; Roychoudhury, P.; Subudhi, P.K. Extended-spectrum β-lactamases producing multidrug resistance *Escherichia coli*, *Salmonella* and *Klebsiella pneumoniae* in pig population of Assam and Meghalaya, India. *Vet. World* **2018**, *11*, 868–873. [CrossRef]

27. Faure, S.; Perrin-Guyomard, A.; Delmas, J.-M.; Laurentie, M. Impact of Therapeutic Treatment with β-Lactam on Transfer of the blaCTX-M-9 Resistance Gene from *Salmonella enterica* Serovar Virchow to *Escherichia coli* in Gnotobiotic Rats. *Appl. Environ. Microbiol.* **2009**, *75*, 5523–5528. [CrossRef] [PubMed]

28. Dahl, K.H.; Mater, D.D.G.; Flores, M.J.; Johnsen, P.J.; Midtvedt, T.; Corthier, G.; Sundsfjord, A. Transfer of plasmid and chromosomal glycopeptide resistance determinants occurs more readily in the digestive tract of mice than in vitro and exconjugants can persist stably In Vivo in the absence of glycopeptide selection. *J. Antimicrob. Chemother.* **2007**, *59*, 478–486. [CrossRef] [PubMed]

29. Licht, T.R.; Christensen, B.; Krogfelt, K.A.; Molin, S. Plasmid transfer in the animal intestine and other dynamic bacterial populations: The role of community structure and environment. *Microbiology* **1999**, *145*, 2615–2622. [CrossRef]

30. Su, L.-H.; Chiu, C.-H.; Chu, C.; Wang, M.-H.; Chia, J.-H.; Wu, T.-L. In Vivo Acquisition of Ceftriaxone Resistance in *Salmonella enterica* Serotype Anatum. *Antimicrob. Agents Chemother.* **2003**, *47*, 563–567. [CrossRef]

31. Sunde, M. Resistance to antibiotics in the normal flora of animals. *Vet. Res.* **2001**, *32*, 227–241. [CrossRef]

32. Stromberg, Z.R.; Van Goor, A.; Redweik, G.A.J.; Brand, M.J.W.; Wannemuehler, M.J.; Mellata, M. Pathogenic and non-pathogenic *Escherichia coli* colonization and host inflammatory response in a defined microbiota mouse model. *Dis. Model. Mech.* **2018**, *11*. [CrossRef]

33. Animal and Plant Quarantine Agency (APQA). *Establishment of Antimicrobial Resistance Surveillance System for Livestock, 2018*; Ministry of Agriculture, Food and Rural Affairs: Gimcheon, Korea; Sejong, Korea, 2019.

Article

Antinematode Activity of Abomasum Bacterial Culture Filtrates against *Haemonchus contortus* in Small Ruminants

Asfa Nazish [1], Fozia [2], Baharullah Khattak [1], Taj Ali Khan [1,3], Ijaz Ahmad [4,*], Riaz Ullah [4,*], Ahmed Bari [4], Majid M. Asmari [5], Hafiz M. Mahmood [6], Muhammad Sohaib [7], Ahmad El Askary [8], Attalla F. El-kott [9,10] and Mohamed M. Abdel-Daim [11]

[1] Department of Microbiology, Kohat University of Science & Technology, Kohat 26000, Pakistan; asfanazish28@gmail.com (A.N.); baharkk75@gmail.com (B.K.); microbiologist63@yahoo.com (T.A.K.)
[2] Boichemistry Department, KMU Institute of Medical Sciences, Kohat 26000, Pakistan; drfoziazeb@yahoo.com
[3] Institute of Pathology and Diagnostic Medicine, Khyber Medical University, Peshawar 25160, Pakistan
[4] Department of Chemistry, Kohat University of Science & Technology, Kohat 26000, Pakistan; abari@ksu.edu.sa
[5] Department of Pharmaceutical Chemistry, College of Pharmacy, P.O. Box 2457, King Saud University, Riyadh 11451, Saudi Arabia; masmary@windowslive.com
[6] Department of Pharmacology, College of Pharmacy, King Saud University, Riyadh 11451, Saudi Arabia; harshad@ksu.edu.sa
[7] Department of Soil Science, College of Food and Agriculture Sciences, King Saud University, P.O. Box 2460, Riyadh 11451, Saudi Arabia; msohaib@ksu.edu.sa
[8] Department of Clinical Laboratory Sciences, College of Applied Medical Sciences, Taif University, P.O. Box 11099, Taif 21944, Saudi Arabia; a.elaskary@tu.edu.sa
[9] Biology Department, Faculty of Science, King Khalid University, Abha 61421, Saudi Arabia; elkottaf@kku.edu.sa
[10] Zoology Department, College of Science, Damanhour University, Damanhour 22511, Egypt
[11] Pharmacology Department, Faculty of Veterinary Medicine, Suez Canal University, Ismailia 41522, Egypt; abdeldaim.m@vet.suez.edu.eg
* Correspondence: drijaz_chem@yahoo.com (I.A.); rullah@ksu.edu.sa (R.U.)

Citation: Nazish, A.; Fozia; Khattak, B.; Ali Khan, T.; Ahmad, I.; Ullah, R.; Bari, A.; Asmari, M.M.; Mahmood, H.M.; Sohaib, M.; et al. Antinematode Activity of Abomasum Bacterial Culture Filtrates against *Haemonchus contortus* in Small Ruminants. *Animals* **2021**, *11*, 1843. https://doi.org/10.3390/ani11061843

Academic Editors: Paola Roncada and Bruno Tilocca

Received: 3 May 2021
Accepted: 14 June 2021
Published: 21 June 2021

Publisher's Note: MDPI stays neutral with regard to jurisdictional claims in published maps and institutional affiliations.

Simple Summary: *Haemonchus contortus* is an important gastrointestinal nematode parasite of the tropical and sub-tropical regions that cause haemonchosis in small ruminants like goats and sheep. It causes low production, reduced growth and may cause death of the infected animals. Due to the resistance development and environmental issues, the use of anthelmintics can be replaced with biological control, which is an environment friendly alternative. In the present study, three bacteria viz; *Comamonas testosteroni*, *C. jiangduensis* and *Pseudomonas weihenstephanesis* showed significant effect on nematode mortality and egg hatch inhibition. It was also observed that the anthelmintic activity of these bacteria was dose dependent, where 100% bacterial metabolite concentration showed the highest activity. It is suggested that these bacteria may included in the integrated nematode management.

Abstract: Haemonchosis is a parasitic disease of small ruminants that adversely affects livestock production. *Haemonchus contortus* is one of the most prevalent nematode parasites that infect the abomasum of small ruminants. This parasite reduces milk production, overall growth and sometimes causes the death of the infected animals. The evaluation of the biocontrol potential of some abomasum bacterial isolates against *H. contortus* is investigated in this study. Out of which, three isolates—*Comamonas testosteroni*, *Comamonas jiangduensis*, *Pseudomonas weihenstephanesis*—show significant effect against the nematode L3, adult, and egg hatch inhibition assays. Various concentrations of metabolites from these bacteria are prepared and applied in different treatments compared with control. In the case of adult mortality assay, 50% metabolites of *C. testosteroni* and *P. weihenstephanesis* show 46% adult mortality, whereas *C. jiangduensis* shows 40% mortality. It is observed that decreasing the concentration of bacterial metabolite, lowers nematode mortality. The minimum nematode mortality rate is recorded at the lowest filtrates concentration of all the bacterial isolates. The same trend is observed in egg hatch inhibition assay, where the higher concentration of bacterial culture filtrates shows 100% inhibition of *H. contortus* egg. It is concluded that the effect of bacterial culture

---

filtrates against *H. contortus* is dose-dependent for their activity against nematode L3, adult, and inhibition of egg hatchment.

Keywords: small ruminants; *H. contortus*; abomasum; fecal samples; bacterial culture filtrates

1. Introduction

Gastrointestinal parasites are considered as the main cause of economic losses in the livestock sector. Among gastrointestinal parasitic infections, haemonchosis is important and dominant that greatly destroys livestock production, particularly the small ruminants [1]. This disease is caused by three species of the genus Haemonchus, namely, *H. similis*, *H. placei*, and *H. contortus*. Among these, *H. contortus* is one of the most prevalent nematode parasites that infect the abomasum of small ruminants worldwide [2]. It is commonly known as a red stomach worm, the wire worm, or the barber's pole worm. It belongs to phylum Nematoda, family Trichostrongyloidae, class Secernentea, and the order Strongylida. The highly susceptible part of ruminant's stomach to *H. contortus* is the abomasum, in which adult worms are present. This parasite causes low production, decreased growth, lower body weight, and sometimes, cause the death of the infected host. This parasite is most prevalent in Africa; however, many cases have been reported in North America as well [3].

Haemonchus contortus mostly affects young animals, having hypo immunological response, showing low resistance to the parasite [4]. Primary symptoms of haemonchosis-include pallor, anemia, edema, ill thrift, lethargy, and depression, which may cause sudden death in acute infection. Another prominent symptom of haemonchosis is the accumulation of fluid in the submandibular tissue, a phenomenon commonly called "bottle jaw" [5]. When the L3 larvae resume development in spring, the threat of haemonchosis increases. The young and pregnant or lactating mothers are highly susceptible to *H. contortus*, because of their low immunogenic response against the parasite infection [6]. A heavy infection (20,000–30,000 worms) of Haemonchus species can kill sheep and goats very quickly [7]. Haemonchosis can be diagnosed based upon the characteristic clinical signs of anemia, low Packed Cell Volume (PCV), pale mucous membranes dehydration, weakness, retarded growth, and edema. [8].

Haemonchus contortus is distributed throughout the world, where warm and humid climate prevails; hence, haemonchosis is a major threat in tropical and subtropical regions. [9]. Haemonchus species are prevalent in Pakistan and are reported almost from every district of Pakistan by different researchers with varying percent prevalence [10–12]. In Khyber Pakhtunkhwa and central Punjab, 72% prevalence of *H. contortus* was reported, while other researchers recorded its prevalence with varying percentages in many districts of Pakistan [13–15].

It is not advisable to eliminate the parasites from livestock, but to keep the population under a threshold, in a sustainable state [16]. To control the nematode parasites, different management practices, such as the use of chemical anthelmintics, sanitation, vaccination, various plant extracts, and biological control, are in common practice [17]. The helminths infection in man and animals is mostly treated by chemotherapy. Gastrointestinal nematode infections can be managed by using chemical anthelmintics, which are used as prophylactic measures. Due to the over-use of chemical anthelmintics, they reduce their effectiveness and emerge resistance in nematodes [18]. The *H. contortus* infection can also be effectively treated with the wire particles of copper oxide (COWP) and copper sulfate ($CuSO_4$). Using copper oxide wire particles 2.5 g to 5 g in sheep, *H. contortus* eggs number was significantly reduced. These anthelmintics have been found to reduce the parasite population in small ruminants with low resistance development. Presently, it is used along with chemical antihelmintics to combat resistance development in *H. contortus* [19].

Limited information available regarding the anthelminthic activity of bacteria antagonists against parasitic nematodes. Some species, including *Bacillus* sp., have been reported to have nematocidal activity. A soil bacterium, *Bacillus thuringienisis*, is widely used as a biological control agent against different pathogens because of its low mammalian toxicity and the species specificity by particular endotoxin groups [20]. *Bacillus* sp. Produce a variety of toxic proteins that are vigorous against different parasites [21]. A number of toxins produced during vegetative growth show lethal activity; however, this lethal activity mainly results from the production of delta endotoxin that is synthesized during sporulation [22,23]. The production of other factors that might contribute to the toxic effects observed, such as proteases, chitinase, exotoxins, and lipases [24].

Little is known about the microbial diversity in the abomasum of the sheep, an important site of nematode infection, and the correlation between the microbial diversity and GIN resistance. Thus, this work aimed to isolate bacteria from the abomasum of sheep and goats to test the anthelmintic activity of metabolites of these bacterial isolates against *H. contortus*.

2. Materials and Methods

The present research work was carried out in the Department of Microbiology, Kohat University of Science and Technology Kohat. All the processes were performed in the aseptic environment.

2.1. Collection of Abomasal Content and Fecal Samples

Abomasum content, as shown in Figure 1a–c, was collected from slaughtered goats and sheep at a slaughterhouse in Kohat, Pakistan. A total of 50 samples of the abomasum and 50 samples of feces were taken from goats and sheep, and the samples were brought to Microbiology Laboratory for further processing.

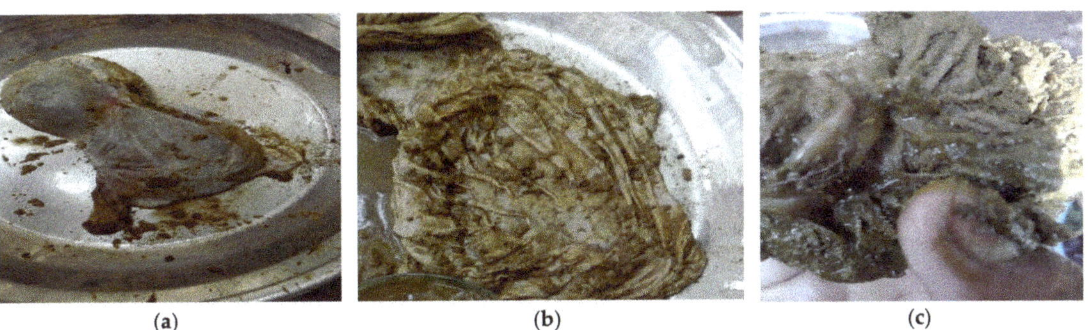

(a) (b) (c)

Figure 1. (**a**) Abomasum, (**b**) opened of abomasum, (**c**) adult *Haemonchus contortus* in abomasum.

2.2. Isolation and Identification of Bacteria from Abomasal Contents

Bacteria were isolated by serial dilution method, taking abomasal fluid in 100 mL of distilled water. After that, 1 mL from the suspension was taken and put in a test tube containing 9 mL of distilled water. Different bacterial dilutions 10^{-1}–10^{-6} and sterile distilled water (control) were used. Streaks were made from 10^{-3}, 10^{-4}, 10^{-5}, and 10^{-6} bacterial dilutions on petri dishes containing nutrient agar. The plates were incubated at 37 °C for 24 h. The bacterial isolates were subcultured and identified by colony morphology and biochemical tests, and finally, through genomic DNA sequencing [25].

2.2.1. Colony Morphology

Size, shape, colony consistency, margins, and elevation are included in the colony morphology.

2.2.2. Gram Staining

Smear from each bacterial isolate was organized on a glass slide, and it was heat-fixed by passing it over the flame. Crystal violet was added drop by drop on the smear and left for 60 s. The slide was washed with tap water. For 45 s, iodine was overlaid on the slide and washed with distilled water. The slide was washed for 10 to 15 s with a decolorizer (alcohol). Smear was rinsed for few seconds with distilled water and counter-stained with safranin and air-dried. Deep violet or purple color emerged for Gram positive bacteria, while Gram negative bacteria appeared purple or red.

2.2.3. Biochemical Characterization of the Bacterial Isolates

For the identification of bacterial isolates, various biochemical tests were performed, such as Oxidase test, Indole test, Sugar fermenter test, Motility test, Catalase test.

Catalase Test

A drop of hydrogen peroxide was added to a slide. A loop filled with each of the bacterial isolates was mixed into the droplet. Bubble formation indicated a positive catalase test.

Oxidase Test

A smooth filter paper was put on a petri plate, and a drop of oxidase reagent was added. Bacterial culture was stretched on the droplet of the oxidase reagent, using an inoculating loop. The formation of the reagent's dark purple color confirmed the oxidase positive test [26].

Indole Test

Bacterial colonies were inoculated into individual tubes of 2 mL tryptone water, incubated at 37 °C for 24 h, and tested for indole production with Kovac's reagent. If the reagent showed a cherry red color layer, then it confirms the positive test.

Sugar fermenter

The sugar fermentation test was performed by inoculating a loop full of a nutrient broth culture of the organisms into the tubes containing different sugar media (five basic sugars, such as dextrose, sucrose, lactose, maltose, and mannitol) and incubated for 24 h at 37 °C. The sugar fermenter was shown by a color change from reddish to yellow and the formation of gas bubbles in the inverted test tubes.

Motility Test

One drop of bacterial culture, grown on nutrient broth, was placed on the coverslip. The same was placed inverted over around the concave depression of the hanging drop slide to make hanging drop preparation and sealed with Vaseline to prevent airflow and evaporation of the fluid. The hanging drop slide was then examined under $100\times$ objective of a compound microscope using immersion oil. The motile and nonmotile organisms were identified by observing motility with to and from the movement of bacteria [27].

2.3. Isolation of Haemonchus Contortus Adults and Larvae from Abomasum

The abomasum of the freshly slaughtered goats and sheep were taken to pick adult *H. contortus* and larvae in the slaughterhouse of Kohat district, Pakistan. The worms were collected from the abomasum by washing with distilled water, and it was transferred into Phosphate Buffer Saline (PBS) with pH 7.4. The PBS was prepared by dissolving 8 g NaCl, 0.2 g KCl, 1.44 g Na_2HPO_4, and 0.24 g of KH_2PO_4 in 800 mL of distilled water, and distilled water was added to adjust the volume to one litter.

2.4. Isolation of H. contortus Eggs from Faecal Sample

In a petri plate (Figure 2), about 4 g of feces were taken, and 60 mL floatation liquid was poured into the container by mixing feces carefully with a moving device. The resultant fecal suspension was poured over a filter of 200 μm into another vessel. The fecal contents were poured into a test tube. Then the test tube was topped gently with the interruption, a curved meniscus was formed at the top of the tube and placed a coverslip carefully. After

20 min, the droplet of liquid sticks to it. Under 10× and 40× of the compound microscope, the slides were observed [28].

Figure 2. Fecal Samples of Sheep and Goats Containing *Haemonchus contortus* Eggs.

2.4.1. Estimation of Eggs in the Fecal Samples

The eggs of *H. contortus* in the animal's fecal samples were calculated by using the McMaster counting chamber method [29]. In this procedure, Fresh feces of 3 g were taken and mix them in 42 mL of saturated sodium chloride solution. Through a tea filter or strainer, passed the suspension three times. Then both compartment of the McMaster counting chamber was filled with suspension and wait for 3–5 min and observed the McMaster chambers under the 10× of a light microscope.

Eggs seen in 1st and 2nd chambers were calculated as:

$$(\text{Eggs seen in chamber 1} + \text{Eggs seen in chamber 2}) \times 50 = \text{Eggs per Gram (EPG)}$$

The nematode eggs were taken from the feces of sheep and goats. Eggs from the highest EPG were isolated by dissolving 5 g of feces in 10 mL of distill water. The diluted feces were filtered through a 100 mesh and transferred into another flask. Then the saturated salt solution was added into filtrate in the threefold volume of the flask. On top of the container, placed a clear plastic sheet; hence, the surface of the solution could touch the plastic sheet and placed it for 60 min. The eggs were adhered to its lower surface, and it was carefully washed away by tap water in a clean container. By removing the upper layer of the water, the eggs were settled down at the end of the container after one hour [30].

2.4.2. Culture Filtrate Collection from the Bacterial Isolates

The nutrient broth, containing 0.5% Peptone and 0.3% beef extract/yeast extract in distilled water, was autoclaved, and inoculated with bacterial isolates, and incubated in a shaking incubator at 37 °C for 7 days. The broth culture was centrifuged at 1000 rpm for 15 min. The pellet was discarded. With the help of Whatman filter paper, the supernatant was filtered and again filter through Miller HA syringe filters (pore size = 0.45 μm). The secondary metabolites were obtained. To check the presence of bacterial cells in metabolites, it was again inoculated on nutrient agar plates at 37 °C for 24 h. Extract of metabolites were mixed with phosphate buffer saline and distilled water to make different concentrations.

The extracts were considered as 100% concentrated, while the bacterial extracts with PBS and anthelmintic agent were considered as a positive control (PC), whereas only PBS and worms were considered as a negative control (NC).

2.5. In Vitro Bioassays

Various concentrations of the bacterial culture filtrates were used to evaluate their anthelmintic effect on adult and larval mortality and egg hatch inhibition of *H. contortus*, by using the standard techniques as per the protocol of Kotze [20].

2.5.1. Adult Mortality Assay (AMA)

Adult worms were taken from the abomasum of freshly slaughtered sheep and goats to perform the adult mortality assay (AMA). Culture filtrate, obtained from three bacterial isolates Abomasum Bacteria with Pinkish Colony (ABP), Abomasum Bacteria with Yellow Colony (ABY), and Abomasum Bacteria with Creamy White Colony (ABCW), were diluted in PBS, as 100%, 50%, 25%, 12.5%, and 6.25% in 48 well plates, into which five adults' nematodes were transferred into each well. The treatments were repeated five times. The bacterial culture filtrates were taken alone with worms, were considered as a 100% concentrated. For negative control, PBS with worms was used, whereas, for positive control, levasole (AgriLabs) 25 µg/mL in PBS with worms was used. Data on the adult nematode mortality were taken after every hour of the treatment, until all the worms were found dead in control. The treated worms were placed in warm PBS and observed their possible motility [31].

The percent nematode mortality was calculated by the following formula:

$$\text{Percentage mortality} = \text{P test}/\text{P total} \times 100$$

P test: number of dead worms
P total: number of dead worms + number of live worms

2.5.2. Larval Mortality Assay (LMA)

Larvae (L3) of *H. contortus* were taken from the abomasum of freshly slaughtered sheep and goats to perform the Larval Mortality Assay (LMA). Various concentrations of the bacterial metabolites (100%, 50%, 25%, 12.5%, and 6.25%) were prepared in PBS. The concentration was considered as 100% with L3 larvae and with metabolites. Well containing larvae and 2.5 mL phosphate buffer saline were considered as a negative control. The larvae to which an antihelmintic agent of 2.5 mL levasole (25 µg/mL) was added, considered as a positive control. The plates were placed in an incubator for 3 h at 37 °C. Under 10× of a stereo microscope, data on the larvae mortality were taken after three hours of treatments, until all the larvae were found dead in control. The treated worms were placed in warm PBS and observed their possible motility [32].

The percent larvae mortality was calculated as:

$$\text{Percentage mortality of larvae} = \text{P check}/\text{P total}$$

P check: number of dead larvae
P total: number of dead larvae + number of live larvae

2.5.3. Egg Hatch Inhibition Assay

Egg hatch inhibition assay was performed to evaluate the inhibitory effect of the metabolites of the bacterial isolates ABP, ABY, and ABCW. This assay was repeated in triplicate following the protocol given by Coles et al. [33].

Nematode eggs were placed in 15 mL of sterile distilled water, and by the McMaster technique, their quantity was adjusted to 100–200 eggs in 75 µL of water and was added into each well of a 24 well titration plate. Metabolites of each bacterial isolates were added to each well at various concentrations of 100%, 50%, 25%, 12.5%, 6.25%, and 3.125%. Wells with nematode eggs, having no metabolites, were considered as a negative control, while 0.025 mg/mL Oxfendazole (Zenith Pharma, Karachi, Pakistan), in 0.3% DMSO, served as a positive control. The plates were incubated at 37 °C for 24 h. A droplet of Lugol's iodine was added to maintain the process for 24 h. Under 10× of an inverted microscope, the total number of hatched and unhatched eggs were counted. The experiment was replicated five times [34].

The percent egg hatch inhibition was calculated as:

$$\text{Inhibition of eggs} = P\ \text{test}/P\ \text{total} \times 100$$

P test: number of unhatched or hatched eggs.
P total: number of unhatched or hatched eggs + Larvae (L1)

2.6. Genomic DNA Extraction

DNA from the abomasal bacteria were extracted by the standard protocol of the phenol chloroform method. Fresh bacterial broth cultures were centrifuged at 13,000 rpm for 5 min. The supernatant was discarded, and pellets were resuspended in 550 μL of Tris-EDTA buffer with the addition of 30 μL of 10% SDS and 5 μL Proteinase K. Vortexed properly and incubated at 37 °C for 1 h. After incubation, 100 μL of 5 M NaCl and 80 μL of CTAB/NaCl were added and mixed properly and incubated again at 65 °C in a water bath for 10 min. An equal volume of phenol, chloroform, and isoamyl alcohol (25:24:1) was added, mixed properly, and centrifuged at 13,000 rpm for 5 min to purify DNA. The supernatant was transferred to a 1.5 mL fresh Eppendorf tube. To this tube, an equal volume of chloroform and isoamyl alcohol, was added and centrifuged at 13,000 rpm for 5 min. The supernatant was transferred to a 1.5 mL Eppendorf tube, and for the precipitation of DNA, 0.6 volume of chilled isopropanol was added and placed at −20 °C for 20 min. DNA was pelleted by centrifugation at 13,000 rpm for 5 min and washed the pellet with 70% ethanol, and then dried at room temperature. Finally, DNA was dissolved in 50 μL Tris-EDTA buffer overnight incubation at 37 °C. The DNA concentration was measured using a spectrophotometer by taking absorbance at 260 nm and diluted for polymerase chain reaction (PCR).

2.6.1. DNA Confirmation and 16S RNA Amplification

The purified genomic DNA was verified through gel electrophoresis by mixing 4 μL genomic DNA with 2 μL loading dye and then load it in 1% agarose gel. The gel was run in a gel tank for 30 min at 120 volts and observed under UV transilluminator for DNA band. The amplification of the 16S RNA gene, universal primers, as shown in Table 1, were run on genomic DNA samples.

Table 1. Primer name and Sequence.

Primer	Sequence	Type
Forward	5-GGAGGCAGCAGTAGGGAATA-3	16s RNA (Universal)
Reverse	5-TGACGGGCGGTGAGTACAAG-3	16s RNA (Universal)

2.6.2. PCR Conditions

The PCR tubes were put in the thermal cycle. The amplification was performed, following the condition given in Table 2.

Table 2. PCR condition for 16s RNA universal primers.

Initial Denaturation		Denaturation		Annealing		Extension		Final Extension	
Temperature	Time	Temperature	Time	Temperature	Time	Temperature	Time	Temperature	Time
94 °C	5 min	94 °C	1 min	50 °C	40 s	72 °C	90 s	72 °C	5 min

2.6.3. Gel Electrophoresis for PCR Products

The PCR products were confirmed on 1.5% agarose gel in TBE buffer containing 3 μL of ethidium bromide and run for 30 min at 100 voltages and 300 milli Ampere current. The bands were visualized by the Gel Doc system [35].

2.7. Statistical Analysis

The information acquired from the bioassays, i.e., adult nematode mortality, nematode larval mortality, and eggs hatch inhibitions assays, were evaluated by P Test via Statistic version 9.

3. Results

3.1. Isolation and Identification of Bacterial Isolates from Abomasal Contents

Three different colonies were subcultured based on different morphological characteristics.

3.1.1. Colony Morphology

The colony morphology of the bacterial isolates are shown in Table 3. The colonial morphology of the ABP bacterial isolate (Figure 3a appeared as irregular with an entire margin, and its elevation was flat and pinkish in color. Similarly, the colony morphology of ABY bacterial isolate was regular with a filamentous margin, and its elevation was flat and yellow. The colony morphology of the ABCW isolate was irregular in shape and creamy white, as shown in Figure 3b.

Table 3. Colony Morphology and Gram Staining of Bacterial Isolates.

Bacterial Isolates	Colony Morphology			Gram Staining
	Color	Shape	Elevation	
ABP	Pinkish	Irregular	Flat	Gram Negative Rods
ABY	Yellow	Regular	Flat	Gram Negative Rods
ABCW	Creamy Yellow	Irregular	Flat	Gram Negative Rods

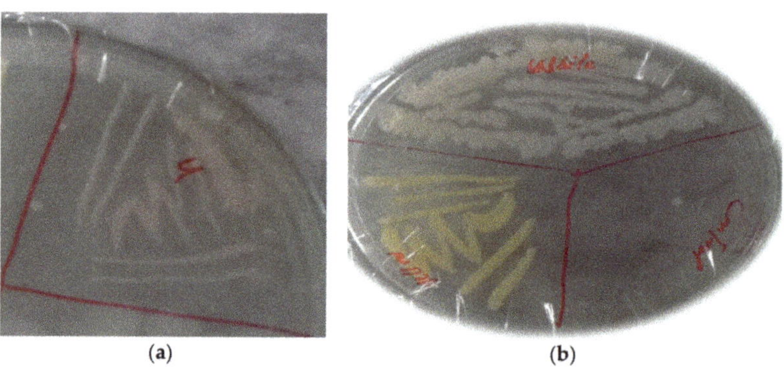

(a)   (b)

Figure 3. (**a**) Bacterial Isolate ABP colony; (**b**) bacterial Isolate ABY and ABCW colony.

3.1.2. Gram Staining

Gram staining revealed that all the bacterial isolates ABP, ABY, and ABCW, were Gram negative rods, as shown in Table 4.

Table 4. Biochemical Tests for Bacteria Identification.

Bacterial Isolates	Oxidase	Indole	Sugar Fermenter	Motility	Catalase
ABP	Negative	Negative	Positive	Motile	Negative
ABY	Positive	Negative	Negative	Non motile	Positive
ABCW	Positive	Negative	Negative	Non motile	Positive

3.1.3. Biochemical Identification

Selected bacterial isolates ABP, ABY, and ABCW were identified by various biochemical tests as given in Table 4. Bacterial isolate ABP was negative for oxidase, catalase, and indole production, while positive for sugar fermenter and amotility test. The biochemical tests showed that the bacterial strain ABP was identified as *Comamonas testosteroni*. Bacterial isolate ABY was positive for catalase, oxidase test, and negative for motility test, indole, and sugar fermenter. Based on biochemical tests, the isolated bacterial strain ABY was reported as *Comamonas jiangduensis*. Bacterial isolate ABCW was positive for catalase and oxidase, while negative for sugar fermenter, motility test, and indole production. Based on the biochemical tests, the bacterial strain ABCW was considered as *Pseudomonas weihenstephanesis*.

3.2. In Vitro Bioassay
3.2.1. Adult Mortality Assay (AMA)

Data regarding the mortality rate of *H. contortus* by various concentrations of bacterial culture filtrates are given in Table 5 and Figure 4. Bacterial isolate *C. testosteroni* caused the highest nematode mortality rate (100%) at 100% metabolite concentration. At 50% metabolite concentration, the nematode mortality was recorded as (46%). The lowest mortality rate (26%) was recorded at 6.25% metabolite concentration. Analysis of the data, regarding the adult nematode mortality by *C. jiangduensis*, showed that the bacterial metabolite had a significant effect on nematode mortality. The highest mortality rate (100%) was recorded at 100% concentration, followed by 50% metabolite concentration, where it was noted as 40%. While the minimum mortality rate (6%) was recorded at the metabolite concentration of 6.25%. As for the nematocidal effect of the bacterial isolate *P. weihenstephanesis* is concerned, the maximum adult nematode mortality rate (100%) was also recorded at 100% bacterial metabolite concentration. At 50% metabolite concentration, the mortality rate was recorded as 46%. The lowest adult nematode mortality of 6% was found at the lowest metabolite concentration, as with the case *C. jiangduensis*. The positive control in all cases showed the maximum activity of adult nematode mortality. While negative control shows no activity in all cases. Interestingly it was noted in all the above cases that the adult nematode mortality was found to be metabolite dosage-dependent.

Table 5. Adult nematodes mortality by metabolite concentrations of the bacterial isolates.

Bacterial Metabolites Conc. (%)	Adult Nematodes Mortality ($n = 15$)			Standard Errors			*p*-Value		
	ABP	ABY	ABCW	ABP	ABY	ABCW	ABP	ABCW	ABY
100	5.0	5.0	5.0	-	-	-	-	-	-
50	2.3	2.0	2.3	0.33	0.57	0.33	0.01	0.01	0.07
25	2.0	1.3	1.6	0.57	0.88	0.66	0.07	0.12	0.26
12.5	1.6	1.0	1.3	0.88	0.57	0.66	0.19	0.18	0.22
6.25	1.3	0.3	0.3	0.33	0.33	0.33	0.05	0.42	0.42
N.C	-	-	-	-	-	-	-	-	-
P.C	5.0	5.0	5.0	-	-	-	-	-	-

Note: *p*-value < 0.05 was considered significantly significant. NC: Negative Control; PC: Positive Control.

3.2.2. Larval Mortality Assay (LMA)

Larval Mortality Assay of *H. contortus*, was carried out on 3rd stage larva (L3) by treating with different concentrations of metabolites extract from *C. testosteroni*, *C. jiangduensis*, and *P. weihenstephanesis* with an exposure time of six hours. The results are given in Table 6 and Figure 5. Analysis of the data revealed that the highest 100% metabolite concentration of *C. testosteroni* caused 100% L3 mortality, followed by 60% with 50% metabolite concentration. The minimum nematode larval mortality (13%) was reported to be caused by the bacterial metabolite at a 6.25% concentration.

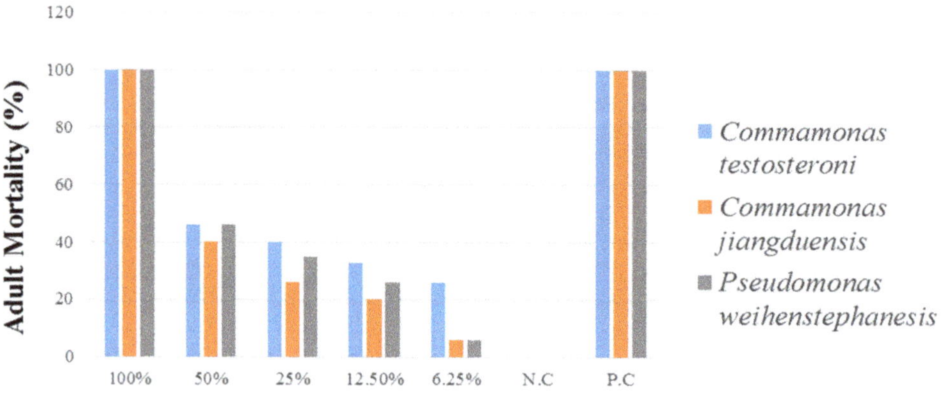

Figure 4. Percent adult nematode mortality by the bacterial isolates, after six hours of treatment.

Table 6. Nematodes larval mortality is caused by various metabolite concentrations of the bacterial isolates.

Con. of Bacterial Metabolites (%)	Mean Nematode Larval Mortality ($n = 15$)			Standard Errors			p-Value		
	ABP	ABY	ABCW	ABP	ABY	ABCW	ABP	ABY	ABCW
100	5.0	5.0	5.0	-	-	-	-	-	-
50	3.0	2.3	2.3	-	0.88	0.33	-	0.11	0.01
25	2.0	2.0	1.3	-	0.57	0.33	-	0.07	0.05
12.5	1.0	1.0	1.3	-	0.57	0.66	-	0.22	0.18
6.25	0.6	0.6	1.0	0.66	0.66	-	0.42	0.42	-
N.C	-	-	-	-	-	-	-	-	-
P.C	5.0	5.0	5.0	-	-	-	-	-	-

Note: p-value < 0.05 was considered significantly significant.

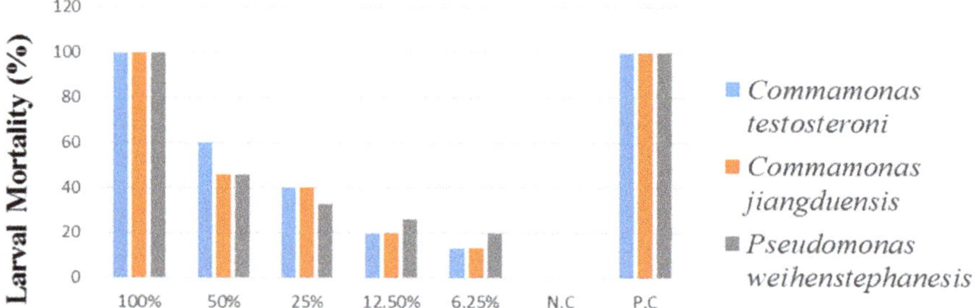

Figure 5. Percent nematode larval mortality by the bacterial isolates after six hours of treatment.

Comamonas jiangduensis showed 100% larval mortality at 100% bacterial metabolite concentration; at a 50% concentration the mortality was recorded as 46%, while the lowest dose (6.25%) of metabolite concentration caused the lowest (13%) mortality of L3 larvae. Analysis of the data showed that *P. weihenstephanesis* with 100% metabolite concentration caused 100% mortality of *H. contortus*' L3 larvae, which was followed by a 46% mortality rate, where 50% bacterial metabolite concentration was used. The lowest concentration of *P. weihenstephanesis* showed the lowest L3 mortality (20%), as in the case with the other bacterial isolates. The 100% concentration of all the bacterial isolates and the positive control showed the maximum L3 larva mortality, while the negative control showed no activity at all.

3.2.3. Egg Hatch Inhibition Assay (EHA)

The results of egg hatch inhibition assay are presented in Figure 6. Analysis of the data showed that all the bacterial culture filtrates showed similar (100%) nematode egg hatch inhibition at 100% metabolite concentration. It was noted that the 50% metabolite concentration of *C. testosteroni* and *C. jiangduensis* inhibited 100% *H. contortus* eggs from hatching, while the same concentration of *P. weihenstephanesis* caused 80% nematode's egg hatch inhibition. It was observed that lowering the bacterial metabolite concentration, lowers the egg hatch inhibition, and at the lowest metabolite concentration of 3.125%, showed the minimum (20%) egg hatch inhibition. The positive control (Oxfendazole) produced complete egg hatch inhibition, even at a very low concentration (0.025 mg/mL), while there was no egg hatch inhibition in the negative control.

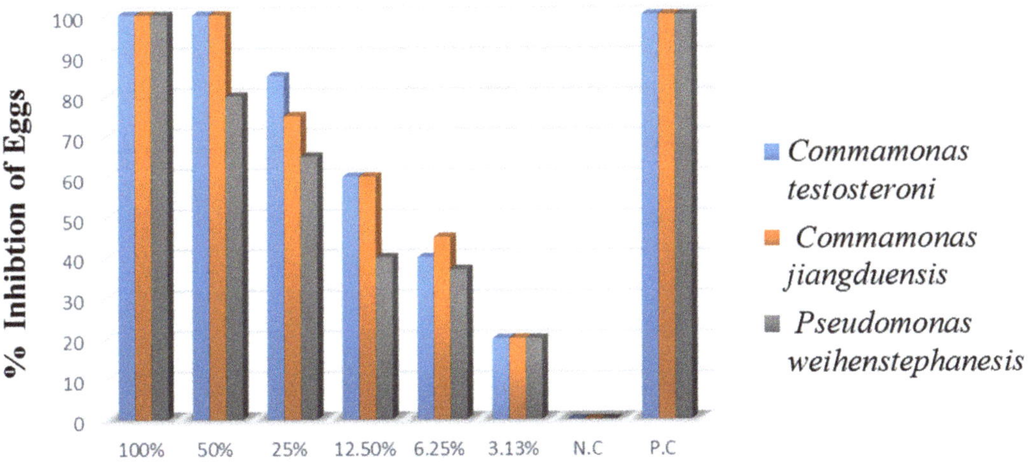

Figure 6. Percent Inhibition of eggs hatching, three days after treatment.

3.3. Genomic DNA Extraction

16S rRNA genes verified the bacterial isolates, and each gene fragment was effectively amplified and sequenced by a polymerase chain reaction from their DNA. On the gel, isolated groups of molecular DNA were noted for separate bacterial isolates. The recognized protein marker below 1 kb size was compared for molecular weight. The pattern of the band, found on agarose gel, revealed that the bacterial isolates ABP, ABY, and ABCW were nearly comparable and about 1000 bp of molecular weight, as shown in the Figure 7. The nucleotide sequences of three isolates were compared with the sequences of nearly linked isolates of 16S rRNA genes. The result revealed that ABP bacterial isolate showed resemblance to *Comamonas testosteroni* was provided by the server and ABY bacterial isolate showed resemblance to *Comamonas jiangduensis*, while ABCW bacterial isolate showed resemblance to *Pseudomonas weihenstephanesis* as provided by the server (Figure 7 and Table 7).

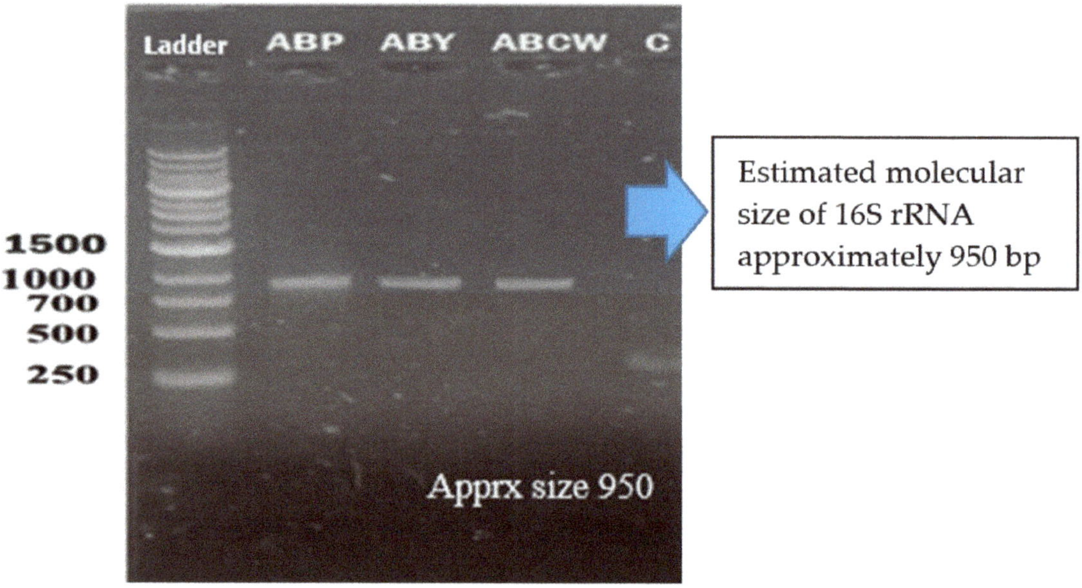

Figure 7. DNA bands of the bacterial isolates.

Table 7. Molecular Identification of bacteria strain.

Strain ID	Number of Nucleotides of 16S rRNA Gene	Closely Related Validly Published Taxa	Similarity (%) of 16S rRNA Sequence with Closely Related Species
ABP	832	*Comamonas testosteroni*	98.2
ABY	798	*Comamonas jiangduensis*	97.06
ABCW	856	*Pseudomonas weihenstephanesis*	96.0

Phylogenetic Analysis

Two forward and two reverse sequences for each sample were aligned using Bionumerics v3.5 (Applied Maths) to obtain a composite sequence. The quality of each sequence trace was visually assessed, and the poor-quality sequence was edited and removed. Organisms were identified for each assay, by comparing consensus sequences to a database library of known 16S rRNA gene sequences in GenBank (http://www.ncbi.nlm.nih.gov/blast/Blast.cgi, accessed on 3 April 2021) by multiple sequence alignment. The bacterial source of the sequence was identified by matching it with a sequence with the highest maximum identity score from the GenBank database. Where more than one bacterial species had the same highest score, all species were recorded in the results (Figures 8 and 9). Sequences with 96% similarity to hits from the GenBank database were of poor quality and were excluded from this study (Figure 10).

The phylogenetic tree was constructed on the origin of 16S rRNA gene sequences for the bacterial isolates using MEGA 6 software. Phylogenetic analysis showed (Figure 11) that ABP was identified as *Comamonas testosteroni* and ABY *Comamonas jiangduensis*.

BLAST ® » **blastn suite** » results for RID-6M6SZC8M01R

Job Title	Nucleotide Sequence ...
RID	6M6SZC8M01R Search expires on 04-06 04:01 am
Program	BLASTN
Database	rRNA_typestrains/16S_ribosomal_RNA
Query ID	lcl|Query_88545
Description	None ...
Molecule type	dna
Query Length	852

Descriptions

Description	Scientific Name	Max Score	Total Score	Query Cover	E value	Per. Ident	Acc. Len	Accession
Comamonas jiangduensis strain YW1 16S ribosomal RNA, partial sequence	Comamonas jiangduensis	1467	1467	98%	0.0	98.10%	1455	NR_109655.1
Limnohabitans planktonicus II-D5 16S ribosomal RNA, partial sequence	Limnohabitans planktonicus II-D5	1461	1461	98%	0.0	97.99%	1486	NR_125541.1
Comamonas aquatilis strain SB30-Chr27-3 16S ribosomal RNA, partial sequence	Comamonas aquatilis	1445	1445	98%	0.0	97.63%	1446	NR_163656.1
Curvibacter delicatus strain NBRC 14919 16S ribosomal RNA, partial sequence	Curvibacter delicatus	1439	1439	98%	0.0	97.51%	1460	NR_113696.1
Comamonas aquatica subsp. rana strain CW-25 16S ribosomal RNA, partial sequence	Comamonas aquatica subsp. rana	1435	1435	98%	0.0	97.51%	1387	NR_165757.1

Figure 8. Sequence similarity of ABP with *Comamonas testosteroni*.

BLAST ® » **blastn suite** » results for RID-6N4B810P013

Job Title	Nucleotide Sequence ...
RID	6N4B810P013 Search expires on 04-06 12:26 pm
Program	BLASTN
Database	rRNA_typestrains/16S_ribosomal_RNA
Query ID	lcl|Query_24761
Description	None ...
Molecule type	dna
Query Length	895

Descriptions

Description	Scientific Name	Max Score	Total Score	Query Cover	E value	Per. Ident	Acc. Len	Accession
Pseudomonas weihenstephanensis strain DSM 29166 16S ribosomal RNA, partial sequence	Pseudomonas weihenstephanensis	826	826	55%	0.0	96.77%	1525	NR_148764.1
Pseudomonas helleri strain DSM 29165 16S ribosomal RNA, partial sequence	Pseudomonas helleri	826	826	55%	0.0	96.77%	1525	NR_148763.1
Pseudomonas endophytica strain BSTT44 16S ribosomal RNA, partial sequence	Pseudomonas endophytica	826	826	55%	0.0	96.77%	1451	NR_136473.1

Figure 9. Sequence similarity of ABY with *Comamonas jiangduensis*.

Description ▾	Scientific Name ▾	Max Score ▾	Total Score ▾	Query Cover ▾	E value ▾	Per. Ident ▾	Acc. Len ▾	Accession
Comamonas testosteroni strain NBRC 14951 16S ribosomal RNA, partial sequence	Comamonas testosteroni	1428	1428	98%	0.0	97.06%	1456	NR_113709.1
Variovorax paradoxus strain DSM 30034 16S ribosomal RNA, partial sequence	Variovorax paradoxus	1428	1428	98%	0.0	97.06%	1483	NR_113329.1
Variovorax paradoxus strain 13-0-1D 16S ribosomal RNA, partial sequence	Variovorax paradoxus	1428	1428	98%	0.0	97.06%	1483	NR_036930.1
Comamonas testosteroni strain KS 0043 16S ribosomal RNA, complete sequence	Comamonas testosteroni	1428	1428	98%	0.0	97.06%	1533	NR_029161.2

Figure 10. Sequence similarity less than 96% of ABCW, and hence, excluded.

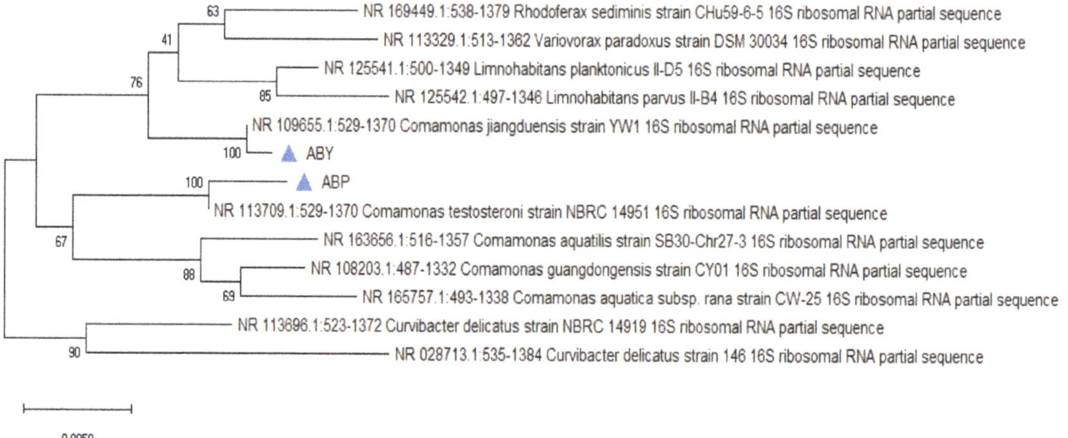

Figure 11. Phylogenetic Tree of *Comamonas testosteroni* (ABP) and *Comamonas jiangduensis* (ABY).

4. Discussion

Haemonchus contortus has a great financial significance causing serious disease and death of cattle and ruminants [2]. Resistance to the available antihelmintic drugs has become a severe threat to livestock production [7]. To decrease the use of chemical anthelmintics, an alternative method is a biocontrol by using bacteria and fungi against the nematode parasites. *Duddingtonia flagrans* is as wide as nematophagous fungi, which is being explored to regulate intestinal nematodes in livestock [32]. *Bacillus thuringiensis* is one of the most commonly used bacterial antagonists in biological control of *H. contortus* that may promote insecticidal crystal proteins, commonly used to control pests and also due to its low mammalian toxicity [23].

The current study tried to explore the bacterial abilities to reduce the population of *H. contortus* at eggs, larval and adult stages. Bacterial isolates were collected from the abomasum of small ruminants. Different bacterial culture filtrate concentrations (100%, 50%, 25%, 12.5%, 6.25%, and 3.125%) were prepared and applied on three life stages of *H. contortus* to observe the mortality of adult, larvae, and egg hatch inhibition. Higher concentration 100% and 50% of *C. testosteroni*, *C. jiangduensis*, and *P. weihenstephanesis* showed 100% nematode eggs hatch inhibition. To our knowledge, these bacterial isolates have never been used against *H. contortus*. Some researchers worked on the effect of *Bacillus thuringiensis* on various life stages of *H. contortus* [30]. Earlier reports of leaf ethyl acetate and methanol extract of *A. squamosa*, *E. prostrata*, *S. torvum*, and *C. roseus* and acetone extract of *T. chebula* showed more consistent results on egg hatch inhibition of *H. contortus* [36].

Previous studies showed the larvicidal effect of various species of Bacillus, such as *Bacillus circulans* (Bcir), *B. thuringienisis* var. osvaldocruzi (Bto), *B. thuringienisis* var. israelensis (Bti), and *B. thuringienisis* var. kurstaki (Btk) on L3 stage of *Haemonchus* sp. among the tested bacteria, *B. circulans* and *B. thuringienisis* var. israelensis showed the best in vitro larvicidal efficiency of 90% and 94%, respectively, against the tested nematodes [37]. In our research studies, *H. contortus* larvae were treated with different bacterial metabolite concentrations, which cause 100% nematode mortality. These results are in line with the earlier reports on the control of nematodes in naturally infected sheep and goats, suggesting that the use of bacteria as an alternative control method for *H. contortus* larvae [21,37].

In the case of egg hatch inhibition assay, the highest bacterial culture filtrates of *C. testosteroni* and *C. jiangduensis* at 100% and 50% bacterial metabolite concentration, resulted in 100% inhibition of *H. contortus* eggs. The extract of the *Annona muricata* plant has also been used to inhibit the eggs hatch and mortality of *H. contortus* larvae and adults [38].

This study has extended the findings by showing that the tested bacteria species are effective against all stages of the nematode parasite. The metabolites of bacterial species *C. testosteroni* and *C. jiangduensis* showed a greater effect than filtrates, obtained from *P. weihenstephanesis*. Similarly, *C. jiangduensis* and *C. testosteroni* showed a higher mortality rate of L3 and adult nematodes as compared to *P. weihenstephanesis*. However, all the bacteria isolates showed a similar impact on eggs hatch inhibition. The positive control in all cases showed the maximum larva mortality, while the negative control exhibited no nematocidal activity at all. It was concluded that the larval and adult nematode mortality, as well as the nematode egg hatch inhibition, have a positive correlation with the doses or concentration of the metabolites, extracted from the bacterial isolates.

5. Conclusion

The present study was conducted to know the antihelmintic capabilities of metabolites extracted from abomasum bacteria *Comamonas testosteroni*, *C. jiangduensis*, and *Pseudomonas weihenstephanesis* against *H. contortus* eggs, larvae, and adults. It was noted that increasing the concentration of bacterial culture filtrates, increased nematode mortality. The same trend was observed in egg hatch inhibition assay bacterial culture filtrates. The effect of bacterial culture filtrates against *H. contortus* was found as dose-dependent. However, further in vivo bacterial culture filtrates investigation is recommended to seethe anthelmintic activity against various developmental stages of *H. contortus*.

Author Contributions: Data curation, B.K., R.U., I.A., A.B.; Formal analysis, F., I.A., T.A.K., A.E.A., A.F.E.-k., M.M.A.-D.; Funding acquisition, R.U., A.B., M.S., A.E.A., A.F.E.-k., M.M.A.-D.; Investigation, A.N., B.K., F., M.S., H.M.M.; Methodology, T.A.K.; Project administration, I.A., B.K., T.A.K., M.M.A.; Writing—review & editing, R.U., B.K., A.E.A., A.F.E.-k., M.M.A., M.S., B.K. All authors have read and agreed to the published version of the manuscript.

Funding: The project financially supported by Taif University under Researchers Supporting Project number (TURSP-2020/82), Taif University, Taif, Saudi Arabia and the deanship of Scientific Research at King Khalid University, Abha, KSA, under grant number R.G.P.2/122/42.

Institutional Review Board Statement: Not applicable.

Informed Consent Statement: Not applicable.

Data Availability Statement: All available data incorporated in the MS.

Acknowledgments: The authors acknowledge and greatly thankful for the financial support of Taif University Researchers Supporting Project number (TURSP-2020/82), Taif University, Taif, Saudi Arabia. This research was funded by the deanship of Scientific Research at King Khalid University, Abha, KSA, under grant number R.G.P.2/122/42.

Conflicts of Interest: The authors declare no conflict of interest. The funders had no role in the design of the study; in the collection, analyses, or interpretation of data; in the writing of the manuscript, or in the decision to publish the results.

References

1. Fthenakis, G.C.; Papadopoulos, E. Impact of parasitism in goat productions. *Small Rumin. Res.* **2018**, *163*, 21–23. [CrossRef]
2. Arsenopoulos, K.V.; Fthenakis, G.C.; Katsarou, E.I.; Papadopoulos, E. Haemonchosis: A Challenging Parasitic Infection of Sheep and Goats. *Animals* **2021**, *11*, 363. [CrossRef]
3. Hoberg, E.P.; Lichtenfels, J.R.; Gibbons, L. Phylogeny for species of Haemonchus (Nematoda: Trichostrongyloidea): Considerations of their evolutionary history and global biogeography among Camelidae and Pecora (Artiodactyla). *J. Parasitol.* **2004**, *90*, 1085–1103. [CrossRef]
4. Manton, V.J.A.; Peacock, R.; Poynter, D.; Silverman, P.H.; Terry, R.J. The influence of age on naturally acquired resistance to *Haemonchus contortus* in lambs. *Res. Vet. Sci.* **1962**, *3*, 308–314. [CrossRef]
5. Besier, R.B.; Kahn, L.P.; Sargison, N.D.; Van Wyk, J.A. The pathophysiology, ecology and epidemiology of *Haemonchus contortus* infection in small ruminants. *Adv. Parasitol.* **2016**, *93*, 95–143.
6. Selemon, M. Review on control of *Haemonchous contortus* in sheep and goat. *J. Vet. Med. Res.* **2018**, *5*, 1139–1145.
7. Houdijk, J.G.M. Influence of periparturient nutritional demand on resistance to parasites in livestock. *Parasite Immunol.* **2008**, *30*, 113–121. [CrossRef] [PubMed]
8. Naeem, M.; Iqbal, Z.; Roohi, N. Ovine haemonchosis: A review. *Trop. Anim. Health Prod.* **2021**, *53*, 19. [CrossRef]
9. Lashari, M.H.; Zahida, T.; Akhtar, M.S.; Chaudhary, M.S.; Nuzhat, S. Prevalence of *Haemonchus contortus* in local goats of D.G. Khan. *World J. Pharm. Pharmachol. Sci.* **2015**, *4*, 190–196.
10. Lashari, M.H.; Tasawar, Z. Prevalence of some gastrointestinal parasites in sheep in southern Punjab, Pakistan. *Pak. Vet. J.* **2011**, *31*, 295–298.
11. Tasawar, Z.; Ahmad, S.; Lashari, M.H.; Chaudhary, S.H. Prevalence of *Haemonchus contortus* in sheep at research centre for conservation of sahiwal cattle (RCCSC) Jehangira bad District Khanewal, Punjab and Pakistan. *Pak. J. Zool.* **2010**, *42*, 735–739.
12. Akhter, N.; Arijo, A.; Phulan, M.; Iqbal, Z.; Mirbahar, K. Prevalence of gastrointestinal nematodes in goats in Hyderabad and adjoining areas. *Pak. Vet. J.* **2011**, *31*, 287–290.
13. Nabi, H.; Saeed, K.; Shah, S.R.; Rashid, M.I.; Akbar, H.; Shehzad, W. Epidemiological study of gastrointestinal nematodes of goats in district Swat, Khyber Pakhtunkhwa, Pakistan. *Sci. Int.* **2014**, *26*, 283–286.
14. Asif, M.; Azeem, S.; Asif, S.; Nazir, S. Prevalence of gastrointestinal parasites of sheep and goats in and around Rawalpindi and Islamabad, Pakistan. *J. Vet. Anim. Sci.* **2008**, *1*, 14–17.
15. Delgado-Núñez, E.J.; Zamilpa, A.; González-Cortazar, M.; Olmedo-Juárez, A.; Cardoso-Taketa, A.; Sánchez-Mendoza, E.; Tapia-Maruri, D.; Salinas-Sánchez, D.O.; Mendoza-de Gives, P. Isorhamnetin: A Nematocidal Flavonoid from *Prosopis laevigata* Leaves Against *Haemonchus contortus* Eggs and Larvae. *Biomolecules* **2020**, *10*, 773. [CrossRef]
16. Castagna, F.; Britti, D.; Oliverio, M.; Bosco, A.; Bonacci, S.; Iriti, G.; Ragusa, M.; Musolino, V.; Rinaldi, L.; Palma, E.; et al. In Vitro Anthelminthic Efficacy of Aqueous Pomegranate (*Punica granatum* L.) Extracts against Gastrointestinal Nematodes of Sheep. *Pathogens* **2020**, *9*, 1063. [CrossRef] [PubMed]
17. Larsen, M. Prospects for controlling animal parasitic nematodes by predacious micro fungi. *Parasitology* **2000**, *120*, 121–131. [CrossRef]
18. Cotter, J.L.; Van Burgel, A.; Besier, R.B. Anthelmintic resistance in nematodes of beef cattle in south-west Western Australia. *Vet. Parasitol.* **2015**, *207*, 276–284. [CrossRef]
19. Mackey, J. *Comparing the Efficacy of Copper Oxide Wire Particles and Copper Sulfate on Haemonchus contortus in Goats*; Murray State University: Murray, KY, USA, 2018.
20. Kotze, A.; O'grady, J.; Gough, J.; Pearson, R.; Bagnall, N.; Kemp, D.; Akhurst, R. Toxicity of *Bacillus thuringiensis* to parasitic and free-living life-stages of nematode parasites of livestock. *Int. J. Parasitol.* **2005**, *35*, 1013–1022. [CrossRef] [PubMed]
21. O'Grady, J.; Akhurst, R.J.; Kotze, A.C. The requirement for early exposure of *Haemonchus contortus* larvae to *Bacillus thuringiensis* for effective inhibition of larval development. *Vet. Parasitol.* **2007**, *150*, 97–103. [CrossRef]
22. Estruch, J.J.; Warren, G.W.; Mullinis, M.A.; Nye, G.J.; Craig, J.A.; Koziel, M.G. A novel *Bacillus thuringiensis* vegetative insecticidal protein with a wide spectrum of activities against lepidopteran insects. *Proc. Natl. Acad. Sci. USA* **1996**, *93*, 5389–5394. [CrossRef] [PubMed]
23. Yu, C.-G.; Mullins, M.A.; Warren, G.W.; Koziel, M.G.; Estruch, J.J. The *Bacillus thuringiensis* vegetative insecticidal protein Vip3 A lyses midgut epithelium cells of susceptible insects. *Appl. Environ. Microbiol.* **1997**, *63*, 532–536. [CrossRef] [PubMed]
24. Lövgren, A.; Zhang, M.; Engström, A.; Dalhammar, G.; Landén, R. Molecular characterization of immune inhibitor A, a secreted virulence protease from *Bacillus thuringiensis*. *Mol. Microbiol.* **1990**, *4*, 2137–2146. [CrossRef]
25. Orden, J.A.; Ruiz-Santa-Quiteria, J.A.; Blanco, M.; Blanco, J.E.; Mora, A.; Cid, D.; González, E.A.; Blanco, J.; De la Fuente, R. Prevalence and characterization of Vero cytotoxin-producing *Escherichia coli* isolated from diarrhoeic and healthy sheep and goats. *Epidemiol. Infect.* **2003**, *130*, 313–321. [CrossRef] [PubMed]
26. Farshad, S.; Norouzi, F.; Aminshahidi, M.; Heidari, B.; Alborzi, A. Two cases of bacteremia due to an unusual pathogen, *Comamonas testosteroni* in Iran and a review literature. *J. Infect. Dev. Ctries.* **2012**, *6*, 521–525. [CrossRef]

27. Purkayastha, M.; Khan, M.S.R.; Alam, M.; Siddique, M.P.; Begum, F.; Mondal, T.; Choudhury, S. Cultural and biochemical characterization of sheep *Escherichia coli* isolated from in and around Bau campus. *Bangladesh J. Vet. Med.* **2010**, *8*, 51–55. [CrossRef]
28. Soulbsy, E.J.L. *Helminths, Arthropods and Protozoa of Domesticated Animals*; Bailliere Tindall: London, UK, 1982; Volume 7, pp. 248–250, 766–770.
29. Dryden, M.W.; Payne, P.A.; Ridley, R.; Smith, V. Comparison of Common Fecal Flotation Techniques for the Recovery of Parasitic Eggs and Oocysts. *Vet. Ther.* **2005**, *6*, 15–28.
30. Sharma, L.D.; Bagha, H.S.; Srivastava, P.S. In Vitro antihelmintic screening of indigenous medicinal plants against *Haemonchus contortus* of sheep and goats. *Indian J. Anim. Res.* **1971**, *5*, 33–38.
31. Varady, M.; Cudekova, P.; Corba, J. In Vitro detection of benzimidazole resistance in *Haemonchus contortus*: Egg hatch test versus larval development test. *Vet. Parasitol.* **2007**, *149*, 104–110. [CrossRef] [PubMed]
32. Hubert, J.; Kerbouef, D. A micro larval developmental assay for the detection of anthelmintic resistance in sheep nematodes. *Vet. Rec.* **1992**, *130*, 442–446. [CrossRef] [PubMed]
33. Coles, G.C.; Bauer, C.; Borgsteede, F.H.M.; Geerts, S.; Klei, T.R.; Taylor, M.A.; Waller, P.J. World Association for the Advancement of Veterinary Parasitology (WAAVP) methods for the detection of antihelmintic resistance in nematodes of veterinary importance. *Vet. Parasitol.* **1992**, *44*, 35–44. [CrossRef]
34. Khattak, B.; Safi, A.U.R.; Sindhu, Z.U.D.; Attaullah, M.; Jamal, Q.; Khan, T.A.; Hussain, M.; Anjum, S.I.; Israr, M.; Khan, I.A. Biological control of *Haemonchus contortus* by fungal antagonists in small ruminants. *Appl. Ecol. Environ. Res.* **2018**, *16*, 5825–5835. [CrossRef]
35. Nazki, S.; Wani, S.A.; Parveen, R.; Ahangar, S.A.; Kashoo, Z.A.; Hamid, S.; Dar, Z.A.; Dar, T.A.; Dar, P.A. Isolation, molecular characterization and prevalence of *Clostridium perfringens* in sheep and goats of Kashmir Himalayas, India. *Vet. World* **2017**, *10*, 1501. [CrossRef]
36. Nikolaou, S.; Gasser, R. Prospects for exploring molecular developmental processes in *Haemonchus contortus*. *Int. J. Parasitol.* **2003**, *36*, 859–868. [CrossRef] [PubMed]
37. Sinott, M.C.; Cunha Filho, N.A.; Castro, L.L.D.; Lorenzon, L.B.; Pinto, N.B.; Capella, G.A.; Leite, F.P.L. Bacillus spp. toxicity against *Haemonchus contortus* larvae in sheep fecal cultures. *Exp. Parasitol.* **2012**, *132*, 103–108. [CrossRef]
38. Ferreira, L.E.; Castro, P.M.N.; Chagas, A.C.S.; França, S.C.; Beleboni, R.O. In vitro anthelmintic activity of aqueous leaf extract of *Annona muricata* L.(Annonaceae) against *Haemonchus contortus* from sheep. *Exp. Parasitol.* **2013**, *134*, 327–332. [CrossRef]

Article

Environmental *Streptococcus uberis* Associated with Clinical Mastitis in Dairy Cows: Virulence Traits, Antimicrobial and Biocide Resistance, and Epidemiological Typing

Norhan K. Abd El-Aziz [1,*], Ahmed M. Ammar [1], Hend M. El Damaty [2,*], Rehab A. Abd Elkader [3], Hosam A. Saad [4], Waleed El-Kazzaz [5] and Eman Khalifa [6]

[1] Department of Microbiology, Faculty of Veterinary Medicine, Zagazig University, Zagazig 44511, Egypt; prof.ahmedammar_2000@yahoo.com
[2] Department of Animal Medicine, Infectious Diseases, Faculty of Veterinary Medicine, Zagazig University, Zagazig 44511, Egypt
[3] Belbies Veterinary Organization, Ministry of Agriculture, Belbies 44974, Egypt; alrehablab01@gmail.com
[4] Department of Chemistry, College of Science, Taif University, P.O. Box 11099, Taif 21944, Saudi Arabia; h.saad@tu.edu.sa
[5] Molecular Microbiology Lab., Botany and Microbiology Department, Faculty of Science, Suez Canal University, Ismailia 41522, Egypt; walid_elkazaz@science.suez.edu.eg
[6] Department of Microbiology, Faculty of Veterinary Medicine, Matrouh University, Matrouh 51511, Egypt; khalifa.eman@alexu.edu.eg
* Correspondence: norhan_vet@hotmail.com or nourhan_vet@zu.edu.eg (N.K.A.E.-A.); hendvet11@yahoo.com or hmsaad@zu.edu.eg (H.M.E.D.)

Citation: Abd El-Aziz, N.K.; Ammar, A.M.; El Damaty, H.M.; Abd Elkader, R.A.; Saad, H.A.; El-Kazzaz, W.; Khalifa, E. Environmental *Streptococcus uberis* Associated with Clinical Mastitis in Dairy Cows: Virulence Traits, Antimicrobial and Biocide Resistance, and Epidemiological Typing. *Animals* **2021**, *11*, 1849. https://doi.org/10.3390/ani11071849

Academic Editors: Paola Roncada and Bruno Tilocca

Received: 9 May 2021
Accepted: 16 June 2021
Published: 22 June 2021

Publisher's Note: MDPI stays neutral with regard to jurisdictional claims in published maps and institutional affiliations.

Simple Summary: Mastitis remains one of dairy cattle's most perplexing and expensive diseases. This study is the first to look into the virulence traits, antimicrobial and biocide resistance, and epidemiological typing of *Streptococcus uberis* (*S. uberis*) isolated from bovine clinical mastitis in dairy farms of diverse hygienic interventions in Egypt. The overall *S. uberis* infection rate was 20.59%; all were multidrug-resistant (MDR). The *sua* gene was the most frequent virulence gene (42.02%), followed by *pauA* (40.57%), *cfu* (21.73%), *skc* (20.28%), and *opp* (11.59%). The *erm*(B) gene serves as the predominant antimicrobial-resistant gene (75.36%), followed by *fexA* (52.63%) and *tet*(M), *blaZ*, and *aac(6′)aph(2″)* genes (46.38% each). Of note, 79.71% of *S. uberis* isolates carried *qac* genes; among them, 55 (79.71%), 54 (78.26%), and 13 (18.84%) harbored *qacED1*, *qacC/D*, and *qacA/B* genes, respectively. Restriction fragment length polymorphism–polymerase chain reaction (RFLP–PCR) indicated that all analyzed isolates were *S. uberis* type I by their unique RFLP pattern. This study shows a significant variation in the occurrence of virulent *S. uberis* in dairy cows with clinical mastitis regarding the prospective hygienic concerns. Furthermore, MDR coupled with the existence of biocide resistance genes indicates the importance of *S. uberis* surveillance and the prudent use of antimicrobials in veterinary clinical medicine to avoid the dissemination of antimicrobial resistance.

Abstract: Mastitis remains a serious problem for dairy animals. The misappropriation of antimicrobial agents helps accelerate resistance, which poses a serious challenge in controlling environmental *S. uberis* infection. Here, we study the virulence attributes, antimicrobial and biocide resistance, and epidemiological typing of *S. uberis* recovered from bovine clinical mastitis in dairy farms of diverse hygienic interventions in Egypt. The overall *S. uberis* infection rate was 20.59%; all were multidrug-resistant (MDR). The *sua* gene was the most frequent virulence gene (42.02%), followed by *pauA* (40.57%), *cfu* (21.73%), *skc* (20.28%), and *opp* (11.59%). The *erm*(B) gene served as the predominant antimicrobial-resistant gene (75.36%), followed by *fexA* (52.63%) and *tet*(M), *blaZ*, and *aac(6′)aph(2″)* genes (46.38% each). Of note, 79.71%, 78.26%, and 18.84% of *S. uberis* isolates harbored *qacED1*, *qacC/D*, and *qacA/B* genes, respectively. All analyzed isolates were *S. uberis* type I by their unique RFLP–PCR pattern. In conclusion, the sustained presence of *pauA* and *sua* genes throughout the investigated farms contributes to a better understanding of the bacterium's pathogenicity. Furthermore, MDR coupled with the existence of biocide resistance genes indicates the importance of *S. uberis* surveillance and the prudent use of antimicrobials in veterinary clinical medicine to avoid the dissemination of antimicrobial resistance.

Keywords: antimicrobial resistance; biocides; dairy cows; intramammary infections; RFLP–PCR; *Streptococcus uberis*; virulence

1. Introduction

Mastitis is a significant concern affecting dairy animals worldwide, causing great losses to breeders and impacting the country's national income [1,2]. Environmental streptococci, notably *Streptococcus uberis* (*S. uberis*), are among the main contributing agents of mastitis in many countries and have increased their significance for udder health in recent decades [3]. This pathogen is not obligatorily adapted to the udder but is ubiquitous as it is considered an environment-associated straw bedding and pasture pathogen [4]. Since *S. uberis* is the prime pathogen in a dairy herd, frequent antimicrobial treatments and several environmental factors favor the development of this form of mastitis [3].

Streptococcus uberis has previously been categorized into two distinct types, I and II; both were isolated from bovine mastitis cases, the latter being reclassified as *Streptococcus parauberis* (*S. parauberis*) [5]. It is impossible to differentiate between *S. uberis* and *S. parauberis* using phenotypic methods [6]. However, *S. uberis* isolates were verified by 16S rRNA gene restriction fragment length polymorphism (RFLP) using the *Hha*I restriction endonuclease for further identification of the *S. uberis* genetic variation [7].

Despite the economic effect of the high prevalence of environmental streptococci in dairy herds, virulence factors related to the pathogenicity of *S. uberis* are not well characterized; these comprise a significant existential threat to the implementation of control strategies [8]. Various potentially virulence factors were identified for *S. uberis*, among these, *sua*, *cfu*, *opp*, *skc*, and *pauA*, that play prominent roles in the adherence and early colonization of bovine mammary epithelial cells [9–11].

Antimicrobial resistance is one of the world's leading threats to human and animal health [12]. It appears to have an extreme occurrence among streptococcal isolates of mastitis in Egypt [13] and *S. uberis* in many countries [4,14,15]. However, this susceptibility can vary from one region to another. Even within the same region, it is necessary to monitor the pathogens' resistance to the antimicrobials used in the treatment of mastitis in various areas [16]. In Egypt, most bovine mastitis studies have focused on the inclusion of *Escherichia coli*, *Staphylococcus aureus*, *Streptococcus agalactiae*, and, infrequently, *S. uberis* [13,17,18].

Disinfectants based on quaternary ammonium compounds (QACs) have a wide range of veterinary medicine implementations and are critical in controlling animal diseases. They are widely used worldwide, which can contribute to bacterial resistance [19].

In Egypt, there has been no exploration of the existence of *S. uberis*-associated virulence genes in mastitic dairy cows and the plausible allocation of virulence dynamics in the distinct hygiene measures applied in dairy farms. Moreover, there are few studies on *S. uberis* resistance to antimicrobials as well as biocides. Therefore, this study was designed to explore the following points: (i) ascertaining the infection rate of *S. uberis* in dairy cows and the hygiene correlation with its abundance, (ii) detecting the most prospective virulence-associated, antimicrobial and antiseptic resistance genes in environmental *S. uberis* isolates using conventional PCR, and (iii) determining the genotypic variation among the virulent *S. uberis* isolates using RFLP–PCR.

2. Materials and Methods

2.1. Lactating Cows and Husbandry Practices

The lactating cows under study were chosen from dairy farms of three distinct hygienic interventions in Alexandria (A) and Sharkia Governorates (B), as well as some individual smallholder cases in different villages of Sharkia Governorates (C and D), over a year, from July 2017 to August 2018. The udder of each lactating cow was screened for recurrent clinical mastitis. Hygienic interventions were based entirely on the following criteria: (i) periodic monitoring of mastitis by an indirect field check during the lactation season, such as the California mastitis test, (ii) pre-milking procedures, such as udder washing and pre- and post-milking teat dipping with antimicrobial dip; and (iii) dry period treatment after the last lactation, bedding materials, and environmental hygiene as well as balanced food. In the first farm (A; $n = 75$), lactating cows were milked three times daily through a computerized system using pre- and post-teat dipping. This farm followed the standard routine management, vaccination program, and control measures against infectious diseases with the implementation of all hygienic measures. On the second farm (B; $n = 50$), cows were milked three times daily using a machinery system with post-milking teat dipping and fair, moderate hygienic measures; the cows were placed in straw-bedding barns. The cows of the third farm (C; $n = 120$) and smallholder cows (D; $n = 90$; reared by local farmers in the villages of Sharkia Governorate) were grazed; thus, infection with *S. uberis* from environmental pasture reservoirs was expected. These animals lived in unhygienic environments and were fed on low nutrient rations. The cows were milked twice daily by hand, and there was no disinfection during the milking process.

2.2. Milk Sampling and Isolation of S. uberis

Three hundred and thirty-five milk samples were collected aseptically, just before treatment, from the affected mammary quarters that had clinical signs of abnormal secretions, containing clots or flakes, with udders showing inflammatory symptoms, with or without systemic reaction appearing on the cows. These samples were placed in sterile screw-capped test tubes, kept in an insulated icebox, then transported to the laboratory for further bacteriological and molecular investigations. Bacteriological analysis of milk samples was carried out following conventional protocols [20]. A milk sample loopful was plated onto Edward's agar medium (Oxoid, Hampshire, UK) and incubated at 37 °C for 24 h. A single, well-isolated colony was subcultured onto a blood agar base (Oxoid, Hampshire, UK) enriched with 7% sterile defibrinated sheep blood and incubated aerobically at 37 °C for 24–48 h. The bacterial isolates were described based on their classic morphological and hemolytic characteristics. Suspected streptococci isolates microscopically appeared as Gram-positive cocci, either in long or short chains. Standard biochemical tests, including catalase, sodium hippurate, and esculin hydrolysis, were carried out [21]. A Christie, Atkins, and Munch-Petersen (CAMP) test was applied [22]. Growth in the presence of 6.5% NaCl at 10 or 45 °C and pH 9.6, combined with resistance to bile salt, was investigated [23].

2.3. Antimicrobial Susceptibility Testing

The Kirby-Bauer disc diffusion test was used to determine the antimicrobial susceptibilities of *S. uberis* isolates [24]. Commercial discs with the following antimicrobials (Oxoid, Hampshire, England, UK), commonly used in veterinary practices or for public health issues, were selected to perform the antibiogram: penicillin (10 IU), ampicillin (10 µg), amoxicillin (25 µg), amoxicillin–clavulanic acid (20/10 µg), cloxacillin (1 µg), cefoperazone (75 µg), ceftriaxone (30 µg), cephalexin (30 µg), imipenem (10 µg), tetracycline (30 µg), clindamycin (2 µg), erythromycin (15 µg), streptomycin (10 µg), neomycin (30 µg), gentamycin (10 µg), kanamycin (30 µg), novobiocin (30 µg), ciprofloxacin (5 µg), chloramphenicol (30 µg), and trimethoprim–sulphamethoxazole (23.75/1.25 µg). The interpretive criteria used for categorizing an isolate as sensitive or resistant to an antimicrobial agent are established in the Clinical and Laboratory Standards Institute guidelines [25]. Isolates showing resistance to at least three different antimicrobial classes are categorized as multidrug-resistant (MDR) [26]. MAR indices were estimated for each antimicrobial and isolate [27].

2.4. DNA Extraction and Molecular Identification of S. uberis

Streptococcus uberis isolates were cultured in tryptone soya broth (TSB, Oxoid, Hampshire, England, UK) at 37 °C for 24 h. Bacterial DNA was extracted using a QIAamp DNA mini kit (Qiagen GmbH, Hilden, Germany), as recommended by the manufacturer. PCR amplification of the *tuf* gene of *Streptococci* species [28] and the 16S rRNA gene of

S. uberis [29] was performed using the oligonucleotide primer pairs listed in Table S1 to confirm the conventional bacteriological identification.

2.5. PCR Amplifications of Virulence Attributes and Antimicrobial and Biocide Resistance Genes

Virulence genes for *S. uberis*, *cfu* (encoding for CAMP factor), *opp* (oligopeptide binding protein), *sua* (*S. uberis* adhesion molecule), *pauA* (plasminogen activator), and *skc* (streptokinase activator), were investigated [9,11,30]. The occurrence of antimicrobial resistance genes, conferring resistance to penicillins (*blaZ*), phenicols (*fexA*), aminoglycosides (*aac(6′)aph(2″)*), tetracyclines (*tet*(M), *tet*(O), *tet*(L) and *tet*(K)), macrolides (*erm*(A), *erm*(B) and *erm*(C)), sulfonamide (*sul1*), and trimethoprim (*dfrA*) was examined [31–40]. Moreover, PCR targeting *qac* genes, *qacA/B*, *qacC/D*, and *qacED1*, conferring a high level of resistance to antiseptics, was applied [41,42]. Oligonucleotide primer sets and thermal cycling profiles are described in Table S1. The amplification reaction for each gene was conducted with a final volume of 25 µL of the following reaction mixture: 12.5 µL Dream*Taq* Green PCR Master Mix (2X) (Thermo Fisher Scientific, Waltham, MA, USA), 1 µL of each primer (20 pmole), 2 µL template DNA, and 8.5 µL water nuclease-free in a programmable thermal cycler PTC-100 TM (MJ Research Inc., Waltham, MA, USA). *S. uberis* ATCC® 27958™ was used as a reference strain. PCR products were analyzed by agarose gel electrophoresis, stained with ethidium bromide (0.5 µg/mL), and visualized using an ultraviolet transilluminator (Spectroline, Wesbury, Meadville, PA, USA).

2.6. PCR–RFLP

Epidemiological typing of recovered *S. uberis* isolates was then performed using *Hha*I restriction endonuclease (Thermo Fisher Scientific, Waltham, MA, USA), as described previously [7]. Aliquots of the amplified restriction endonuclease-digested fragments were electrophoresed on 0.5 µg/mL ethidium bromide (Sigma-Aldrich, Chemie GmbH, Schnelldorf, Germany) stained agarose gel with a 100 bp standard DNA molecular weight ladder (Fermentas, Inc., Hanover, NH, USA). The numbers of DNA fragments and their sizes in base pairs were then assessed using Pro-Score/RFLP software version 2.39 (DNA ProScan, Inc.; Nashville, TN, USA).

2.7. Bioinformatics and Statistical Analysis

The overall distribution of the antimicrobial resistance phenotypes, virulence-associated genes, and antimicrobial and biocide resistance genes in *S. uberis* isolates was visualized using a heatmap. The clustering pattern of the isolates and various features were determined by hierarchical clustering dendrogram [43]. These analyses were done using R software (version 3.4.2, R Foundation for Statistical Computing, Vienna, Austria), package pheatmap. To estimate the similarity among *S. uberis* isolates from various farms, the binary distances were calculated among isolates based on the presence or absence of the four studied features (virulence, resistance phenotype/genes, and biocide resistance genes). This analysis was done using the functions *dist* and *hlcust* in the R environment. Correlation analyses were done on the raw data after data conversion to binary outcomes (1 = feature presence, 0 = feature absence). The correlation was estimated on a scale from +1 to −1. The significance of the correlation was assessed at a significance level of 0.05. The variables that have similar occurrences in all isolates were excluded from this analysis. The correlation analyses and visualization were done using R packages *corrplot*, *heatmaply*, *hmisc*, and *ggpubr* [44–46]. Fisher's exact two-tailed test [47] was used to study the infection rates of *S. uberis* among farms of varying hygiene interventions and their antimicrobial resistance; *p*- values < 0.05 were statistically significant.

3. Results

3.1. Infection Rate and Characterization of S. uberis in Clinically Mastitic Dairy Cows

The overall infection rate of *S. uberis* was 20.59% (69/335), which significantly ($p < 0.05$) differed between farms, being 8.8% (11/125) in animals living in farms with adequately applied hygiene measures and 27.61% (58/210) in animals living on low hygiene, hand machine farms and smallholders. On Edward's media, *S. uberis* isolates appeared as colorless dewdrop-like, pinpoint rounded colonies. Phenotypic characteristics of the isolates denoted Gram-positive cocci, arranged mainly in chains and, sometimes, in diplococci. They showed β or γ hemolytic colonies on blood agar media. CAMP-like hemolytic activities were determined, together with beta-toxin-producing *Staphylococcus aureus*, on sheep blood agar in 60 out of 69 (86.9%) *S. uberis* isolates. Biochemically, *S. uberis* isolates were catalase-test-negative, whereas all isolates were positive for sodium hippurate and bile-esculin hydrolyses tests. The isolates fail to grow on MacConkey's agar, media containing 6.5% NaCl, or at 45 °C, which is characteristic for *S. uberis*.

3.2. Antimicrobial Resistance Patterns of S. uberis Isolates

The antimicrobial resistance of *S. uberis* isolates ($n = 69$) was validated against 21 antimicrobials of 12 chemotherapeutic classes. As shown in Table 1 and Figure 1, *S. uberis* exhibited 100% resistance to cloxacillin, ceftriaxone, cephalexin, clindamycin, and novobiocin. Moreover, high levels of resistance were reported for ampicillin (89.85%), streptomycin (86.96%), penicillin (79.71%), and erythromycin (73.91%). On the other hand, kanamycin (30.43%), cefoperazone (26.04%), ciprofloxacin (21.74 %), and gentamycin (20.28%) showed the lowest resistance levels, and none of the isolates exhibited imipenem resistance. Of note, all *S. uberis* isolates were MDR, with MAR indices ranged from 0.38–0.81, whereas the MAR indices for tested antimicrobials were up to 0.048. Statistical analysis revealed a significant variation in the resistance levels of *S. uberis* isolates to various antimicrobial agents ($p < 0.05$).

Table 1. Antimicrobial resistance of *S. uberis* isolated from lactating cows with clinical mastitis.

Antimicrobial Class	AMA	No. of Resistant Isolates (%)	MAR Index	Fisher Exact *p*-Value *
Beta-lactams	CX	69 (100.00)	0.048	NE
	AMP	62 (89.85)	0.043	0.007
	AX	48 (69.57)	0.033	<0.001
	P	55 (79.71)	0.038	<0.001
Beta-lactamase inhibitor	AMC	24 (34.78)	0.017	<0.001
Cephalosporins	CRO	69 (100.00)	0.048	NE
	CFP	18 (26.09)	0.012	<0.001
	CL	69 (100.00)	0.048	NE
	FEP	69 (100.00)	0.048	NE
Non-beta Lactams (Carbapenems)	IPM	0 (00.00)	0.00	<0.001
Lincomycins	DA	69 (100.00)	0.048	NE
Fluoroquinlones	CIP	15 (21.74)	0.010	<0.001
Tetracyclines	TE	45 (65.22)	0.031	<0.001
Macrolides	E	51 (73.91)	0.035	0.001
Aminoglycosides	S	60 (86.96)	0.041	0.002
	GEN	14 (20.28)	0.009	<0.001
	NEO	28 (40.57)	0.019	<0.001
	K	21 (30.43)	0.014	<0.001
Phenicols	C	38 (55.07)	0.026	<0.001
Aminocoumarins	NV	69 (100.00)	0.048	NE
Sulfonamides	SXT	33 (47.83)	0.023	<0.001

MAR, multiple antibiotic resistance index; AX, amoxicillin; AMC, amoxicillin–clavulanic acid; CX, cloxacillin; CRO, ceftriaxone; CFP, cefoperazone; CL, cephalexin; FEP, cefepime; IPM, imipenem; S, streptomycin; DA, clindamycin; CIP, ciprofloxacin; TE, tetracycline; E, erythromycin; NEO, neomycin; P, penicillin; NV, novobiocin; AMP, ampicillin; GEN, gentamycin; C, chloramphenicol; K, kanamycin; SXT, trimethoprim–sulfamethoxazole; NE, not estimated. * *p*-value < 0.01 was considered highly statistically significant.

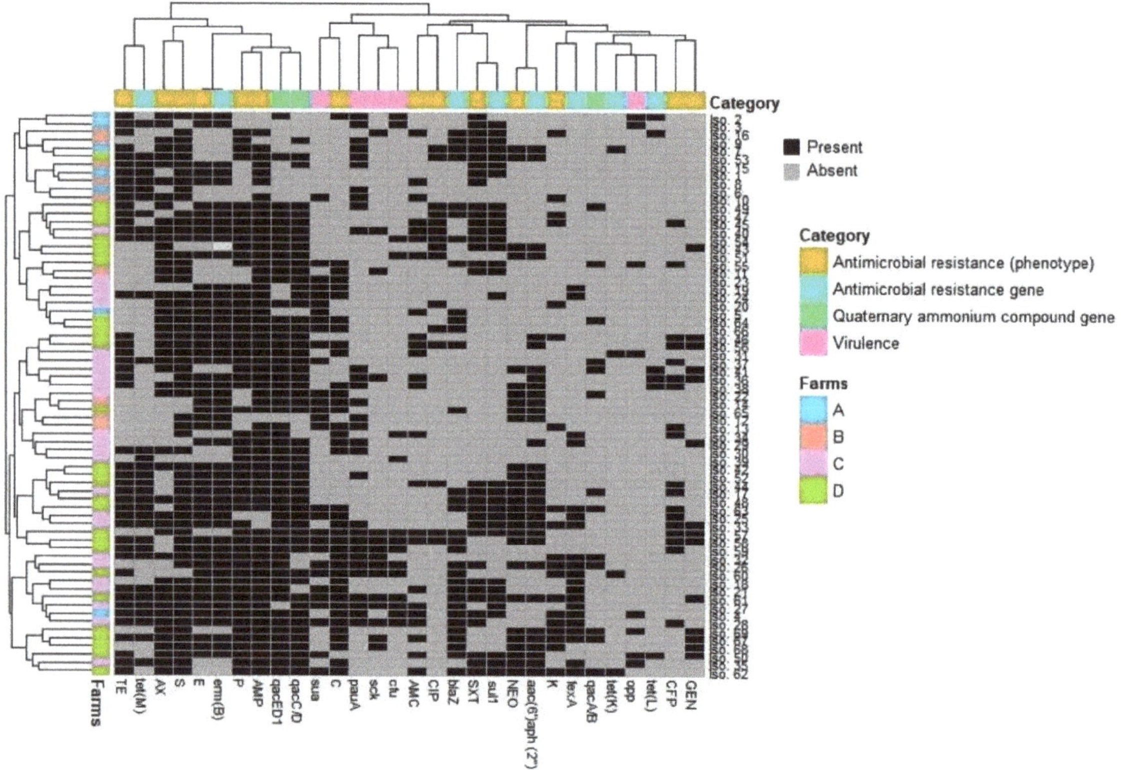

Figure 1. Overall occurrence and clustering of *S. uberis* isolates (*n* = 69) in the investigated farms, their virulence attributes, and antimicrobial and biocide resistance patterns. The heatmap shows the occurrence of features in all isolates. The dendrogram indicates the hierarchical clustering of features and isolates. Different farms and feature categories are color-coded, as shown in the label. GEN, gentamycin; CFP, cefoperazone; K, kanamycin; NEO, neomycin; SXT, trimethoprim-sulfamethoxazole; CIP, ciprofloxacin; AMC, amoxicillin–clavulanic acid; C, chloramphenicol; AMP, ampicillin; P, penicillin; E, erythromycin; S, streptomycin; AX, amoxicillin; TE, tetracycline.

3.3. Molecular Characteristics and Virulence Gene Profiling of S. uberis

Conventional identification of *S. uberis* isolates (*n* = 69) was confirmed by PCR-based amplification of the genus-specific *tuf* gene (DNA fragment ~196 bp). Further, the 16S rRNA gene identified *S. uberis* at the species level (DNA product ~854 bp). *S. uberis* isolates were tested by PCR for the existence of five major genes potentially involved in virulence (Table 2). The most frequent gene was *sua* (42.02%), followed by *pauA* (40.57%), *cfu* (21.73%), and *skc* (20.28%). In contrast, the *opp* gene was detected with a low percentage (11.59%). The frequency of putative virulence gene patterns among *S. uberis* isolates is summarized in Table 3. Most of the examined isolates (58/69; 84.06%) harbored at least one virulence gene. Moreover, 11 of 69 isolates (15.94%) possessed simultaneously 3 to 4 virulence-associated genes, and 7 (10.14%) *S. uberis* isolates carried 2 different virulence-related genes. The most frequent virulence gene pattern was *sua* + *pauA* + *skc* + *cfu*, which was observed in 8 of 69 isolates (11.59%) from 2 different herds (C and D) of low hygiene measures.

Table 2. Virulence traits and antimicrobial and biocide resistance profiles of *S. uberis* (*n* = 69) isolated from dairy cows of different hygiene interventions.

Isolate No.	Herd	Virulence Pattern	Antimicrobial Resistance Profile Phenotype	Resistance Genes	Biocide Resistance Genes
1	A	ND	CX, CRO, CL, FEP, S, DA, TE, E, SXT, AMP, NV	*tet*(M), *erm*(B), *sul1*	ND
2		*pauA, cfu, opp*	AX, CX, CRO, CL, FEP, S, DA, TE, E, SXT, AMP, NV, C	*tet*(L), *erm*(B)	*qacED1*
3		*pauA, cfu, opp*	AX, CX, CRO, CL, FEP, DA, TE, E, SXT, NV	*tet*(M), *erm*(B), *sul1*	ND
4		*pauA, cfu, opp*	AX, AMC, CX, CRO, CL, FEP, S, DA, TE, E, SXT, P, AMP, NV, K, C	*blaZ, fexA, tet*(M), *sul1*	*qacED1, qacC/D*
5		*sua*	AX, AMC, CX, CRO, CL, FEP, S, DA, E, P, AMP, NV	*blaZ, erm*(B)	ND
6		*pauA*	CX, CRO, CL, FEP, S, DA, TE, NV	*tet*(M)	ND
7		*pauA*	AX, CX, CRO, CL, FEP, S, DA, CIP, TE, SXT, P, AMP, NV, NEO	*blaZ, aac(6′)aph(2″), tet*(K), *sul1*	ND
8	B	ND	AX, AMC, CX, CRO, CL, FEP, S, DA, TE, E, SXT, AMP, NV	*tet*(M), *erm*(B)	ND
9		*pauA*	AX, CX, CRO, CL, FEP, S, DA, SXT, P, NV	*blaZ, sul1*	*qacC/D*
10		*sua, pauA*	AX, AMC, CX, CRO, CL, FEP, S, DA, TE, AMP, NV, K	*tet*(M)	ND
11		*skc*	AX, CX, CRO, CL, FEP, S, DA, SXT, AMP, NV, C	*sul1*	*qacED1, qacC/D*
12		*sua, pauA*	CX, CRO, CL, FEP, S, DA, E, NV, NEO	*aac(6′)aph(2″), erm*(B)	ND
13		*sua, pauA*	CX, CRO, CFP, CL, FEP, S, DA, E, AMP, NV, K, C	*erm*(B)	*qacED1*
14		*sua, pauA*	CX, CRO, CL, FEP, DA, E, AMP, NV, NEO, C	*aac(6′)aph(2″), erm*(B)	*qacED1, qacC/D*
15		*pauA*	AX, CX, CRO, CL, FEP, S, DA, TE, E, SXT, AMP, NV	*tet*(M), *erm*(B), *sul1*	*qacC/D*
16		*skc*	AMC, CX, CRO, CL, FEP, S, DA, TE, SXT, P, AMP, NV, K	*blaZ, tet*(L), *sul1*	ND
17	C	ND	CX, CRO, CFP, CL, FEP, S, DA, TE, E, SXT, P, AMP, NV, NEO	*blaZ, aac(6′)aph(2″), tet*(M), *erm*(B), *sul1*	*qacA/B, qacED1, qacC/D*
18		ND	AX, CX, CRO, CL, FEP, S, DA, E, SXT, P, AMP, NV, C	*blaZ, fexA, erm*(B), *sul1*	*qacED1, qacC/D*
19		*sua*	CX, CRO, CL, FEP, DA, E, P, AMP, NV, C	*fexA, erm*(B)	*qacC/D*
20		*sua*	AX, CX, CRO, CL, FEP, S, DA, CIP, E, P, AMP, NV, K	*erm*(B)	*qacED1, qacC/D*
21		*sua*	AX, AMC, CX, CRO, CL, FEP, S, DA, TE, E, SXT, P, AMP, NV, C	*blaZ, fexA, tet*(M), *erm*(B), *sul1*	*qacED1*
22		*sua*	AX, CX, CRO, CL, FEP, S, DA, E, AMP, NV, NEO, C	*aac(6′)aph(2″), erm*(B)	*qacA/B, qacED1, qacC/D*
23		*sua*	AX, CX, CRO, CL, FEP, S, DA, AMP, NV, C	ND	*qacED1, qacC/D*
24		*sua*	AX, CX, CRO, CL, FEP, S, DA, TE, E, P, AMP, NV, C	*fexA, tet*(M), *erm*(B), *sul1*	*qacC/D*
25		*sua*	AX, CX, CRO, CFP, CL, FEP, S, DA, TE, E, SXT, NV, NEO, C	*fexA, aac(6′)aph(2″), tet*(M), *erm*(B), *sul1*	*qacED1, qacC/D*
26		*sua, pauA, skc, cfu*	CX, CRO, CL, FEP, S, DA, E, SXT, P, AMP, NV, NEO, K, C	*blaZ, fexA, aac(6′)aph(2″), erm*(B)	*qacA/B, qacED1*
27		*sua, pauA, skc, cfu*	AX, AMC, CX, CRO, CL, FEP, S, DA, TE, E, P, AMP, NV, C	*blaZ, fexA, tet*(M), *erm*(B), *sul1*	*qacC/D*
28		*sua, pauA, skc, cfu*	AX, CX, CRO, CFP, CL, FEP, S, DA, TE, E, SXT, P, AMP, NV, K, C	*blaZ, fexA, tet*(M), *erm*(B)	*qacED1, qacC/D*
29		*pauA*	CX, CRO, CL, FEP, DA, E, P, AMP, NV, GEN, C	*fexA, aac(6′)aph(2″), erm*(B)	*qacED1, qacC/D*
30		*pauA*	AX, CX, CRO, CL, FEP, S, DA, TE, AMP, NV, K, C	*tet*(M)	*qacED1, qacC/D*
31		*opp*	AX, AMC, CX, CRO, CL, FEP, S, DA, TE, E, P, AMP, NV, C	*tet*(K), *erm*(B)	*qacED1, qacC/D*

Table 2. *Cont.*

Isolate No.	Herd	Virulence Pattern	Antimicrobial Resistance Profile		Biocide Resistance Genes
			Phenotype	**Resistance Genes**	
32		*pauA, skc*	AX, CX, CRO, CL, FEP, S, DA, TE, E, P, AMP, NV, K, C	*fexA, tet*(M), *erm*(B)	*qacA/B, qacED1*
33		*pauA*	AX, CX, CRO, CFP, CL, FEP, S, DA, TE, E, SXT, P, NV, GEN, NEO, K, C	*fexA, aac(6′)aph(2″), tet*(M), *erm*(B), *sul1*	*qacED1, qacC/D*
34		*cfu*	AMC, CX, CRO, CFP, CL, FEP, S, DA, P, AMP, NV, C	*fexA*	*qacED1*
35		*opp*	AX, CX, CRO, CL, FEP, S, DA, TE, SXT, P, AMP, NV, NEO, C	*fexA, aac(6′)aph(2″), tet*(M), *sul1*	*qacED1, qacC/D*
36		*pauA, skc*	AMC, CX, CRO, CFP, CL, FEP, S, DA, TE, E, P, AMP, NV, GEN	*aac(6′)aph(2″), tet*(L), *erm*(B)	*qacED1, qacC/D*
37		*sua*	AX, CX, CRO, CFP, CL, FEP, S, DA, TE, E, P, AMP, NV	*tet*(M), *erm*(B)	*qacA/B, qacED1, qacC/D*
38		*pauA*	AX, AMC, CX, CRO, CFP, CL, FEP, S, DA, TE, E, P, AMP, NV, NEO, C	*aac(6′)aph(2″), tet*(L), *erm*(B)	*qacED1, qacC/D*
39		*cfu*	CX, CRO, CL, FEP, DA, P, AMP, NV	*tet*(M)	*qacED1, qacC/D*
40		*pauA, skc*	AX, AMC, CX, CRO, CL, FEP, S, DA, CIP, TE, E, SXT, P, AMP, NV	*tet*(M), *erm*(B), *sul1*	*qacED1, qacC/D*
41		*pauA*	AX, CX, CRO, CL, FEP, S, DA, TE, P, AMP, NV, GEN, NEO	*aac(6′)aph(2″), tet*(L)	*qacA/B, qacED1, qacC/D*
42	D	ND	AX, CX, CRO, CL, FEP, S, DA, TE, E, P, AMP, NV, NEO	*aac(6′)aph(2″), tet*(M), *erm*(B)	*qacED1, qacC/D*
43		ND	AX, CX, CRO, CL, FEP, DA, CIP, SXT, P, AMP, NV, GEN, NEO	*aac(6′)aph(2″), sul1*	*qacED1, qacC/D*
44		ND	AMC, CX, CRO, CFP, CL, FEP, S, DA, CIP, TE, E, SXT, P, AMP, NV, NEO	*aac(6′)aph(2″), tet*(M), *erm*(B), *sul1*	*qacED1, qacC/D*
45		ND	AX, AMC, CX, CRO, CFP, CL, FEP, S, DA, CIP, TE, E, SXT, P, AMP, NV, K	*erm*(B), *sul1*	*qacED1, qacC/D*
46		ND	AX, AMC, CX, CRO, CFP, CL, FEP, S, DA, TE, E, P, AMP, NV, GEN, K	*aac(6′)aph(2″), erm*(B)	*qacED1, qacC/D*
47		ND	CX, CRO, CL, FEP, S, DA, CIP, TE, E, SXT, P, AMP, NV, K	*blaZ, tet*(M), *erm*(B), *sul1*	*qacED1, qacC/D*
48		ND	AX, CX, CRO, CL, FEP, S, DA, TE, E, SXT, P, AMP, NV, NEO	*blaZ, aac(6′)aph(2″), tet*(M), *erm*(B), *sul1*	*qacED1, qacC/D*
49		*pauA*	AX, CX, CRO, CL, FEP, DA, CIP, TE, E, SXT, P, AMP, NV	*blaZ, erm*(B), *sul1*	*qacA/B, qacED1, qacC/D*
50		*sua, opp*	AX, AMC, CX, CRO, CL, FEP, S, DA, TE, E, SXT, P, AMP, NV, GEN, NEO	*blaZ, aac(6′)aph(2″), tet*(L), *erm*(B), *sul1*	*qacED1, qacC/D*
51		*cfu*	AX, AMC, CX, CRO, CL, FEP, S, DA, CIP, E, P, AMP, NV, NEO	*aac(6′)aph(2″), erm*(B)	*qacED1, qacC/D*
52		*pauA*	CX, CRO, CL, FEP, S, DA, TE, E, P, AMP, NV, NEO	*aac(6′)aph(2″), tet*(M), *erm*(B)	*qacED1, qacC/D*
53		*pauA*	AX, CX, CRO, CL, FEP, S, DA, CIP, TE, SXT, P, AMP, NV, NEO	*blaZ, aac(6′)aph(2″), tet*(M), *sul1*	*qacED1, qacC/D*
54		*cfu*	AX, AMC, CX, CRO, CL, FEP, S, DA, CIP, TE, E, SXT, P, AMP, NV	*blaZ, tet*(M), *erm*(B), *sul1*	*qacED1, qacC/D*
55		*sua, opp*	AX, CX, CRO, CFP, CL, FEP, S, DA, SXT, P, AMP, NV, C	*blaZ, sul1*	*qacA/B, qacED1, qacC/D*
56		*sua*	AX, AMC, CX, CRO, CFP, CL, FEP, S, DA, CIP, TE, E, P, AMP, NV, GEN, C	*blaZ, aac(6′)aph(2″), erm*(B)	*qacED1, qacC/D*
57		*sua, pauA, skc, cfu*	AX, AMC, CX, CRO, CFP, CL, FEP, S, DA, CIP, SXT, P, AMP, NV, GEN, NEO, C	*blaZ, aac(6′)aph(2″), sul1*	*qacED1, qacC/D*
58		*sua, pauA, skc, cfu*	AMC, CX, CRO, CFP, CL, FEP, S, DA, CIP, TE, E, P, AMP, NV, GEN, NEO, C	*blaZ, aac(6′)aph(2″), tet*(M), *erm*(B)	*qacED1, qacC/D*

Table 2. *Cont.*

Isolate No.	Herd	Virulence Pattern	Antimicrobial Resistance Profile		Biocide Resistance Genes
			Phenotype	**Resistance Genes**	
59		*sua, pauA, skc, cfu*	CX, CRO, CFP, CL, FEP, S, DA, TE, E, P, AMP, NV, C	*blaZ, tet*(M), *erm*(B)	*qacED1, qacC/D*
60		*sua, pauA, skc, cfu*	CX, CRO, CL, FEP, DA, TE, E, P, AMP, NV, K, C	*blaZ, fexA, tet*(K), *erm*(B)	*qacED1, qacC/D*
61		*sua, pauA, skc, cfu*	AX, CX, CRO, CL, FEP, S, DA, TE, E, SXT, P, AMP, NV, GEN, NEO, K, C	*blaZ, fexA, aac(6′)aph(2″), tet*(M), *erm*(B), *sul1*	*qacED1, qacC/D*
62		*sua*	AX, AMC, CX, CRO, CL, FEP, S, DA, TE, SXT, P, AMP, NV, NEO, K, C	*blaZ, fexA, aac(6′)aph(2″), tet*(K), *sul1*	*qacA/B, qacED1, qacC/D*
63		*sua*	CX, CRO, CFP, CL, FEP, S, DA, TE, E, SXT, P, AMP, NV, NEO, K, C	*blaZ, fexA, aac(6′)aph(2″), tet*(M), *erm*(B), *sul1*	*qacA/B, qacED1, qacC/D*
64		*sua*	AX, CX, CRO, CL, FEP, S, DA, E, P, AMP, NV, NEO, C	*blaZ, erm*(B)	*qacA/B, qacED1, qacC/D*
65		*sua*	CX, CRO, CL, FEP, DA, E, P, AMP, NV, C	*blaZ, aac(6′)aph(2″), erm*(B)	*qacED1, qacC/D*
66		*sua*	AX, CX, CRO, CL, FEP, S, DA, CIP, E, P, AMP, NV, C	*blaZ, erm*(B)	*qacED1, qacC/D*
67		*skc*	AX, AMC, CX, CRO, CL, FEP, S, DA, TE, E, P, AMP, NV, GEN, NEO, K, C	*blaZ, fexA, aac(6′)aph(2″), tet*(M), *erm*(B)	*qacA/B, qacED1, qacC/D*
68		*skc*	AX, CX, CRO, CL, FEP, S, DA, E, P, AMP, NV, GEN, NEO, K, C	*blaZ, aac(6′)aph(2″), erm*(B)	*qacED1, qacC/D*
69		*opp*	AX, CX, CRO, CL, FEP, S, DA, P, AMP, NV, GEN, NEO, K, C	*blaZ, fexA, aac(6′)aph(2″),*	*qacA/B, qacED1, qacC/D*

AX, amoxicillin; AMC, amoxicillin–clavulanic acid; CX, cloxacillin; CRO, ceftriaxone; CFP, cefoperazone; CL, cephalexin; FEP, cefepime; IPM, imipenem; DA, clindamycin; CIP, ciprofloxacin; TE, tetracycline; E, erythromycin; NEO, neomycin; P, penicillin; NV, novobiocin; AMP, ampicillin; S, streptomycin; GEN, gentamycin; C, chloramphenicol; K, kanamycin; SXT, trimethoprim-sulfamethoxazole; ND, not detected.

Table 3. Virulence gene profiles of *S. uberis* isolated from lactating cattle experience clinical mastitis.

Molecular Pathotype	Virulence Genes	No. of *S. uberis* Isolates (%)	Farms
I	*sua, pauA, skc, cfu*	8 (11.59)	C, D
II	*pauA, cfu, opp*	3 (4.35)	A
III	*sua, pauA*	4 (5.8)	B
IV	*pauA, skc*	3 (4.35)	C
V	*sua, opp*	2 (2.9)	D
VI	*sua*	15 (21.74)	A, C, D
VII	*pauA*	10 (14.49)	A, B, C, D
VIII	*skc*	3 (4.35)	B, D
IX	*cfu*	4 (5.8)	C, D
X	*opp*	3 (4.35)	C, D

3.4. Detection of Antimicrobial Resistance Genes in S. uberis Isolates

The detection of antimicrobial resistance genes confirmed the phenotypic resistance patterns of the respective *S. uberis* isolates (Table 2). As presented in Figure 1, the erythromycin resistance gene *erm*(B) was the most prevalent among the analyzed isolates (75.36%). However, *erm*(C) and *erm*(A) genes were not amplified in either erythromycin-resistant or erythromycin-susceptible *S. uberis* isolates. The most frequent tetracycline resistance gene was *tet*(M) (46.38%), whereas *tet*(L) and *tet*(K) genes were recorded in lower frequencies (8.7 and 5.8%, respectively), and the *tet*(O) gene was not detected in any of the tested *S. uberis* isolates. The *blaZ* and *aac(6′)aph(2″)* genes, conferring resistance to penicillins and aminoglycosides, respectively, were similarly found in 32 out of

69 examined isolates (46.38% each). Furthermore, the *fexA* gene, conferring resistance to chloramphenicol, was detected in 20 *S. uberis* isolates (28.99%). The sulfonamide resistance gene, *sul1*, was found in 31 (44.93%) *S. uberis* isolates, but the trimethoprim *dfrA* gene was not detected in any analyzed isolate.

3.5. Biocide Resistance Genes and Biocide–Antimicrobial Cross-Resistance

Biocide resistance profiling showed that 55 out of 69 *S. uberis* isolates (79.71%) carried *qac* genes; among them, 55 (79.71%), 54 (78.26%), and 13 (18.84%) exhibited resistance to *qacED1*, *qacC/D*, and *qacA/B*, respectively. Biocide resistance gene combinations were detected among the isolates; 3 gene combinations were found in 11 (15.94%) isolates, and 2 combinations, either *qacED1* + *qacC/D* (38/69, 55.07%) or *qacED1* + *qacA/B* (2/69, 2.9%), were also reported (Table 2). The selective pressure employed by exposure to biocides may be associated with increasing antimicrobial resistance. As shown in Figure 2 and Table S2, significant ($p < 0.05$) positive correlations ($r = 0.01–0.43$) between QAC tolerance and resistance to various antimicrobials indicate the pervasive occurrence of multi-drug efflux pumps. However, non-significant ($p > 0.05$) negative correlations were observed between the existence of *qac* genes and resistance to certain antimicrobials such as amoxicillin–clavulanate ($r = -0.01, -0.1$ and -0.05), tetracycline ($r = -0.06, -0.03$ and -0.01), and trimethoprim-sulfamethoxazole ($r = -0.02, -0.01$ and -0.05) for *qacED1*, *qacC/D*, and *qacA/B* genes, respectively.

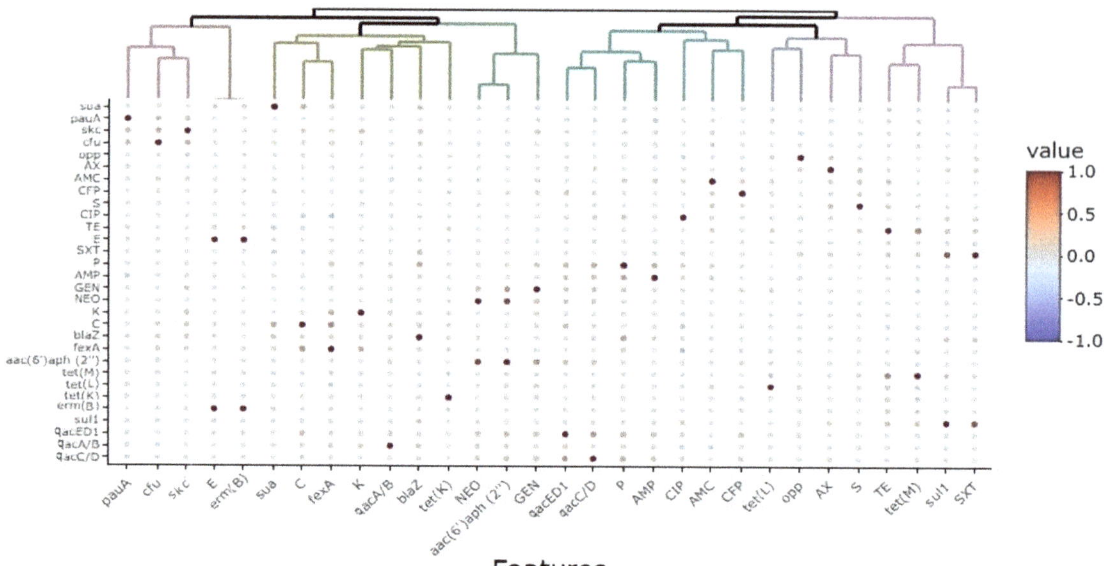

Figure 2. Correlations among various features in *S. uberis* isolates ($n = 69$) from various farms. The color scale represents the correlation coefficient (R) on a scale from +1 to −1 (+1 is the highest positive correlation, and −1 is the highest negative correlation). AX, amoxicillin; AMC, amoxicillin–clavulanic acid; CFP, cefoperazone; S, streptomycin; CIP, ciprofloxacin; TE, tetracycline; E, erythromycin; SXT, trimethoprim-sulfamethoxazole; P, penicillin; AMP, ampicillin; GEN, gentamycin; NEO, neomycin; K, kanamycin; C, chloramphenicol.

3.6. Typing of Virulent S. uberis Isolates Using RFLP–PCR

Phenotypically, *S. uberis* type I and *S. parauberis* (*S. uberis* type II) isolates had similar cultural, morphological, and biochemical characteristics and could not be differentiated by conventional methods. Therefore, RFLP–PCR analysis of the 16S rRNA gene was used to characterize them, and the results indicated that all isolates ($n = 69$) were indeed *S. uberis* (*S. uberis* type I) by their unique RFLP pattern (Figure S1).

3.7. Association between the Existence of Virulence Traits, Antimicrobial and Biocide Resistance, and Hygienic Interventions for Dairy Cows

As shown in Table 2, the occurrence of virulence-associated and antimicrobial- and QAC resistance genes was distributed among *S. uberis* isolates ($n = 69$) over all the investigated dairy farms. However, the simultaneous existence of four virulence genes (*sua* + *pauA* + *skc* + *cfu*, 11.59%), more than four antimicrobial resistance genes (17.39%), and the three tested *qac* genes (*qacA/B* + *qacED1* + *qacC/D*, 15.94%) was reported only in C and D dairy herds. However, nine (13.04%) *S. uberis* isolates recovered from the C and D dairy herds did not carry any QAC resistance genes. Overall, as presented in Figure 3, the four studied features (virulence, resistance phenotype/genes, and *qac* genes) were prominent in the dairy herds with moderate and low hygiene measures (C and D, respectively).

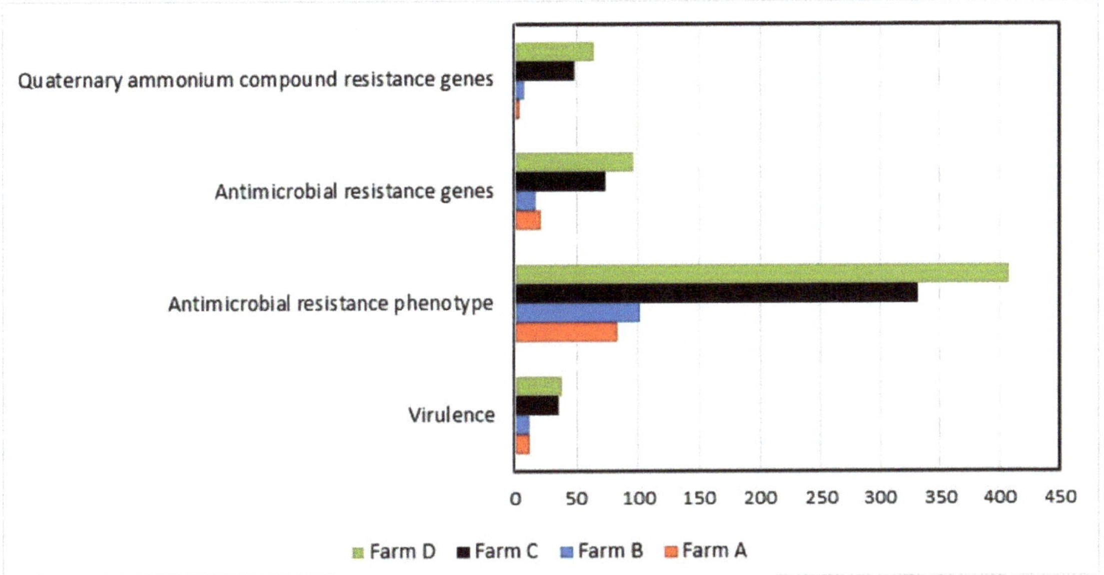

Figure 3. Differences among farms are shown in the term of studied features. Each horizontal bar represents the overall number of isolates (*x*-axis) showing a certain feature (including redundancy). Farms are shown in different colors. Farm D possessed the highest number of isolates harboring the studied features compared to other farms.

Figure 4 and Table S3 demonstrate that *S. uberis* isolates ($n = 69$) had a low-to-moderate diversity (Euclidean distance = 0.11–0.73) among the investigated dairy herds. The dendrogram analysis (Figure 5) classified the isolates into four clusters (1, 2, 3, and 4). A close relatedness was noticed among certain *S. uberis* isolates from different dairy herds. Most *S. uberis* isolated from C and D dairy herds were closely related and gathered in clusters 1 and 2. In addition, *S. uberis* of A and B dairy herds were clustered closely in cluster 4. Few isolates of the four dairy herds clustered together in cluster 3.

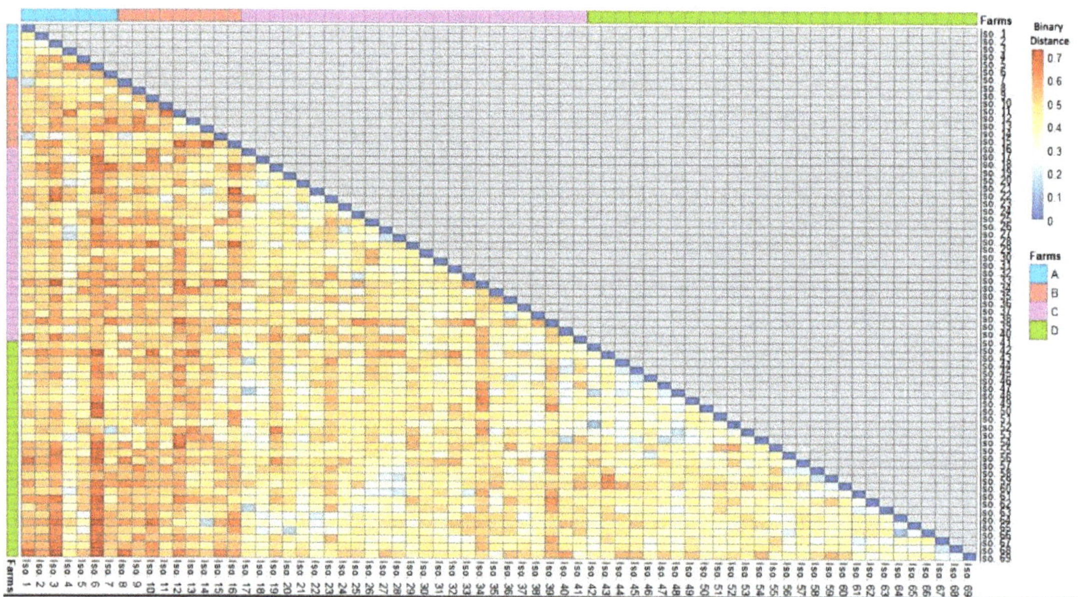

Figure 4. A heatmap showed the binary distances among *S. uberis* isolates based on the presence or absence of the four studied features (Scheme 0.7).

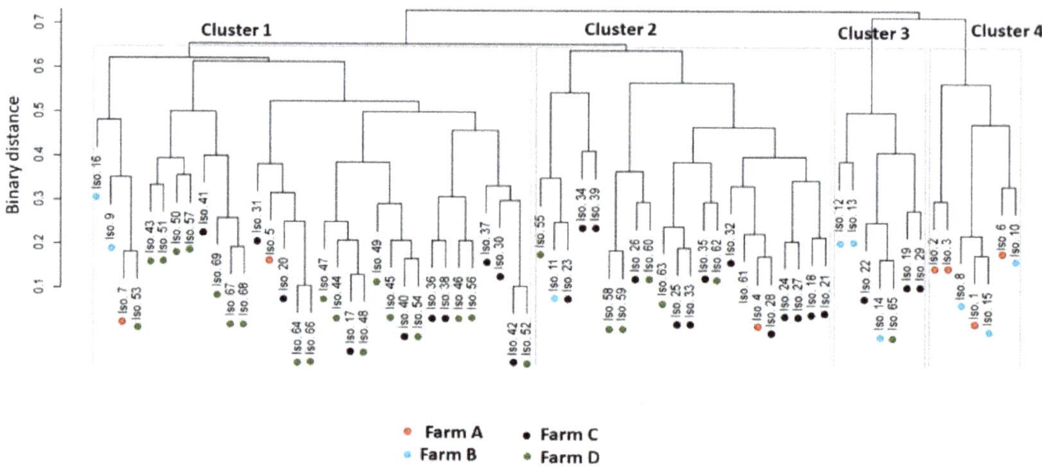

Figure 5. Hierarchical clustering dendrogram showing the relatedness (closeness) of various isolates (shown as numbers) from different farms (shown as colored dots) based on all the feature categories. All isolates were classified into four clusters.

4. Discussion

Mastitis remains a critical problem for dairy animals, causing drastic losses during lactation seasons. Such losses are attributed mainly to decreased milk yield, lower milk quality, and higher treatment and control costs [48]. *S. uberis* is a well-known pathogen that causes bovine intramammary infections worldwide. Nonetheless, there are scant epidemiological data on *S. uberis* isolated from lactating cows in Egypt, especially in the smallholder production system, despite the fact that this extensive system type is the most common traditional livestock farming system among Egyptian farmers [49,50].

In the current study, the overall infection rate of *S. uberis* in dairy cows of different parity, showing gross signs of clinical mastitis associated with or without systemic reactions, was 20.59%, which is nearly similar to a previously published work (23.5%) [51]. Comparatively lower infection rates (6.3%, 9.3%, and 11.8%, respectively) of *S. uberis* isolated from mastitic cattle were previously reported by several studies [52–54]. Higher rates of *S. uberis* infection (55.38 and 33%, respectively) were reported in previous studies [55,56]. Our findings may represent a potential hygiene deficiency that has a significant role in the occurrence of environmental *S. uberis* mastitis [57].

Animals with adequate hygiene during milking (A and B dairy herds) had a lower prevalence of infection (11/125; 8.8%) than those with poor hygiene (C and D) during the milking process (58/210; 27.61%). The predominance of the microorganisms varies according to the handling practices of the animals and the hygiene conditions during milking [58]. The lower infection rate may be attributed to good management practices such as the milkers' hygiene, sanitization of the milking machine, healthy udder environment, dry period treatment, and the control of other predisposing diseases. Meanwhile, the higher infection rate (herds C, D) may be ascribed to a group of shared breeding factors where the dairy cattle live, including bad habitats, unbalanced food, terrible drafts, and the lack of pre-milking procedures. These conditions play a role in rendering the udder more susceptible to intramammary infections [59]. Furthermore, variations in the microenvironments and management practices between the different hygiene features applied to farms can influence the existence of the disease.

Herein, 21 antimicrobials of 12 antimicrobial classes were chosen to be tested, considering their availability for the intramammary treatment of clinical mastitis. In addition, we monitored penicillin, phenicol, aminoglycoside, tetracycline, sulfonamide, and trimethoprim resistance phenotypes/genotypes among the bovine mastitis *S. uberis* isolates.

Penicillin is widely used in the treatment of clinical bovine mastitis. The proportions of ampicillin-, penicillin-, and amoxicillin-resistant isolates in this study were high (89.85%, 79.71%, and 69.57%, respectively). Our finding strongly supported the previous results of Haenni et al. [60], who described a shift toward penicillin resistance among a subpopulation of *S. uberis* isolates. Additionally, they identified the presence of resistance-associated mutations among isolates considered intermediately susceptible to penicillin. Here, the *blaZ* gene conferring resistance to penicillin was found in 46.38% of the examined isolates, indicating an alarming level of potential resistance in bovine mastitis. This finding conflicts with the claim that environmental streptococci are still susceptible to β-lactam-active substances [61]. Previous studies [62,63] have documented that penicillin is effective against streptococci isolates with percentages of 92% and 96%, respectively. Moreover, Minst and coauthors [64] noticed the absence of penicillin and ampicillin resistance, suggesting that β-lactam antibiotics should remain the drug of choice for treating streptococcal mastitis. In this study, a low level of gentamicin resistance (20.28%) against *S. uberis* isolates was observed. On the contrary, an earlier study [65] reported that up to 93% of streptococci were resistant to gentamicin. Additionally, Rato et al. [66] stated that most *S. uberis* isolates (80%) were resistant to gentamycin. Nevertheless, our results are comparable to a lower rate of gentamicin-resistant *S. uberis* in a previous German study [67]. On the other hand, the *S. uberis* resistance rate to streptomycin was 96%, precluding its use in the treatment of bovine mastitis, which is consistent with a previous study [66]. The aminoglycoside resistance gene, *aac(6′)aph(2″)*, was detected here within a reasonable rate (46.38%), which provides evidence suggesting that it confers resistance to a broad spectrum of aminoglycosides in Gram-positive bacteria, including streptococci [68], whereas the *aac(6′)-Ib* gene confers resistance to tobramycin, kanamycin, and amikacin in Gram-negative bacteria [69].

Our results showed that *S. uberis* is highly resistant to erythromycin (73.91%), which is higher than previous reports from France (21%) [70], Argentina (27.6%) [71], and northwestern China (31.2%) [15]. This explains that the *erm*(B) gene is the most prevalent among the analyzed isolates (75.36%). Furthermore, 65.22% of *S. uberis* isolates displayed resistance against tetracycline, mainly due to the inclusion of the *tet*(M) gene (46.38%) in most resistant isolates, which is nearly similar to a previous report (60%) [66]. However, the levels in our findings were lower than that previously described in a previous research (81.3%) [15]. Another study performed on *S. uberis* isolates from dairy cattle with clinical mastitis found results lower than ours for tetracycline resistance (18.1%) [72]. High tetracycline resistance levels may be attributed to their widespread use in treating numerous cattle infections for several years, proposing that tetracyclines, quinolones, and aminoglycosides should be avoided for the treatment of streptococcal mastitis. Differences in the susceptibility patterns among various studies could be due to different antimicrobial use in farms or countries, which could be a consequence of antimicrobial overuse for treating clinical mastitis or for growth promotion purposes in dairy herds [14], thus resulting in the inclusion of drug-resistant bacteria even in raw milk [73].

In the present investigation, conventional PCR allowed the amplification of virulence-associated genes of *S. uberis*, namely, *sua*, *skc*, *cfu*, *pauA*, and *opp*, each represented by a single band to their respective base pairs in the corresponding region of the DNA marker. The detection of genes encoding virulence factors could explain a possible association in the pathogenesis of mammary infections. The CAMP gene (*cfu*) is recorded here with a percentage of 21.73%; nearly similar results have been reported (25%) [74]. However, previous studies have reported high frequencies of the *cfu* gene in *S. uberis* isolates: 76.9% [11], 55.5% [53], and 46.1% [52]. On the contrary, a lower *cfu* percentage (3.8%) was reported in a previous research article [7]. The results suggest that this gene might not be the only gene related to the expression of the CAMP reaction. The *opp* gene was found in 11.59% of the examined *S. uberis* isolates. Previous studies have described a higher percentage of *opp* in *S. uberis* isolates: 64.1% [11] and 22.2% [53]. In contrast, an earlier study reported that the *opp* gene could not be amplified from all the strains, suggesting this gene may not be the only one responsible for the growth of *S. uberis* in milk [75]. The *pauA* gene was found in 40.57% of the examined *S. uberis* isolates. On the contrary, Ward et al. [76] reported that expression of *pauA* is not essential for infection of the mammary gland, as none of the examined isolates harbored the *pauA* gene from mastitic cows in an experimental study. In the same way, previous reports [10,11] found the *pauA* gene in *S. uberis* isolates with a higher percentage (94.9% and 61.5%, respectively). The streptokinase gene (*skc*) was detected at a percentage of 20.28%. A higher result was recorded by Shome and coauthors [52], who reported the *skc* gene in *S. uberis* strains at an incidence of 100%. The *sua* gene was recorded in our research at 42.02%. Nearly similar results were obtained previously (38.5%) [52]. In contrast, higher rates (97.8% and 83.3%, respectively) have been previously recorded [10,11].

Quaternary ammonium compounds (QACs) are amongst the most frequently used disinfectants. They are known to hinder the activity of a broad spectrum of microorganisms. They can disrupt the microbial cell wall, resulting in the leaking of the cytoplasm out of the cells [77]. Regrettably, the prevailing usage of QAC-based antiseptics in animal husbandry may result in bacterial resistance. In this study, QAC resistance genes were examined in *S. uberis* isolates. The *qacED1* (79.71%) and *qacC/D* (78.26%) genes were found more frequently than *qacA/B* (18.84%). A paucity of data is currently available regarding the extent of QAC resistance genes in environmental streptococcal mastitis in Egypt. In a previous study in Egypt, all examined *S. uberis* isolates from bovine mastitis showed 100% phenotypic resistance to QACs (TH4; concentration = 0.25%) [78]. However, there are no Egyptian reports on QAC resistance in *S. uberis* isolates at the genetic level. The selective pressure employed by exposure to biocides has been concomitant with increasing resistance to antimicrobial agents. It has been documented that biocides and antimicrobial agents may share joint target sites and be situated together in mobile elements, resulting in co-resistance [79]. In addition, chromosomal efflux pumps may be involved in antimicrobial and biocide resistance due to their non-specific mechanism [80]. Inconsistent with a previous investigation on biocide resistance in *Staphylococcus aureus* [81], we report a weak or moderate genetic correlation between the existence of QAC and antimicrobial resistance genes.

5. Conclusions

The present investigation is the first to look into the virulence attributes and genotypic resistance to antimicrobials and biocides in *S. uberis* isolates from bovine clinical mastitis in Egypt. It adds to our knowledge of the high diversity of *S. uberis* and its occurrence in relation to prospective hygienic concerns. The sustained presence of *pauA* and *sua* genes throughout the investigated farms contributes to a better understanding of the pathogenicity of the bacterium, which provides the need to use such virulence factors as potential constituents of a vaccine against *S. uberis*. The co-existence of MDR and biocide resistance indicates the importance of *S. uberis* surveillance and the prudent use of antimicrobials and antiseptics in veterinary clinical medicine to avoid the dissemination of resistance.

Supplementary Materials: The followings are available online at https://www.mdpi.com/article/10.3390/ani11071849/s1, Figure S1: RFLP–PCR analysis of the 16S rRNA gene of representative *S. uberis* isolates showed unique patterns, Table S1: Oligonucleotide primer sequences used for PCR assays, Table S2: Significance and correlations among pairs of the four features under study, Table S3: Calculated binary distances among isolates based on the presence or absence of the four studied features.

Author Contributions: Validation, A.M.A., E.K. and R.A.A.E.; formal analysis, N.K.A.E.-A., H.M.E.D. and E.K.; investigation, N.K.A.E.-A., H.M.E.D., A.M.A., R.A.A.E., W.E.-K. and E.K.; data curation, N.K.A.E.-A. and H.M.E.D.; writing—original draft preparation, N.K.A.E.-A. and H.M.E.D.; writing— review and editing, N.K.A.E.-A., A.M.A., W.E.-K. and E.K.; project administration, H.A.S. and W.E.-K.; funding acquisition, H.A.S. and W.E.-K. All authors have read and agreed to the published version of the manuscript.

Funding: This research received no external funding.

Institutional Review Board Statement: Ethical review and approval were waived for this study as the study does not involve human cases or animal interventions. In place, recruitment of dairy cows into the work was done in consultation with veterinarians, and written informed consent was obtained from dairy farm owners for the collection of milk samples from mastitic cows.

Informed Consent Statement: Informed consent was obtained from all subjects involved in the study.

Acknowledgments: We appreciate and thank Taif University for financial support for the Taif University Researchers Supporting Project (TURSP-2020/07), Taif University, Taif, Saudi Arabia. We thank the staff members of the Faculty of Veterinary Medicine, Zagazig University, Egypt, for their technical support during this study.

Conflicts of Interest: The authors declare no conflict of interest.

References

1. Collenburg, L.; Beyersdorf, N.; Wiese, T.; Arenz, C.; Saied, E.M.; Becker-Flegler, K.A.; Schneider-Schaulies, S.; Avota, E. The Activity of the Neutral Sphingomyelinase Is Important in T Cell Recruitment and Directional Migration. *Front. Immunol.* **2017**, *8*. [CrossRef]
2. Abebe, R.; Hatiya, H.; Abera, M.; Megersa, B.; Asmare, K. Bovine mastitis: Prevalence, risk factors and isolation of *Staphylococcus aureus* in dairy herds at Hawassa milk shed, South Ethiopia. *BMC Vet. Res.* **2016**, *12*, 270. [CrossRef]
3. Lone, M.A.; Hülsmeier, A.J.; Saied, E.M.; Karsai, G.; Arenz, C.; von Eckardstein, A.; Hornemann, T. Subunit Composition of the Mammalian Serine-Palmitoyltransferase Defines the Spectrum of Straight and Methyl-Branched Long-Chain Bases. *Proc. Natl. Acad. Sci. USA* **2020**, *117*, 15591–15598. [CrossRef] [PubMed]
4. Käppeli, N.; Morach, M.; Zurfluh, K.; Corti, S.; Nüesch-Inderbinen, M.; Stephan, R. Sequence types and antimicrobial resistance profiles of *Streptococcus uberis* isolated from bovine mastitis. *Front. Vet. Sci.* **2019**, *6*, 234. [CrossRef]
5. Vasiliauskaité-Brooks, I.; Healey, R.D.; Rochaix, P.; Saint-Paul, J.; Sounier, R.; Grison, C.; Waltrich-Augusto, T.; Fortier, M.; Hoh, F.; Saied, E.M.; et al. Structure of a Human Intramembrane Ceramidase Explains Enzymatic Dysfunction Found in Leukodystrophy. *Nat. Commun.* **2018**, *9*, 5437. [CrossRef] [PubMed]

6. Facklam, R. What happened to the streptococci: Overview of taxonomic and nomenclature changes. *Clin. Microbiol. Rev.* **2002**, *15*, 613–630. [CrossRef] [PubMed]

7. Khan, I.U.; Hassan, A.A.; Abdulmawjood, A.; Lanimler, C.; Wolter, W.; Zschock, M. Identification and epidemiological characterization of *Streptococcus uberis* isolated from bovine mastitis using conventional methods. *J. Vet. Sci.* **2003**, *4*, 213–224. [CrossRef]

8. Almeida, R.A.; Luther, D.A.; Park, H.M.; Oliver, S.P. Identification, isolation and partial characterization of a novel *Streptococcus uberis* adhesion molecule (SUAM). *Vet. Microbiol.* **2006**, *115*, 183–191. [CrossRef]

9. Smith, A.J.; Kitt, A.J.; Ward, P.N.; Leigh, J.A. Isolation and characterization of a mutant strain of *Streptococcus uberis*, which fails to utilize a plasmin derived beta-casein peptide for the acquisition of methionine. *J. Appl. Microbiol.* **2002**, *93*, 631–639. [CrossRef]

10. Ward, P.N.; Leigh, J.A. Genetic analysis of *Streptococcus uberis* plasminogen activators. *Indian J. Med. Res.* **2004**, *119*, 136–140.

11. Reinoso, E.B.; Lasagno, M.C.; Dieser, S.A.; Odierno, L.M. Distribution of virulence-associated genes in *Streptococcus uberis* isolated from bovine mastitis. *FEMS Microbiol. Lett.* **2011**, *318*, 183–188. [CrossRef]

12. Palma, E.; Tilocca, B.; Roncada, P. Antimicrobial resistance in veterinary medicine: An overview. *Int. J. Mol. Sci.* **2020**, *21*, 1914. [CrossRef] [PubMed]

13. Abd El-Aziz, N.K.; Ammar, A.M.; El-Naenaeey, E.Y.M.; El Damaty, H.M.; Elazazy, A.A.; Hefny, A.A.; Shaker, A.; Eldesoukey, I.E. Antimicrobial and antibiofilm potentials of cinnamon oil and silver nanoparticles against *Streptococcus agalactiae* isolated from bovine mastitis: New avenues for countering resistance. *BMC Vet. Res.* **2021**, *17*, 1–14. [CrossRef]

14. Tomazi, T.; Freu, G.; Alves, B.G.; de Souza Filho, A.F.; Heinemann, M.B.; Veiga Dos Santos, M. Genotyping and antimicrobial resistance of *Streptococcus uberis* isolated from bovine clinical mastitis. *PLoS ONE* **2019**, *14*, e0223719. [CrossRef] [PubMed]

15. Zhang, H.; Yang, F.; Li, X.-P.; Luo, J.-Y.; Wang, L.; Zhou, Y.-L.; Yan, Y.; Wang, X.-R.; Li, H.S. Detection of antimicrobial resistance and virulence-related genes in *Streptococcus uberis* and *Streptococcus parauberis* isolated from clinical bovine mastitis cases in northwestern China. *J. Integr. Agric.* **2020**, *19*, 2784–2791. [CrossRef]

16. Pitkälä, A.; Koort, J.; Björkroth, J. Identification and antimicrobial resistance of *Streptococcus uberis* and *Streptococcus parauberis* isolated from bovine milk samples. *J. Dairy Sci.* **2008**, *91*, 4075–4081. [CrossRef]

17. Roussel, P.; Porcherie, A.; Répérant-Ferter, M.; Cunha, P.; Gitton, C.; Rainard, P.; Germon, P. *Escherichia coli* mastitis strains: In Vitro phenotypes and severity of infection In Vivo. *PLoS ONE* **2017**, *12*, e0178285. [CrossRef]

18. Abd El-Aziz, N.K.; Abd El-Hamid, M.I.; Bendary, M.M.; El-Azazy, A.A.; Ammar, A.M. Existence of vancomycin resistance among methicillin resistant *S. aureus* recovered from animal and human sources in Egypt. *Slov. Vet. Res.* **2018**, *55*, 221–230.

19. Bjorland, J.; Steinum, T.; Kvitle, B.; Waage, S.; Sunde, M.; Heir, E. Widespread distribution of disinfectant resistance genes among staphylococci of bovine and caprine origin in Norway. *J. Clin. Microbiol.* **2005**, *43*, 4363–4368. [CrossRef]

20. Quinn, P.J.; Carter, M.E.; Markey, B.; Carter, G.R. *Clinical Veterinary Microbiology*; Mosby: London, UK, 1999; pp. 21–66.

21. Hardie, J.M. Genus *Streptococcus*. In *Bergey's Manual of Systematic Bacteriology*; Sneath, P.H.A., Ed.; Williams and Wilkins: Baltimore, MD, USA, 1986; Volume 2, pp. 1043–1071.

22. Christie, R.; Atkins, N.E.; Munch-Petersen, E. A note on a lytic phenomenon shown by group B streptococci. *Aust. J. Exp. Biol. Med. Sci.* **1944**, *22*, 197–200. [CrossRef]

23. Domig, K.J.; Mayer, H.K.; Kneifel, W. Methods used for the isolation, enumeration, characterization and identification of *Enterococcus* spp. 2. Pheno- and genotypic criteria. *Int. J. Food Microbiol.* **2003**, *88*, 165–188. [CrossRef]

24. Bauer, A.W.; Kirby, W.M.; Sherris, J.C.; Turck, M. Antibiotic susceptibility testing by a standard single disc method. *Am. J. Clin. Pathol.* **1966**, *45*, 493–496. [CrossRef]

25. Clinical and Laboratory Standards Institute. *Performance Standard for Antimicrobial Susceptibility Testing: Nineteenth Informational Supplement M100*; CLSI: Wayne, PA, USA, 2019.

26. Magiorakos, A.P.; Srinivasan, A.; Carey, R.B.; Carmeli, Y.; Falagas, M.E.; Giske, C.G.; Harbarth, S.; Hindler, J.F.; Kahlmeter, G.; Olsson-Liljequist, B.; et al. Multidrug-resistant, extensively drug-resistant and pandrug-resistant bacteria: An international expert proposal for interim standard definitions for acquired resistance. *Clin. Microbiol. Infect.* **2012**, *18*, 268–281. [CrossRef] [PubMed]

27. Tambekar, D.; Dhanorkar, D.; Gulhane, S.; Khandelwal, V.; Dudhane, M. Antibacterial susceptibility of some urinary tract pathogens to commonly used antibiotics. *Afr. J. Biotechnol.* **2006**, *5*, 1562–1565.

28. Picard, F.J.; Ke, D.; Boudreau, D.K.; Boissinot, M.; Huletsky, A.; Richard, D.; Ouellette, M.; Roy, P.H.; Bergeron, M.G. Use of *tuf* sequences for genus-specific PCR detection and phylogenetic analysis of 28 *Streptococcal* species. *J. Clin. Microbiol.* **2004**, *42*, 3686–3695. [CrossRef]

29. Nithin Prabhu, K.; Isloor, S.; Hegde, R.; Rathnamma, D.; Veeregowda, B.M.; Narasimha Murthy, H.; Shome, R.; Suryanarayana, V.V.S. Development of polymerase chain reaction for detection of predominant streptococcal isolates causing subclinical bovine mastitis. *Indian J. Biotechnol.* **2013**, *12*, 208–212.

30. Nithin Prabhu, K.; Isloor, S.K.; Hegde, R.; Suryanarayana, W. Standardization of PCR and phylogenetic analysis for predominant streptococcal species isolated from subclinical mastitis. In Proceedings of the International Symposium on "Role of Biotechnology in Conserving Biodiversity and Livestock Development for Food Security and Poverty Alleviation" and XVII Annual Convention of Indian Society of Veterinary Immunology & Biotechnology (ISVIB), Bikaner, India, 23 March 2010; Volume 50, p. 47.

31. Vesterholm-Nielsen, M.; Ølholm Larsen, M.; Elmerdahl Olsen, J.; Møller Aarestrup, E. Occurrence of the *blaZ* Gene in penicillin resistant *Staphylococcus aureus* isolated from bovine mastitis in Denmark. *Acta Vet. Scand.* **1999**, *40*, 279–286. [CrossRef] [PubMed]

32. Kehrenberg, C.; Schwarz, S. Distribution of florfenicol resistance genes *fexA* and *cfr* among chloramphenicol resistant *Staphylococcus* isolates. *Antimicrob. Agents Chemother.* **2006**, *50*, 1156–1163. [CrossRef] [PubMed]

33. Duran, N.; Ozer, B.; Duran, G.G.; Onlen, Y.; Demir, C. Antibiotic resistance genes & susceptibility patterns in staphylococci. *Indian J. Med. Res.* **2012**, *135*, 389–396.
34. Ng, L.K.; Martin, I.; Alfa, M.; Mulvey, M. Multiplex PCR for the detection of tetracycline resistant genes. *Mol. Cell Probes.* **2001**, *15*, 209–215. [CrossRef]
35. Malhotra-Kumar, S.; Lammens, C.; Piessens, J.; Goossens, H. Multiplex PCR for simultaneous detection of macrolide and tetracycline resistance determinants in streptococci. *Antimicrob. Agents Chemother.* **2005**, *49*, 4798–4800. [CrossRef]
36. Lina, G.; Quaglia, A.; Reverdy, M.E.; Leclercq, R.; Vandenesch, F.; Etienne, J. Distribution of genes encoding resistance to macrolides, lincosamides, and streptogramins among staphylococci. *Antimicrob. Agents Chemother.* **1999**, *43*, 1062–1066. [CrossRef] [PubMed]
37. Schlegelova, J.; Vlkova, H.; Babak, V.; Holasova, M.; Jaglic, Z.; Stosova, T.; Sauer, P. Resistance to erythromycin of *Staphylococcus* spp. isolates from the food chain. *Vet. Med.* **2008**, *53*, 307–314. [CrossRef]
38. Jensen, L.B.; Frimodt-Møller, N.; Aarestrup, F.M. Presence of *erm* gene classes in gram-positive bacteria of animal and human origin in Denmark. *FEMS Microbiol. Lett.* **1999**, *170*, 151–158. [CrossRef] [PubMed]
39. Ibekwe, A.M.; Murinda, S.E.; Graves, A.K. Genetic diversity and antimicrobial resistance of *Escherichia coli* from human and animal sources uncovers multiple resistances from human sources. *PLoS ONE* **2011**, *6*, e20819. [CrossRef] [PubMed]
40. Grape, M.; Motakefi, A.; Pavuluri, S.; Kahlmeter, G. Standard and real-time multiplex PCR methods for detection of trimethoprim resistance *dfr* genes in large collections of bacteria. *Clin. Microbiol. Infect.* **2007**, *13*, 1112–1118. [CrossRef]
41. Noguchi, N.; Suwa, J.; Narui, K.; Sasatsu, M.; Ito, T.; Hiramatsu, K.; Song, J. Susceptibilities to antiseptic agents and distribution of antiseptic-resistance genes *qacA/B* and *smr* of methicillin-resistant *Staphylococcus aureus* isolated in Asia during 1998 and 1999. *J. Med. Microbiol.* **2005**, *54*, 557–565. [CrossRef] [PubMed]
42. Chuanchuen, R.; Khemtong, S.; Padungtod, P. Occurrence of *qacE/qacEDelta1* genes and their correlation with class 1 integrons in *Salmonella enterica* isolates from poultry and swine. *Southeast Asian J. Trop. Med. Public Health.* **2007**, *38*, 855–862.
43. Kolde, R. Package 'Pheatmap': Pretty Heat Map. 2018. Available online: https://rdrr.io/cran/pheatmap/ (accessed on 12 May 2020).
44. Friendly, M. Corrgrams: Exploratory displays for correlation matrices. *Am. Stat.* **2002**, *56*, 316–324. [CrossRef]
45. Galili, T.; O'Callaghan, A.; Sidi, J.; Sievert, C. Heatmaply: An R package for creating interactive cluster heatmaps for online publishing. *Bioinformatics* **2018**, *34*, 1600–1602. [CrossRef]
46. Harrel, F.E., Jr. Package 'Hmisc'. Available online: https://cran.r-project.org/web/packages/Hmisc/Hmisc.pdf (accessed on 4 May 2020).
47. SAS. *SAS Statistics User's Guide. Statistical Analytical System*, 5th rev. SAS ed.; Institute Inc.: Cary, NC, USA, 2012.
48. Palanivel, K.M.; Suresh, R.V.; Jayakumar, R.; Ganesan, P.I.; Dhanapalan, P. Retrospective study of sub-clinical mastitis in buffaloes. *Indian J. Vet. Med.* **2008**, *28*, 34–36.
49. El Damaty, H.M.; Mahmmod, Y.S.; Gouda, S.M.; Sobhy, N.M. Epidemiological and ultrasonographic investigation of bovine fascioliasis in smallholder production system in Eastern Nile Delta of Egypt. *Prev. Vet. Med.* **2018**, *158*, 35–42. [CrossRef]
50. El Damaty, H.M.; Fawzi, E.M.; Neamat-Allah, A.N.F.; Elsohaby, I.; Abdallah, A.; Farag, G.K.; El-Shazly, Y.A.; Mahmmod, Y.S. Characterization of foot and mouth disease virus serotype SAT-2 in swamp water buffaloes (*Bubalus bubalis*) under the Egyptian smallholder production system. *Animals* **2021**, *11*, 1697. [CrossRef]
51. Bradley, A.J.; Leach, K.A.; Breen, J.E.; Green, L.E.; Green, M.J. Survey of the incidence and etiology of mastitis in dairy' farms in England and Wales. *Vet. Rec.* **2007**, *160*, 253–257. [CrossRef]
52. Shome, B.R.; Bhuvana, M.; Mitra, S.D.; Krithiga, N.; Shome, R.; Velu, D.; Banerjee, A.; Barbuddhe, S.B.; Prabhudas, K.; Rahman, H. Molecular characterization of *Streptococcus agalactiae* and *Streptococcus uberis* isolates from bovine milk. *Trop. Anim. Health Prod.* **2012**, *44*, 1981–1992. [CrossRef]
53. Eldesouky, I.E.; Refae, M.A.; Nada, H.S.; Elnaby, G.R.H. Molecular detection of *Streptococcus* species isolated from cows with mastitis. *World Vet. J.* **2016**, *6*, 193–202. [CrossRef]
54. Amin, B.; Deneke, Y.; Abdela, N. Bovine mastitis: Prevalence, risk factors, and isolation of *Streptococcus* species from small holders dairy farms in and around Haramaya town, Eastern Ethiopia. *Glob. J. Med. Res.* **2017**, *17*, 27–38.
55. Amosun, E.A.; Ajuwape, A.T.P.; Adetosoye, A.I. Bovine streptococcal mastitis in Southwest and Northern States of Nigeria. *Afr. J. Biomed. Res.* **2010**, *13*, 33–37.
56. Vasil', M. Etiology, course, and reduction of incidence of environmental mastitis in the herd of dairy cows. *Slovak J. Anim. Sci.* **2009**, *3*, 136–144.
57. Abozaid, A.A.; El Balkemy, F.A.; El Damaty, H.M. Impact of risk factors on the prevalence of mastitis in dairy cattle. *Zag. Vet. J.* **2013**, *41*, 162–168. [CrossRef]
58. Jayarao, B.M.; Pillai, S.R.; Sawant, A.A.; Wolfgang, D.R.; Hegde, N.V. Guidelines for monitoring bulk tank milk somatic cell and bacterial counts. *J. Dairy Sci.* **2004**, *87*, 3561–3573. [CrossRef]
59. Kivaria, F.M.; Noordhuizen, J.P.T.M.; Msamia, H.M. Risk factors associated with the incidence rate of clinical mastitis in smallholder dairy cows in the Dar es Salaam region of Tanzania. *Vet. J.* **2007**, *173*, 623–629. [CrossRef] [PubMed]
60. Haenni, M.; Galofaro, L.; Ythier, M.; Giddey, M.; Majcherczyk, P.; Moreillon, P.; Madec, J.Y. Penicillin-binding-protein gene alterations in *Streptococcus uberis* isolates presenting decreased susceptibility to penicillin. *Antimicrob. Agents Chemother.* **2010**, *54*, 1140–1145. [CrossRef]

61. Cameron, M.; Saab, M.; Heider, L.; McClure, J.T.; Rodriguez-Lecompte, J.C.; Sanchez, J. Antimicrobial susceptibility patterns of environmental streptococci recovered from bovine milk samples in the Maritime Provinces of Canada. *Front. Vet. Sci.* **2016**, *3*, 1–14. [CrossRef]

62. Erskine, R.J.; Walker, R.D.; Bolin, C.A.; Bartlett, P.C.; White, D.G. Trends in antibacterial susceptibility of mastitis pathogens during a seven-year period. *J. Dairy Sci.* **2002**, *85*, 1111–1118. [CrossRef]

63. Gianneechini, R.E.; Concha, C.; Franklin, A. Antimicrobial susceptibility of udder pathogens isolated from dairy herds in the West Littoral region of Uruguay. *Acta Vet. Scand.* **2002**, *43*, 31–41. [CrossRef] [PubMed]

64. Minst, K.; Märtlbauer, E.; Miller, T.; Meyer, C. Short communication: *Streptococcus* species isolated from mastitis milk samples in Germany and their resistance to antimicrobial agents. *J. Dairy Sci.* **2012**, *95*, 6957–6962. [CrossRef] [PubMed]

65. Schröder, A.; Hoedemacker, M.; Klein, G. Susceptibility of mastitis pathogens in northern Germany. *Berl. Munch. Tierarztl. Wochenschr.* **2005**, *118*, 393–398. [PubMed]

66. Rato, M.G.; Bexiga, R.; Florindo, C.; Cavaco, L.M.; Vilela, C.L.; Santos-Sanches, I. Antimicrobial resistance and molecular epidemiology of streptococci from bovine mastitis. *Vet. Microbiol.* **2013**, *161*, 286–294. [CrossRef] [PubMed]

67. Kaspar, H. Results of the antimicrobial agent susceptibility study raised in a representative, cross-sectional monitoring study on a national basis. *Int. J. Med. Microbiol.* **2006**, *296*, 69–79. [CrossRef]

68. Lollai, S.A.; Ziccheddu, M.; Duprè, I.; Piras, D. Characterization of resistance to tetracyclines and aminoglycosides of sheep mastitis pathogens: Study of the effect of gene content on resistance. *J. Appl. Microbiol.* **2016**, *121*, 941–951. [CrossRef]

69. Abd El-Aziz, N.K.; Gharib, A.A. Coexistence of plasmid-mediated quinolone resistance determinants and AmpC-beta- lactamases in *Escherichia coli* strains in Egypt. *Cell Mol. Biol.* **2015**, *61*, 29–35. [PubMed]

70. Guérin-Faublée, V.; Tardy, F.; Bouveron, C.; Carret, G. Antimicrobial susceptibility of *Streptococcus* species isolated from clinical mastitis in dairy cows. *Int. J. Antimicrob. Agents* **2002**, *19*, 219–226. [CrossRef]

71. Denamiel, G.; Llorente, P.; Carabella, M.; Rebuelto, M.; Gen-tilini, E. Anti-microbial susceptibility of *Streptococcus* spp. isolated from bovine mastitis in Argentina. *J. Vet. Med. B Infect. Dis. Vet. Public Health* **2005**, *52*, 125–128. [CrossRef] [PubMed]

72. Boireau, C.; Cazeau, G.; Jarrige, N.; Calavas, D.; Madec, J.Y.; Leblond, A.; Haenni, M.; Gay, É. Antimicrobial resistance in bacteria isolated from mastitis in dairy cattle in France, 2006–2016. *J. Dairy Sci.* **2018**, *101*, 9451–9462. [CrossRef]

73. Abd El-Aziz, N.K.; Ammar, A.M.; Hamdy, M.M.; Gobouri, A.A.; Azab, E.; Sewid, A.H. First report of aacC5-aadA7_4 gene cassette array and phage tail tape measure protein on class 1 integrons of *Campylobacter* species isolated from animal and human sources in Egypt. *Animals* **2020**, *10*, 2067. [CrossRef]

74. Lasagno, M.C.; Reinoso, E.B.; Dieser, S.A.; Calvinho, L.F.; Buzzola, F.; Vissio, C.; Bogni, C.; Odierno, L.M. Phenotypic and genotypic characterization of *Streptococcus uberis* isolated from bovine subclinical mastitis in Argentinean dairy farms. *Rev. Argent. Microbiol.* **2011**, *43*, 212–217.

75. Zadoks, R.N.; Schukken, Y.H.; Wiedmann, M. Multilocus sequence typing of *Streptococcus uberis* provides sensitive and epidemi-ologically relevant subtype information and reveals positive selection in the virulence gene *pauA*. *J. Clin. Microbiol.* **2005**, *43*, 2407–2417. [CrossRef] [PubMed]

76. Ward, P.N.; Field, T.R.; Rapier, C.D.; Leigh, J.A. The activation of bovine plasminogen by *pauA* is not required for virulence of *Streptococcus uberis*. *Infect. Immun.* **2003**, *71*, 7193–7196. [CrossRef]

77. Bragg, R.; Jansen, A.; Coetzee, M.; van der Westhuizen, W.; Boucher, C. Bacterial resistance to quaternary ammonium compounds (qac) disinfectants. In *Infectious Diseases and Nanomedicine II. Advances in Experimental Medicine and Biology*; Adhikari, R., Thapa, S., Eds.; Springer: New Delhi, India, 2014; Volume 808.

78. Haggag, Y.N.; Nossair, M.A.; Mansour, A.M.; Abd El Rahman, A.H. Streptococci in dairy farms: Isolation, antibiogram pattern and disinfectant sensitivity. *Alex. J. Vet. Sci.* **2018**, *59*, 85–92. [CrossRef]

79. Levy, S.B.; Marshall, B. Antibacterial resistance worldwide: Causes, challenges and responses. *Nat. Med.* **2004**, *10*, S122–S129. [CrossRef] [PubMed]

80. Cambau, E.; Guillard, T. Antimicrobials that affect the synthesis and conformation of nucleic acids. *Rev. Sci. Tech.* **2012**, *31*, 77–87. [CrossRef] [PubMed]

81. Oggioni, M.R.; Coelho, J.R.; Furi, L.; Knight, D.R.; Viti, C.; Orefici, G.; Martinez, J.L.; Freitas, A.T.; Coque, T.M.; Morrissey, I.; et al. Significant differences characterise the correlation coefficients between biocide and antibiotic susceptibility profiles in *Staphylococcus aureus*. *Curr. Pharm. Des.* **2015**, *21*, 2054–2057. [CrossRef] [PubMed]

Article

Characteristics of High-Level Aminoglycoside-Resistant *Enterococcus faecalis* Isolated from Bulk Tank Milk in Korea

Hyo Jung Kang [1], Sunghyun Yoon [1,2], Koeun Kim [1] and Young Ju Lee [1,*]

[1]   College of Veterinary Medicine & Zoonoses Research Institute, Kyungpook National University, Daegu 41566, Korea; sa01083@knu.ac.kr (H.J.K.); goguma0707@gmail.com (S.Y.); kke02062@gmail.com (K.K.)
[2]   Division of Microbiology, National Center for Toxicological Research, U.S. Food and Drug Administration, Jefferson, AR 72079, USA
*   Correspondence: youngju@knu.ac.kr; Tel.: +82-53-950-7793

Simple Summary: Aminoglycosides are used to treat various infections in veterinary and human medicine. However, with the emergence of high-level aminoglycoside resistance in human and food-producing animals, the synergism of aminoglycosides with beta-lactam or glycopeptide is being threatened. Moreover, the environmental mastitis-causing agent, enterococci, has emerged as a cause of nosocomial infection due to its antimicrobial resistance. Therefore, the purpose of this study was to investigate the characteristics of high-level aminoglycoside-resistant *Enterococcus faecalis* isolated from bulk tank milk in Korea. It showed that 185 (61.5%) isolates out of 301 were high-level aminoglycoside resistant, while 149 isolates were multidrug resistant.

Abstract: Enterococci, which are considered environmental mastitis-causing pathogens, have easily acquired aminoglycoside-resistant genes that encode various aminoglycoside-modifying enzymes (AME). Therefore, this study was conducted to compare the distribution of high-level aminoglycoside-resistant (HLAR) and multidrug-resistant (MDR) *Enterococcus faecalis* (*E. faecalis*) bacteria isolated from bulk tank milk in four dairy companies in Korea. Moreover, it analyzed the characteristics of their antimicrobial resistance genes and virulence factors. Among the 301 *E. faecalis* bacteria studied, 185 (61.5%) showed HLAR with no significant differences among the dairy companies. Furthermore, 129 (69.7%) of the 185 HLAR *E. faecalis* showed MDR without significant differences among companies. In contrast, HLAR *E. faecalis* from companies A, B, and C were significantly higher in resistance to the four classes than those in company D, which had the highest MDR ability against the three antimicrobial classes ($p < 0.05$). In addition, in the distribution of AME genes, 72 (38.9%) and 36 (19.5%) of the isolates carried both *aac(6′)Ie-aph(2″)-Ia* and *ant(6)-Ia* genes, and the *ant (6)-Ia* gene alone, respectively, with significant differences among the companies ($p < 0.05$). In the distribution of virulence genes, the *ace* (99.5%), *efa A* (98.9%), and *cad 1* (98.4%) genes were significantly prevalent ($p < 0.05$). Thus, our results support that an advanced management program by companies is required to minimize the dissemination of antimicrobial resistance and virulence factors.

Keywords: high-level aminoglycoside resistance; *Enterococcus faecalis*; bulk tank milk

Citation: Kang, H.J.; Yoon, S.; Kim, K.; Lee, Y.J. Characteristics of High-Level Aminoglycoside-Resistant *Enterococcus faecalis* Isolated from Bulk Tank Milk in Korea. *Animals* **2021**, *11*, 1724. https://doi.org/10.3390/ani11061724

Academic Editors: Paola Roncada and Bruno Tilocca

Received: 20 May 2021
Accepted: 7 June 2021
Published: 9 June 2021

Publisher's Note: MDPI stays neutral with regard to jurisdictional claims in published maps and institutional affiliations.

1. Introduction

Aminoglycosides are antimicrobials, including gentamicin, streptomycin, and kanamycin, that inhibit bacterial protein synthesis [1]. In particular, aminoglycosides are used in the treatment of aerobic Gram-negative bacilli infections, and with broad-spectrum beta-lactam for severe infections [2]. However, the emergence of resistance to aminoglycosides has continuously been reported among isolates from humans and food-producing animals. This resistance has also been associated with exposure to the commonly used agents [3,4].

Enterococci have increasingly emerged as a cause of serious nosocomial infection in humans and have also been considered environmental mastitis-causing pathogens in veterinary medicine [5,6]. Although synergic combinations of penicillin or a glycopeptide with an aminoglycoside have been used for treating such infections, enterococci have easily acquired aminoglycoside-resistant genes that encode various aminoglycoside-modifying enzymes (AME). These acquired genes cause high resistance to aminoglycosides [7]. In particular, high-level resistance to aminoglycosides can abolish the synergic effect between commercially available aminoglycosides and cell-wall active agents, such as beta-lactams or glycopeptides [8].

Although high-level aminoglycoside-resistant (HLAR) enterococci were first reported in the 1980s in humans [9], and have been described in several studies investigating antimicrobial resistance profiles in raw milk or dairy products worldwide [10,11], there have been no comprehensive surveys to date on the characteristics of HLAR enterococci obtained from raw milk or dairy products in Korea. Hence, this study was conducted to compare the distribution of HLAR and multidrug-resistant (MDR) *Enterococcus faecalis* (*E. faecalis*) isolated from bulk tank milk in four major dairy companies in Korea. Moreover, it analyzed the characteristics of the antimicrobial resistance genes and virulence factors of the bacterial strains of interest.

2. Materials and Methods

2.1. Bacterial Isolation

A total of 1584 batches of bulk tank milk samples from 395 farms belonging to four dairy companies in Korea were collected twice during the summer and winter each (July–December 2019). Then, 50 mL of milk samples were aseptically collected from each bulk sample and sent to the laboratory at 4 °C. For the isolation and identification of *E. faecalis*, 1 mL of the milk sample was cultured in 9 mL buffered peptone water (BPW; BD Biosciences, San Jose, CA, USA); the pre-enriched BPW was then mixed with an Enterococcosel broth (BD Biosciences) at a 1:10 ratio, and incubated at 37 °C for 18–24 h. Furthermore, each medium was streaked onto an Enterococcosel agar (BD Biosciences), and confirmation of *E. faecalis* was performed using PCR, as described previously [12]. Among the isolates showing the same antimicrobial susceptibility patterns from the same origin, only one isolate was chosen for this study. As a result, *E. faecalis* isolates were tested.

2.2. Antimicrobial Susceptibility Testing

According to the guidelines of the Clinical and Laboratory Standards Institute (CLSI, 2019) [13], the disk diffusion method was performed for all *E. faecalis* isolates against 11 antimicrobial agents (BD Bioscience, Sparks, MD, USA). These 11 antimicrobial agents are: ampicillin (AM, 10 µg), chloramphenicol (C, 30 µg), ciprofloxacin (CIP, 5 µg), doxycycline (DOX, 30 µg), erythromycin (E, 15 µg), high-level gentamicin (G, 120 µg), penicillin (P, 10 units), rifampin (RA, 5 µg), high-level streptomycin (S, 300 µg), tetracycline (TE, 30 µg), and vancomycin (VA, 30 µg). MDR was defined as acquired resistance to at least one agent of the three or more antimicrobial classes [14].

2.3. Detection of HLAR Enterococci

The standard agar dilution method conducted on brain heart infusion agar was used to determine the minimum inhibitory concentration values for G and S, with a concentration range of 256–2048 µg/mL (serial 2-fold dilutions). Moreover, breakpoints for high-level G and S were set at ≥500 and ≥2000 µg/mL, respectively, following the CLSI guidelines (CLSI, 2019) [13].

2.4. Detection of Antimicrobial Resistance and Virulence Genes

The presence of genes conferring resistance to aminoglycosides (*aac (6″)Ie-aph(2″)-Ia, ant (6)-Ia, aph(2″)-Ic*, and *aph (2″)-Id*), macrolide (*erm A, erm B*, and *mef*), oxazolidinone (*optr A* and *poxt A*), phenicols (*cat A, cat B, cfr*, and *fex A*), and tetracyclines (*tet L, tet M*, and *tet O*) were investigated using PCR, as described previously [15–21]. Genes encoding virulence factors such as collagen-binding protein (*ace*), aggregation substance (*asa 1*), pheromone cAD1 precursor lipoprotein (*cad 1*), cytolysin (*cyl A activator*), *E. faecalis* endocarditis antigen (*efa A*), enterococcal surface protein (*esp*), and gelatinase (*gel E*) were also detected, as described previously [22,23]. The primers used in this study are presented in Table 1.

Table 1. Primer sequences used for this study.

Locus	Target Gene	Sequence (5′-3′)	Size (pb)	Reference
Aminoglycoside-modifying enzymes	*aac(6″)Ie-aph(2″)-1la*	F: CAGAGCCTTGGGAAGATGAAG R: CCTCGTGTAATTCATGTTCTGGC	348	[17]
	ant(6)-Ia	F: ACTGGCTTAATCAATTTGGG R: GCCTTTCCGCCACCTCACCG	597	[15]
	aph(2″)-Ic	F: CCACAATGATAATGACTCAGTTCCC R: CCACAGCTTCCGATAGCAAGAG	444	[17]
	aph(2″)-Id	F: GTGGTTTTTACAGGAATGCCATC R: CCCTCTTCATACCAATCCATATAACC	641	[17]
Macrolide resistance	*ermA*	F: TAACATCAGTACGGATATTG R: AGTCTACACTTGGCTTAGG	200	[19]
	ermB	F: CCGAACACTAGGGTTGCTC R: ATCTGGAACATCTGTGGTATG	139	[19]
	mef	F: AGTATCATTAATCACTAGTGC R: TTCTTCTGGTACTAAAAGTGG	348	[19]
Oxazolidinone resistance	*optrA*	F: AGGTGGTCAGCGAACTAA R: ATCAACTGTTCCCATTCA	1395	[20]
	poxtA	F: TCCACAAAGGATGGGTTATG R: ATGCCCGTATTGGTTATCTC	1336	[22]
Phenicol resistance	*catA*	F: GGATATGAAATTTATCCCTC R: CAATCATCTACCCTATGAAT	486	[21]
	catB	F: TGAACACCTGGAACCGCAGAG R: GCCATAGTAAACACCGGAGCA	482	[21]
	cfr	F: TGAAGTATAAAGCAGGTTGGGAGTCA R: ACCATATAATTGACCACAAGCAGC	746	[18]
	fexA	F: GTACTTGTAGGTGCAATTACGGCTGA R: CGCATCTGAGTAGGACATAGCGTC	1272	[18]
Tetracycline resistance	*tetL*	F: ATAAATTGTTTCGGGTCGGTAAT R: AACCAGCCAACTAATGACAATGAT	1077	[16]
	tetM	F: GTTAAATAGTGTTCTTGGAG R: CTAAGATATGGCTCTAACAA	657	[16]
	tetO	F: CAATATCACCAGAGCAGGCT R: TCC CAC TGT TCC ATA TCG TCA	614	[16]
Virulence gene	*ace*	F: GGAATGACCGAGAACGATGGC R: GCTTGATGTTGGCCTGCTTCCG	616	[23]
	asa1	F: CACGCTATTACGAACTATGA R: TAAGAAAGAACATCACCACGA	375	[23]
	cad1	F: TTCCAA AACTACGCACAACA R: CTTTTTCAGCAGCATTCACTAATT	423	[24]
	cylA	F: GACTCGGGGATTGATAGGC R: GCTGCTAAAGCTGCGCTTAC	688	[23]
	efaA	F: CGTGAGAAAGAAATGGAGGA R: CTACTAACACGTCACGAATG	499	[23]
	esp	F: AGATTTCATCTTTGATTCTTG R: AATTGATTCTTTAGCATCTGG	510	[23]
	gelE	F: TATGACAATGCTTTTTGGGAT R: AGATGCACCCGAAATAATATA	213	[23]

2.5. Statistical Analysis

The Statistical Package for Social Science version 25 (IBM SPSS Statistics for Windows, Armonk, NY, USA) was used for statistical analysis. Further, Pearson's chi-square test and Fisher's exact test with Bonferroni correction were used to compare the prevalence of isolates between companies [25]. A p value < 0.05 was considered statistically significant.

3. Results

3.1. Prevalence of MDR and HLAR E. faecalis

The distribution of MDR and HLAR in *E. faecalis* from the bulk tank milk of four dairy companies is presented in Table 2. Among the 301 *E. faecalis* isolates studied, 149 (49.5%) and 185 (61.5%) showed MDR and HLAR, respectively. Moreover, although company D showed the highest prevalence of *E. faecalis*, the prevalence of MDR *E. faecalis* was significantly higher in isolates from company A (61.5%) ($p < 0.05$). However, the prevalence of HLAR *E. faecalis* showed no significant difference among the four dairy companies.

Table 2. Distribution of multidrug-resistant and high-level aminoglycoside-resistant *Enterococcus faecalis* isolates from the bulk tank milk of four dairy companies.

Company (No. of Farms)	No. of *E. faecalis*	No. of MDR [1] (%)	No. of HLAR [2] (%)
A (106)	52	37 (71.2) [a]	36 (69.2)
B (47)	39	20 (51.3) [a,b]	28 (71.8)
C (120)	86	41 (47.7) [b]	54 (62.8)
D (122)	124	51 (41.1) [b]	67 (54.0)
Total (395)	301	149 (49.5)	185 (61.5)

Bulk tank milk samples were collected in summer and winter from each farm. [a,b] Values in a column without the same subscript letter differ significantly ($p < 0.05$). [1] MDR: multidrug resistance. [2] HLAR: high-level aminoglycoside resistance.

3.2. Antimicrobial Resistance of HLAR E. faecalis

The distribution of resistance against nine antimicrobial agents of 185 HLAR *E. faecalis* is presented in Table 3. The results showed that the significantly highest resistance was against TE (93.5%), followed by E (71.9%), then DOX (70.8%). In particular, resistance to DOX also showed significant differences among the dairy companies ($p < 0.05$). However, resistance to AM, CIP, P, RA, and VA was only 0% to 5.9%.

Table 3. Antimicrobial resistance of 185 high-level aminoglycoside-resistant *Enterococcus faecalis* isolates from the bulk tank milk of four dairy companies.

	No. (%) of Antimicrobial-Resistant HLAR *E. faecalis* by Company				
Antimicrobials	A ($n = 36$) *	B ($n = 28$)	C ($n = 54$)	D ($n = 67$)	Total ($n = 185$)
Ampicillin	0 (0.0)	1 (3.6)	0 (0.0)	1 (1.5)	2 (1.1) [A,B]
Chloramphenicol	27 (75.0) [a]	14 (50.0) [a,b]	19 (35.2) [b,c]	15 (22.4) [c]	75 (40.5) [C]
Ciprofloxacin	0 (0.0)	0 (0.0)	2 (3.7)	0 (0.0)	2 (1.1) [A,B]
Doxycycline	24 (66.7) [a,b]	20 (71.4) [a,b]	31 (57.4) [b]	56 (83.6) [a]	131 (70.8) [D]
Erythromycin	30 (83.3)	18 (64.3)	43 (79.6)	42 (62.7)	133 (71.9) [D]
Penicillin	0 (0.0)	0 (0.0)	0 (0.0)	0 (0.0)	0 (0.0) [B]
Rifampin	4 (11.1)	0 (0.0)	5 (9.3)	2 (3.0)	11 (5.9) [A]
Tetracycline	33 (91.7)	25 (89.3)	50 (92.6)	65 (97.0)	173 (93.5) [E]
Vancomycin	0 (0.0)	1 (3.6)	0 (0.0)	0 (0.0)	1 (0.5) [A,B]

* n = No. of high-level aminoglycoside-resistant *Enterococcus faecalis* isolated from bulk tank milk by company. Values with different subscript letters ([a–c]) represent significant differences among farms, while superscript letters ([A–E]) represent the total significant difference ($p < 0.05$).

3.3. Distribution of MDR Patterns

The distribution of MDR isolates among 185 HLAR *E. faecalis* is presented in Figure 1. Although the prevalence of MDR (129 isolates, 69.7%) in HLAR *E. faecalis* showed no significant differences among the dairy companies, HLAR *E. faecalis* from company A showed the highest MDR (80.6%), followed by company D (73.2%), C (62.9%), and B (60.7%), respectively. Likewise, all MDR isolates showed resistance against three to five antimicrobial classes. In particular, HLAR *E. faecalis* from companies A, B, and C were significantly higher in resistance against the four classes than company D, which showed the highest MDR against only three of the antimicrobial classes ($p < 0.05$). Furthermore, MDR to five classes was observed only in HLAR *E. faecalis* from companies A (5.6%) and C (3.7%).

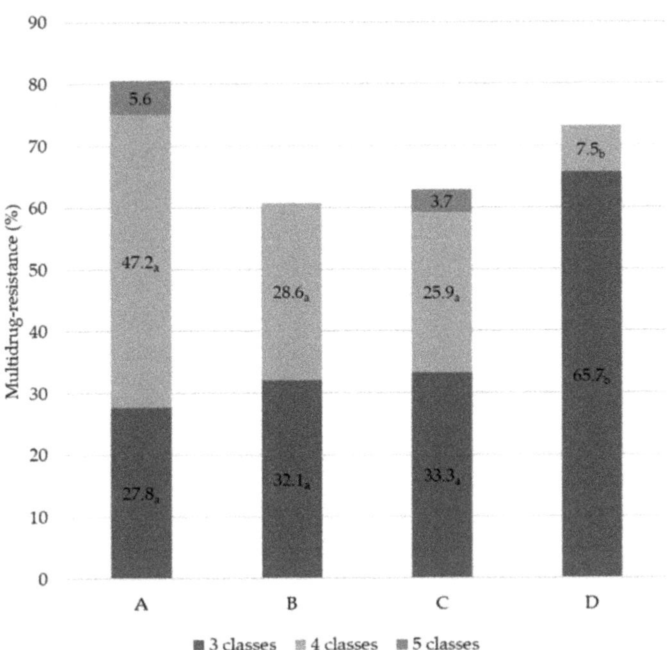

Figure 1. Distribution of multidrug resistance of 185 high-level aminoglycoside-resistant *Enterococcus faecalis* isolated from the bulk tank milk of four dairy companies. Values without the same subscript letter ($_{a,b}$) differ significantly ($p < 0.05$).

3.4. Distribution of Antimicrobial Resistance Genes

The distributions of resistance genes in 185 HLAR *E. faecalis* are presented in Table 4. In the distribution of AME genes, 72 (38.9%) and 36 (19.5%) isolates, respectively, expressed both *aac (6′)Ie-aph(2″)-Ia* and *ant (6)-Ia* genes, and the *ant (6)-Ia* gene alone, with significant differences among the dairy companies ($p < 0.05$). Likewise, for the E resistance genes, the *erm B* gene (71.4%) among the three genes had the highest prevalence ($p < 0.05$), although no significant difference among the dairy companies was observed. Moreover, in the tetracycline resistance genes, the prevalence of both *tet M* and *tet L* genes (46.5%), as well as the *tet M* gene (36.3%) alone, had the highest prevalence with significant differences among the studied dairy companies ($p < 0.05$). Furthermore, the *cat A* and *cfr* genes related to resistance to C were observed among 27 (14.6%) and two (1.1%) isolates, respectively. In contrast, the *optr A* and *poxt A* genes related to resistance to linezolid were observed only in two (1.1%) isolates and one (0.5%) isolate, respectively.

Table 4. Distribution of antimicrobial resistance genes of 185 high-level aminoglycoside-resistant *Enterococcus faecalis* from the bulk tank milk of four dairy companies.

Genes	No. (%) of Isolates with Antimicrobial Resistance Gene(s) by Company				
	A (n = 36) *	**B** (n = 28)	**C** (n = 54)	**D** (n = 67)	**Total** (n = 185)
Aminoglycoside-modifying enzymes					
aac(6′)Ie-aph(2″)-Ia	2 (5.6)	1 (3.6)	6 (11.1)	4 (6.0)	13 (7.0) [A]
ant(6)-Ia	8 (22.2) [a,b]	10 (35.7) [b]	13 (24.1) [a,b]	5 (7.5) [a]	36 (19.5) [B]
aph(2″)-Ic	0 (0.0)	1 (3.6)	1 (1.9)	0 (0.0)	2 (1.1) [A,C]
aph(2″)-Id	1 (2.8)	2 (7.1)	2 (3.7)	8 (11.9)	13 (7.0) [A]
aac(6″)Ie-aph(2″)-Ia, ant(6)-Ia	12 (33.3) [a,b]	6 (21.4) [b]	18 (33.3) [a,b]	36 (53.7) [a]	72 (38.9) [D]
aac(6″)Ie-aph(2″)-Ia, aph(2″)-Id	0 (0.0)	0 (0.0)	1 (1.9)	0 (0.0)	1 (0.5) [C]
aph(2″)-Ic, aph(2″)-Id	0 (0.0)	0 (0.0)	1 (1.9)	0 (0.0)	1 (0.5) [C]
aph(2″)-Ic, ant(6)-Ia	1 (2.8)	0 (0.0)	1 (1.9)	1 (1.5)	3 (1.6) [A,C]
aph(2″)-Id, ant(6)-Ia	1 (2.8)	3 (10.7)	5 (9.3)	3 (4.5)	12 (6.5) [A,C]
aph(2″)-Ic, aph(2″)-Id, ant(6)-Ia	0 (0.0)	0 (0.0)	1 (1.9)	0 (0.0)	1 (0.5) [C]
Macrolides					
ermA	0 (0.0)	1 (3.6)	0 (0.0)	0 (0.0)	1 (0.5) [A]
ermB	29 (80.6)	18 (64.3)	43 (79.6)	42 (62.7)	132 (71.4) [B]
mef	0 (0.0)	0 (0.0)	0 (0.0)	0 (0.0)	0 (0.0) [A]
Oxazolidinones					
optrA	0 (0.0)	1 (3.6)	0 (0.0)	1 (1.5)	2 (1.1) [A]
poxtA	0 (0.0)	0 (0.0)	1 (1.9)	0 (0.0)	1 (0.5) [A]
Phenicols					
catA	11 (30.6) [a]	3 (10.7) [a,b]	8 (14.8) [a,b]	5 (7.5) [b]	27 (14.6) [B]
catB	0 (0.0)	0 (0.0)	0 (0.0)	0 (0.0)	0 (0.0) [A]
cfr	2 (5.6)	0 (0.0)	0 (0.0)	0 (0.0)	2 (1.1) [A]
fexA	0 (0.0)	0 (0.0)	0 (0.0)	0 (0.0)	0 (0.0) [A]
Tetracyclines					
tetL	0 (0.0)	1 (3.6)	3 (5.6)	2 (3.0)	6 (3.2) [A]
tetM	19 (52.8) [a]	3 (10.7) [b]	30 (55.6) [a]	15 (22.4) [b]	67 (36.2) [B]
tetO	0 (0.0)	2 (7.1)	0 (0.0)	0 (0.0)	2 (1.1) [A]
tetM + tetL	11 (30.6) [a]	15 (53.6) [a,b]	16 (29.6) [a]	44 (65.7) [b]	86 (46.5) [B]

* n = No. of high-level aminoglycoside-resistant *Enterococcus faecalis* isolated from bulk tank milk by company. Values with different subscript letters ([a,b]) represent significant differences among farms, while superscript letters ([A–D]) represent the total significant difference ($p < 0.05$).

3.5. Distribution of Virulence Genes

The distributions of virulence genes in 185 HLAR *E. faecalis* are presented in Table 5. The *ace* (99.5%), *efa A* (98.9%), and *cad 1* (98.4%) genes were the prevalent genes ($p < 0.05$), followed by the *gel E* (85.9%), *asa 1* (61.6%), *esp* (12.4%), and *cyl A* (6.5%) genes. However, significant differences among the dairy companies were observed in *efa A*, *cyl A*, and *gel E* genes ($p < 0.05$).

Table 5. Distribution of virulence genes of 185 high-level aminoglycoside-resistant *Enterococcus faecalis* isolated from the bulk tank milk of four dairy companies.

	No. (%) of Isolates with Virulence Gene(s) by Company				
Genes	**A** ($n = 36$)	**B** ($n = 28$)	**C** ($n = 54$)	**D** ($n = 67$)	**Total** ($n = 185$)
ace	36 (100)	27 (96.4)	54 (100)	67 (100)	184 (99.5) [A]
asa1	24 (66.7)	17 (60.7)	28 (51.9)	45 (67.2)	114 (61.6) [B]
cad1	36 (100)	28 (100)	53 (98.1)	65 (97.0)	182 (98.4) [A]
cylA	5 (13.9) [a]	2 (7.1) [a,b]	5 (9.3) [a,b]	0 (0) [b]	12 (6.5) [C]
efaA	36 (100) [a]	26 (92.9) [b]	54 (100) [a]	67 (100) [a]	183 (98.9) [A]
esp	4 (11.1)	4 (14.3)	11 (20.4)	4 (6.0)	23 (12.4) [C]
gelE	31 (86.1) [a,b]	23 (82.1) [a,b]	40 (74.1) [b]	65 (97.0) [a]	159 (85.9) [D]

Values with different subscript letters ([a,b]) represent significant differences among farms, while superscript letters ([A–D]) represent the total significant difference ($p < 0.05$).

4. Discussion

In Korea, five major dairy companies produce 84% of the total milk and dairy products consumed. Confinement housing is used to run most of the farms, which is a primary management system of dairy production [26]. Washburn et al., (2002) [27] reported that confined cows had 1.8 times more clinical mastitis compared with cows on pasture. Therefore, various antimicrobials have been used for treating mastitis every year in Korea [28]. In particular, aminoglycosides, which are used along with cell-wall active agents, are effective for treating serious enterococcus infection [29]. However, enterococci showing high resistance to aminoglycosides has been reported continuously in food-producing animals [30–32]. In this study, 61.5% of the 301 *E. faecalis* isolates from bulk tank milk were HLAR, although no significant difference among the four companies was shown. Chajęcka-Wierzchowska et al., (2020) [10] reported that 30.2% of the ready-to-eat dairy products in Poland showed HLAR to enterococci. Özdemir and Tuncer (2020) [33] also reported that 59 HLAR enterococci were observed in 100 samples of milk and dairy in Turkey. Furthermore, the high prevalence of HLAR enterococci in Korea compared with that in Poland and Turkey is indirect proof of the consistent use of aminoglycosides to treat bacterial infection in Korea.

Interestingly, 69.7% of the HLAR isolates showed MDR in this study. Hegstad et al. (2010) [34] reported that enterococci have a common feature of easily transferring DNA via plasmid or transposons encoding resistance genes. Therefore, the capacity of transference to other bacteria by these plasmids or transposons leads to the spread of various resistance strains, which ultimately results in MDR. The distribution of MDR patterns in HLAR *E. faecalis* showed significant differences among the companies in this study. Likewise, HLAR *E. faecalis* from companies A, B, and C showed the highest prevalence in MDR against the four classes. In contrast, isolates from company D showed the highest prevalence in the three classes. MDR of five classes was observed only in HLAR *E. faecalis* from companies A and C. These results suggest that the critical point for reducing the emergence of resistant bacteria is the management of which and how dairy companies use antimicrobials.

Hollenbeck and Rice (2012) [35] reported that all enterococci possess intrinsic low-level resistance to all aminoglycosides by limiting the uptake of drugs, and this resistance originated from their facultative anaerobic metabolism. However, genes encoding diverse AME acquire high-level resistance to aminoglycosides of enterococci. Thus, it eliminates the synergism of aminoglycosides with cell-wall synthesis interfering agents, such as β-lactams [36–38]. In this study, the combination of both *aac (6′)Ie-aph(2″)-Ia* and *ant (6)-Ia* genes was significantly prevalent ($p < 0.05$). The *ant (6)-Ia* gene is responsible for encoding the ANT (O-adenyltransferase) enzyme that the catalyzes ATP-dependent adenylation of a hydroxyl group, and grant resistance to streptomycin without cross-resistance to other aminoglycosides [39]. In contrast, *aac (6′)Ie-aph(2″)-Ia* encodes the bifunctional enzymes AAC (N-Acetyltransferase) and APH (O-Phosphotransferase), which are responsible for resistance to all types of aminoglycosides, except streptomycin and spectinomycin [40]. Therefore, the high prevalence of HLAR in this study can be related to the distribution of these genes.

In this study, HLAR *E. faecalis* showed the highest resistance to TE (93.5%), followed by E (71.9%). The most prevalent of the different types of antimicrobial resistance genes were *tet M* (82.7%), including a combination of *tet M* and *tet L*, which is related to resistance to tetracyclines, and *erm B* (71.4%), which has shown resistance to macrolides. Although the high distribution of these genes has been continuously reported in enterococci isolated from humans and food-producing animals [41,42], it is important that the *erm B* and *tet* (M) genes can also be transferred easily by conjugative transposons, such as the Tn*916/1545* and Tn*5397* families [43]. Therefore, the horizontal transfer of these genes in enterococci should be continuously monitored in the future. In Korea, tetracyclines as feed additives have been banned since 2009, but a large amount of chlortetracycline calcium, chlortetracycline HCL, oxytetracycline dehydrate, and oxytetracycline HCL have still been used for treating mastitis [44]. Furthermore, E is rarely used in the dairy industry [25,44], but resistance to E is linked to the use of tylosin, which is a macrolide, and is used widely for treating streptococcal mastitis [45]. Therefore, our results support that acquiring resistance genes in food-producing animals was induced by the use of antimicrobials, which also contribute to the burden of growing antimicrobial resistance in humans.

Furthermore, due to the acquisition of virulence genes being directly related to the capability of bacteria to cause illness [46], monitoring virulence genes of enterococci is crucial regarding the public health concerns of dairy products. In this study, HLAR *E. faecalis* showed a high prevalence of genes such as *ace* (99.5%), *efa A* (98.9%), *cad 1* (98.4%), *gel E* (85.9%), and *asa 1* (61.6%). This result was in accordance with the high prevalence of virulence genes of *E. faecalis* from buffalo milk in Brazil [47] and *E. faecalis* from dairy products in Egypt [48]. Although the presence of virulence genes does not increase pathogenicity, virulence factors promote tissue colonization in the hosts. Moreover, the combination of antimicrobial resistance genes and virulence factors in enterococci, which are potential opportunistic pathogens regarding clinical or subclinical mastitis, could be a public health problem. Thus, in this study, the phenotypic and genotypic characteristics of HLAR *E. faecalis* showed significant differences among the dairy companies. However, an advanced management protocol by companies is warranted to minimize the dissemination of antimicrobial resistance and virulence factors.

5. Conclusions

In this comprehensive research on HLAR, *E. faecalis* isolated from the bulk tank milk of four dairy companies in Korea, the distribution of antimicrobial resistance in these bacteria and the genetic characteristics of HLAR *E. faecalis* showed a significant difference among the companies. Therefore, our results suggest that advanced management programs by companies are warranted to minimize the emergence of antimicrobial-resistant bacteria and to reduce the dissemination of these resistance genes and virulence factors.

Author Contributions: Conceptualization, S.Y. and Y.J.L.; Data curation, H.J.K. and S.Y.; Formal analysis, H.J.K., S.Y. and K.K.; Investigation, H.J.K.; Methodology, H.J.K. and Y.J.L.; Software, K.K.; Supervision, Y.J.L.; Writing—original draft, H.J.K.; Writing—review & editing, Y.J.L. All authors have read and agreed to the published version of the manuscript.

Funding: This research received no external funding.

Institutional Review Board Statement: Not applicable.

Data Availability Statement: Not applicable.

Conflicts of Interest: The authors declare no conflict of interest. The funders had no role in the design of the study; in the collection, analyses, or interpretation of the data; in the writing of the manuscript; or in the decision to publish the results.

References

1. Mingeot-Leclercq, M.P.; Glupczynski, Y.; Tulkens, P.M. Aminoglycosides: Activity and resistance. *Antimicrob. Agents Chemother.* **1999**, *43*, 727–737. [CrossRef]
2. Krause, K.M.; Serio, A.W.; Kane, T.R.; Connolly, L.E. Aminoglycosides: An Overview. *Cold Spring Harbor Perspect. Med.* **2016**, 1–18. [CrossRef]
3. Verraes, C.; Van Boxstael, S.; Van Meervenne, E.; Van Coillie, E.; Butaye, P.; Catry, B.; de Schaetzen, M.A.; Van Huffel, X.; Imberechts, H.; Dierick, K.; et al. Antimicrobial resistance in the food chain: A review. *Int. J. Environ. Res. Public Health* **2013**, *10*, 2643–2669. [CrossRef]
4. Economou, V.; Gousia, P. Agriculture and food animals as a source of antimicrobial-resistant bacteria. *Infect. Drug Resist.* **2015**, *8*, 49–61. [CrossRef]
5. Giraffa, G. Enterococci from foods. *FEMS Microbiol. Rev.* **2002**, *26*, 163–171. [CrossRef]
6. Rózańska, H.; Lewtak-Piłat, A.; Kubajka, M.; Weiner, M. Occurrence of enterococci in mastitic cow's milk and their antimicrobial resistance. *J. Vet. Res.* **2019**, *63*, 93–97. [CrossRef]
7. Chow, J.W. Special Section: Antimicrobial Resistance Aminoglycoside Resistance in Enterococci. *Clin. Infect. Dis.* **2000**, *31*, 586–589. [CrossRef] [PubMed]
8. Khodabandeh, M.; Mohammadi, M.; Abdolsalehi, M.R.; Hasannejad-Bibalan, M.; Gholami, M.; Alvandimanesh, A.; Pournajaf, A.; Rajabnia, R. High-Level Aminoglycoside Resistance in Enterococcus Faecalis and Enterococcus Faecium; as a Serious Threat in Hospitals. *Infect. Disord. Drug Targets* **2018**, *20*, 223–228. [CrossRef]
9. Murray, B.E. The life and times of the enterococcus. *Clin. Microbiol. Rev.* **1990**, *3*, 46–65. [CrossRef]
10. Chajęcka-Wierzchowska, W.; Zadernowska, A.; García-Solache, M. Ready-to-eat dairy products as a source of multidrug-resistant Enterococcus strains: Phenotypic and genotypic characteristics. *J. Dairy Sci.* **2020**, *103*, 4068–4077. [CrossRef]
11. Palmeri, M.; Mancuso, I.; Gaglio, R.; Arcuri, L.; Barreca, S.; Barbaccia, P.; Scatassa, M.L. Identification and evaluation of antimicrobial resistance of enterococci isolated from raw ewes' and cows' milk collected in western Sicily: A preliminary investigation. *Ital. J. Food Saf.* **2021**, *9*. [CrossRef]
12. Dutka-Malen, S.; Evers, S.; Courvalin, P. Erratum: Detection of glycopeptide resistance genotypes and identification to the species level of clinically relevant enterococci by PCR. *J. Clin. Microbiol.* **1995**, *33*, 1434. [CrossRef]
13. Clinical and Laboratory Standards Institute (CLSI). *M100 Performance Standards for Antimicrobial Susceptibility Testing*; CLSI: Wayne, PA, USA, 2019; Volume 29.
14. Magiorakos, A.P.; Srinivasan, A.; Carey, R.B.; Carmeli, Y.; Falagas, M.E.; Giske, C.G.; Harbarth, S.; Hindler, J.F.; Kahlmeter, G.; Olsson-Liljequist, B.; et al. Multidrug-resistant, extensively drug-resistant and pandrug-resistant bacteria: An international expert proposal for interim standard definitions for acquired resistance. *Clin. Microbiol. Infect.* **2012**, *18*, 268–281. [CrossRef] [PubMed]
15. Clark, N.C.; Olsvik, Ø.; Swenson, J.M.; Spiegel, C.A.; Tenover, F.C. Detection of a streptomycin/spectinomycin adenylyltransferase gene (aadA) in Enterococcus faecalis. *Antimicrob. Agents Chemother.* **1999**, *43*, 157–160. [CrossRef]
16. Aarestrup, F.M.; Agerso, Y.; Gerner-Smidt, P.; Madsen, M.; Jensen, L.B. Comparison of antimicrobial resistance phenotypes and resistance genes in Enterococcus faecalis and Enterococcus faecium from humans in the community, broilers, and pigs in Denmark. *Diagn. Microbiol. Infect. Dis.* **2000**, *37*, 127–137. [CrossRef]
17. Vakulenko, S.B.; Susan, M.; Donabedian, A.M.V.; Zervos, M.J.; Lerner, S.A.; Chow, J.W. Multiplex PCR for detection of aminoglycoside resistance genes in enterococci. *Antimicrob. Agents Chemother.* **2003**, *47*, 1423–1426. [CrossRef]
18. Kehrenberg, C.; Schwarz, S. Distribution of florfenicol resistance genes fexA and cfr among chloramphenicol-resistant Staphylococcus isolates. *Antimicrob. Agents Chemother.* **2006**, *50*, 1156–1163. [CrossRef]
19. Di Cesare, A.; Luna, G.M.; Vignaroli, C.; Pasquaroli, S.; Tota, S.; Paroncini, P.; Biavasco, F. Aquaculture Can Promote the Presence and Spread of Antibiotic-Resistant Enterococci in Marine Sediments. *PLoS ONE* **2013**, *8*, e62838. [CrossRef]
20. Wang, Y.; Lv, Y.; Cai, J.; Schwarz, S.; Cui, L.; Hu, Z.; Zhang, R.; Li, J.; Zhao, Q.; He, T.; et al. A novel gene, optrA, that confers transferable resistance to oxazolidinones and phenicols and its presence in Enterococcus faecalis and Enterococcus faecium of human and animal origin. *J. Antimicrob. Chemother.* **2015**, *70*, 2182–2190. [CrossRef]
21. Osman, K.M.; Badr, J.; Orabi, A.; Elbehiry, A.; Saad, A.; Ibrahim, M.D.S.; Hanafy, M.H. Poultry as a vector for emerging multidrug resistant Enterococcus spp.: First report of vancomycin (van) and the chloramphenicol–florfenicol (cat-fex-cfr) resistance genes from pigeon and duck faeces. *Microb. Pathog.* **2019**, *128*, 195–205. [CrossRef]
22. Huang, J.; Wang, M.; Gao, Y.; Chen, L.; Wang, L. Emergence of plasmid-mediated oxazolidinone resistance gene poxtA from CC17 Enterococcus faecium of pig origin. *J. Antimicrob. Chemother.* **2019**, *74*, 2524–2530. [CrossRef]
23. Billström, H.; Lund, B.; Sullivan, Å.; Nord, C.E. Virulence and antimicrobial resistance in clinical Enterococcus faecium. *Int. J. Antimicrob. Agents* **2008**, *32*, 374–377. [CrossRef]
24. Song, H.S.; Bae, Y.C.; Jeon, E.J.; Kwon, Y.K.; Joh, S.J. Multiplex PCR analysis of virulence genes and their influence on antibiotic resistance in Enterococcus spp. isolated from broiler chicken. *J. Vet. Sci.* **2019**, *20*, e26. [CrossRef]
25. Yoon, S.; Lee, Y.J. Molecular characteristics of enterococcus faecalis and enterococcus faecium from bulk tank milk in Korea. *Animals* **2021**, *11*, 1–9. [CrossRef] [PubMed]
26. Korea Agro-Fisheries & Food Trade Corporation Food Information Statistics System (ATFIS). Available online: https://www.atfis.or.kr/home/M000000000/index.do# (accessed on 18 May 2021).

27. Washburn, S.P.; White, S.L.; Green, J.T.; Benson, G.A. Reproduction, mastitis, and body condition of seasonally calved holstein and jersey cows in confinement or pasture systems. *J. Dairy Sci.* **2002**, *85*, 105–111. [CrossRef]

28. Animal and Plant Quarantine Agency (APQA). *Antimicrobial Use and Monitoring in Animals and Andimal Products*; APQA: Gimcheon, Korea, 2019.

29. Shahini Shams Abadi, M.; Taji, A.; Salehi, F.; Kazemian, H.; Heidari, H. High-level Gentamicin Resistance among Clinical Isolates of Enterococci in Iran: A Systematic Review and Meta-analysis. *Folia Med.* **2021**, *63*, 15–23. [CrossRef]

30. Jackson, C.R.; Fedorka-Cray, P.J.; Barrett, J.B.; Ladely, S.R. High-level aminoglycoside resistant enterococci isolated from swine. *Epidemiol. Infect.* **2005**, *133*, 367–371. [CrossRef]

31. Riboldi, G.P.; Frazzon, J.; D'Azevedo, P.A.; Frazzon, A.P.G. Antimicrobial resistance profile of Enterococcus spp. isolated from food in southern Brazil. *Braz. J. Microbiol.* **2009**, *40*, 125–128. [CrossRef]

32. Kim, Y.B.; Seo, K.W.; Son, S.H.; Noh, E.B.; Lee, Y.J. Genetic characterization of high-level aminoglycoside-resistant Enterococcus faecalis and Enterococcus faecium isolated from retail chicken meat. *Poult. Sci.* **2019**, *98*, 5981–5988. [CrossRef]

33. Özdemir, R.; Tuncer, Y. Detection of antibiotic resistance profiles and aminoglycoside-modifying enzyme (AME) genes in high-level aminoglycoside-resistant (HLAR) enterococci isolated from raw milk and traditional cheeses in Turkey. *Mol. Biol. Rep.* **2020**, *47*, 1703–1712. [CrossRef]

34. Hegstad, K.; Mikalsen, T.; Coque, T.M.; Werner, G.; Sundsfjord, A. Mobile genetic elements and their contribution to the emergence of antimicrobial resistant Enterococcus faecalis and Enterococcus faecium. *Clin. Microbiol. Infect.* **2010**, *16*, 541–554. [CrossRef]

35. Hollenbeck, B.L.; Rice, L.B. Intrinsic and acquired resistance mechanisms in enterococcus. *Virulence* **2012**, *3*, 421–569. [CrossRef] [PubMed]

36. Shaw, K.J.; Rather, P.N.; Hare, R.S.; Miller, G.H. Molecular genetics of aminoglycoside resistance genes and familial relationships of the aminoglycoside-modifying enzymes. *Microbiol. Rev.* **1993**, *57*, 138–163. [CrossRef]

37. Udo, E.E.; Al-Sweih, N.; John, P.; Jacob, L.E.; Mohanakrishnan, S. Characterization of high-level aminoglycoside-resistant enterococci in Kuwait hospitals. *Microb. Drug Resist.* **2004**, *10*, 139–145. [CrossRef] [PubMed]

38. Bartash, R.; Nori, P. Beta-lactam combination therapy for the treatment of Staphylococcus aureus and Enterococcus species bacteremia: A summary and appraisal of the evidence. *Int. J. Infect. Dis.* **2017**, *63*, 7–12. [CrossRef]

39. Livermore, D.M.; Winstanley, T.G.; Shannon, K.P. Interpretative reading: Recognizing the unusual and inferring resistance mechanisms from resistance phenotypes. *J. Antimicrob. Chemother.* **2001**, *48*, 87–102. [CrossRef] [PubMed]

40. Davies, J.; Wright, G.D. Bacterial resistance to aminoglycoside antibiotics. *Trends Microbiol.* **1997**, *5*, 234–240. [CrossRef]

41. Yang, F.; Zhang, S.; Shang, X.; Wang, X.; Yan, Z.; Li, H.; Li, J. Short communication: Antimicrobial resistance and virulence genes of Enterococcus faecalis isolated from subclinical bovine mastitis cases in China. *J. Dairy Sci.* **2019**, *102*, 140–144. [CrossRef]

42. Choi, J.M.; Woo, G.J. Transfer of Tetracycline Resistance Genes with Aggregation Substance in Food-Borne Enterococcus faecalis. *Curr. Microbiol.* **2015**, *70*, 476–484. [CrossRef]

43. Agersø, Y.; Pedersen, A.G.; Aarestrup, F.M. Identification of Tn5397-like and Tn916-like transposons and diversity of the tetracycline resistance gene tet(M) in enterococci from humans, pigs and poultry. *J. Antimicrob. Chemother.* **2006**, *57*, 832–839. [CrossRef]

44. Ministry of Food and Drug Safety (MFDS). *Processing Standards and Ingredient Specifications for Livestock Products*; NIFDS: Cheongju, Korea, 2018.

45. Gao, X.; Fan, C.; Zhang, Z.; Li, S.; Xu, C.; Zhao, Y.; Han, L.; Zhang, D.; Liu, M. Enterococcal isolates from bovine subclinical and clinical mastitis: Antimicrobial resistance and integron-gene cassette distribution. *Microb. Pathog.* **2019**, *129*, 82–87. [CrossRef] [PubMed]

46. Tatsing Foka, F.E.; Ateba, C.N.; Lourenco, A. Detection of virulence genes in multidrug resistant enterococci isolated from feedlots dairy and beef cattle: Implications for human health and food safety. *BioMed Res. Int.* **2019**, *2019*. [CrossRef] [PubMed]

47. Pereira, R.I.; Prichula, J.; Santesteva, N.A.; d'Azevedo, P.A.; Motta, A.d.S.; Frazzon, A.P.G. Virulence Profiles in Enterococcus spp. Isolated from Raw Buffalo's Milk in South Brazil. *Res. J. Microbiol.* **2017**, *12*, 248–254. [CrossRef]

48. Shafeek, M.; El-Malt, L.; Abdel Hameed, K.; El-Zamkan, M. Some Virulence Genes of Pathogenic Enterococci Isolated from Raw Milk and Some Milk Products. *SVU Int. J. Vet. Sci.* **2018**, *1*, 102–113. [CrossRef]

Article

Epidemiological Cut-Off Values and Multidrug Resistance of *Escherichia coli* Isolated from Domesticated Poultry and Pigs Reared in Mwanza, Tanzania: A Cross-Section Study

Conjester I. Mtemisika [1,2,*], Helmut Nyawale [1], Ronald J. Benju [3], Joseph M. Genchwere [3], Vitus Silago [1], Martha F. Mushi [1], Joseph Mwanga [4], Eveline Konje [4], Mariam M. Mirambo [1] and Stephen E. Mshana [1]

[1] Department of Microbiology and Immunology, Weill Bugando School of Medicine, Catholic University of Health and Allied Sciences, Mwanza P.O. Box 1464, Tanzania; helmutny@yahoo.com (H.N.); vsilago@bugando.ac.tz (V.S.); mushimartha@gmail.com (M.F.M.); mmmirambo@gmail.com (M.M.M.); stephen72mshana@gmail.com (S.E.M.)
[2] Bugando Medical Centre, Molecular Biology Laboratory, Mwanza P.O. Box 1370, Tanzania
[3] Tanzania Veterinary Laboratory Agency, Ministry of Livestock and Fisheries, Mwanza P.O. Box 129, Tanzania; rbenju@yahoo.com (R.J.B.); jgenchwere@yahoo.com (J.M.G.)
[4] Department of Epidemiology, Biostatistics and Behavioral Sciences, School of Public Health, Catholic University of Health and Allied Sciences, Mwanza P.O. Box 1464, Tanzania; jrmwanga@yahoo.co.uk (J.M.); ekonje28@yahoo.com (E.K.)
* Correspondence: conjestermtemisika@yahoo.com

Citation: Mtemisika, C.I.; Nyawale, H.; Benju, R.J.; Genchwere, J.M.; Silago, V.; Mushi, M.F.; Mwanga, J.; Konje, E.; Mirambo, M.M.; Mshana, S.E. Epidemiological Cut-Off Values and Multidrug Resistance of *Escherichia coli* Isolated from Domesticated Poultry and Pigs Reared in Mwanza, Tanzania: A Cross-Section Study. *Animals* **2022**, *12*, 835. https://doi.org/10.3390/ani12070835

Academic Editors: Sangeeta Rao, Paola Roncada and Bruno Tilocca

Received: 25 February 2022
Accepted: 18 March 2022
Published: 25 March 2022

Publisher's Note: MDPI stays neutral with regard to jurisdictional claims in published maps and institutional affiliations.

Simple Summary: The objectives of this study were to determine the prevalence of multidrug resistance phenotypes and the distribution of *Escherichia coli* among poultry and pigs. Laboratory procedures were conducted according to standard operating procedures and international guidelines. Our findings showed that poultry and pigs reared in Mwanza, Tanzania, are colonized with resistant bacterial phenotypes. Further, different populations of intestinal flora, *E. coli*, exist between poultry and pigs.

Abstract: Increasing antimicrobial resistance (AMR) colonizing domesticated animals is a global concern threatening food safety. This study aimed at determining the prevalence of multidrug resistance (MDR) and epidemiological cut-off values (ECVs) of *E. coli* isolated from poultry and pigs in Mwanza, Tanzania. This cross-sectional study was conducted between June and August 2021, involving 297 pigs, 191 broilers, and 203 layers. Rectal and cloacal swabs were collected and processed following standard guidelines. ECVs were determined using normalized resistance interpretation (NRI), a computer software, and descriptive analysis was performed using STATA version 13.0. The overall prevalence of MDR *E. coli* was 63.2%, whereas poultry (87.5% layers and 86.3% broilers) were more colonized than pigs (31.8%) ($p < 0.001$). Based on ECVs of antibiotics tested, *E. coli* from broilers, layers, and pigs exhibited different resistance patterns hence different populations. Exotic breed ($p < 0.001$) and recent antimicrobial use ($p < 0.001$) significantly predicted colonization with MDR *E. coli*. Veterinary officers should implement regulations that prohibit the inappropriate use of antimicrobial agents in livestock keeping.

Keywords: antimicrobial resistance; epidemiological cut-off values; *Escherichia coli*; poultry; pigs

1. Introduction

The use of antimicrobials in livestock to maintain health and promote production is increasing [1], resulting in antimicrobial selection pressure leading to the proliferation of antibiotic-resistant bacteria [2]. Generally, the use of antimicrobials in animals is reported higher in poultry and pigs than in cattle, threatening the safe consumption of poultry and pork and increasing environmental contamination with MDR bacterial strains [1]. Moreover, MDR strains may be transmitted to humans directly via contact with live animals or manure and indirectly via the consumption of contaminated animal products [3,4]. This may result in humans being colonized by multidrug-resistant (MDR) bacteria and subsequently MDR bacterial infections [3].

The increasing unregulated use of antimicrobials in livestock production, particularly in Tanzania, lacks AMR data to create evidence-based standard treatment guidelines for animals [5–7]. In Africa, including Tanzania, studies have documented antimicrobials use among domesticated animals ranging from 77% to 100%, whereas carriage of MDR producing bacteria, particularly Gram-negative bacteria were found to range from 20% to 100% [8,9]. The proportion of MDR strains among *E. coli* isolated from poultry and pigs was 55.2% and 44.8%, respectively, along the Msimbazi basin in Dar es Salaam, Tanzania [10]. However, human antimicrobial susceptibility testing disks and guidelines for interpretation of zones of inhibition used among animal surveillances lack veterinary breakpoints. This practice could over and/or under-report the burden of MDR in livestock [11].

In Mwanza, Tanzania, the prevalence of extended-spectrum β-lactamase (ESBL) among companion and domesticated farm animals was 21.7%. ESBL *E. coli* (93.3%) was predominantly isolated, and pigs were more colonized (33.1%) than other animals [8]. Despite the availability of this information, the prevalence and patterns of MDR Gram-negative bacteria (GNB), notably *E. coli,* is not clearly known as the previous study from the same setting used selective culture media to screen for ESBL producing GNB. The lack of this information may underestimate strategic efforts to prevent the emergence and spreading of MDR bacterial strains among livestock, humans, and environments. Therefore, we designed this study to investigate the prevalence and patterns of MDR *E. coli* and establish epidemiological cut-off values (ECVs) of *E. coli* colonizing poultry and pigs reared in Mwanza, Tanzania. The information from this study will not only facilitate the review of empirical treatment guidelines but also necessitate the implementation of MDR control and preventive measures among poultry and pigs reared in Mwanza, Tanzania. *E. coli* is frequently used as indicator bacteria to monitor trends of antimicrobial resistance (AMR) because it can easily acquire and preserve antibiotics resistance genes from other organisms in the environment and animal populations [12–14]. *E. coli* is also considered a good indicator of the selective pressure imposed by antimicrobial use (AMU) in food animals [12,15,16].

2. Material and Methods

2.1. Ethical Approval

Ethical approval for this study was obtained from the joint CUHAS/BMC research ethics and review committee with certificate number CREC/474/2021. Permission to conduct this study was requested from Livestock and Fisheries authorities. Livestock keepers (farmers) were requested to sign permission forms before sample collection. Unique identification laboratory numbers were used throughout the study.

2.2. Study Design, Population Setting, and Duration

This cross-sectional study was conducted between June and August 2021 among domesticated poultry and pigs reared in 16 wards located in 3 districts in Mwanza, Tanzania, namely, Nyamagana (Nyegezi, Buhongwa, Igoma, Busenga, Kilimahewa, Lumala, Mahina alliance, Malimbe, Mkolani, Nyamongolo, and Mabatini), Ilemela (Buswelu, Kiseke PPF, Nyamanoro, and Pasiansi), and Misungwi (Nyashishi).

2.3. Animals and Farms Selection

Pigs and poultry farms were systematically selected from a list provided by the livestock officers within the study area. Pig farms with pigs aged ≥ 20 weeks and poultry farms with ≥100 poultry were selected. A total of 29 farms (9-pig farms, 9-broiler farms, and 11-layer farms) were visited and enrolled in this study. Poultry farms were selected based on the number of flocks, and 5% of poultry ready to enter the food chain (aged ≥ 12 weeks) in each selected farm were identified for sampling.

For pigs, 10% of pigs per pig pen ready to enter the food chain (aged $\geq$ 20 weeks) in each selected farm were randomly identified and sampled. In all 9 farms, a total of 628 pigs were reared, and sampled pigs in each farm totaled: 1st farm 20 pigs, 2nd farm 31 pigs, 3rd farm 28 pigs, 4th farm 41 pigs, 5th farm 33 pigs, 6th farm 45 pigs, 7th farm 49 pigs, 8th farm 31 pigs, and 9th farm 19 pigs, a total of 297 pigs.

2.4. Sample Collection and Transportation

Fecal samples from the rectum (pigs) or cloaca (chicken) were collected using a sterile cotton swab. Briefly, a sterile swab was gently inserted into the cloaca/rectum and rotated to ensure sufficient sample was collected. Samples were transported to the Microbiology laboratory of the Catholic University of Health and Allied Sciences (CUHAS; Mwanza, Tanzania) in Stuart transport media in a clean, cool box within 8 h following collection.

2.5. Laboratory Procedure

2.5.1. Isolation of E. coli

Swab samples were directly inoculated onto plain MacConkey agar (MCA; HiMedia, India followed by aerobic incubation at 37 °C for 18–24 h. After incubation, in the case of mixed growth, a single colony resembling *E. coli* among morphological predominant similar colonies (deep pinkish, round, mid-sized, and flat) was selected for the purity-plate, its sub-culture onto another plain MCA plate, which was incubated aerobically at 37 °C for 16–20 h, as reported previously [8,10]. Pure growth of presumptive *E. coli* was further identified by in-house prepared biochemical identification tests to species level.

2.5.2. Physiological and Biochemical Identification of E. coli

The presumptive isolates of *E. coli* were preliminarily identified by using conventional in-house prepared physiological and biochemical identification tests, including lactose fermentation, production of CO_2 from sugar fermentation, and sulfur production by triple sugar iron (TSI) agar; sulfur production, indole production, and motility by sulfur-indole-motility (SIM) medium; utilization of sodium citrate as the sole source of carbohydrate by Simmons citrate; and urease production by Christensen's urea agar. Identification tests were interpreted as reported previously [17]. Identified isolates of *E. coli* were subjected to antibiotic susceptibility testing (AST) and phenotypic confirmation of ESBL production.

2.5.3. Antibiotics Susceptibility Testing (AST)

All *E. coli* isolates were tested for antibiotics susceptibility by using the disk diffusion method as reported by Kirby-Bauer [18]. Briefly, isolates were suspended in sterile 0.85% normal saline and adjusted to 0.5 McFarland standard solution. Then, MHA plates were inoculated, and antibiotic disks were seeded within 15 min after inoculation of MHA plates. MHA plates were incubated aerobically at 37 °C for 16–18 h. The interpretations of zones of inhibitions were performed as recommended by the CLSI 29th Edition guidelines [19]. All *E. coli* that showed intermediate susceptibility to the antibiotics tested were regarded as resistant to such particular antibiotics. Antibiotics tested included ciprofloxacin (CIP 5 µg; HiMedia, Mumbai, India), ampicillin (AMP 10 µg; HiMedia, India), tetracycline (TE 30 µg; HiMedia, India), meropenem (MEM 10 µg; HiMedia, India), ceftazidime (CAZ 30 µg; HiMedia, India), gentamicin (CN 10 µg; HiMedia, India), cefepime (FEP 30 µg; HiMedia, India), and trimethoprim-sulfamethoxazole (SXT 25 µg; HiMedia, India).

2.5.4. Screening and Phenotypic Confirmation of ESBL Production

Isolates from plain MCA were sub-cultured on MCA plates which were supplemented with cefotaxime 2 µg/mL (MCA-C) for the screening of ESBL producing *E. coli* (ESBL-EC) as documented previously [20]. Plates were incubated aerobically at 37 °C for 18–24 h. All isolates grown on MCA-C were further confirmed for ESBL production using the phenotypic method, a combination disc method recommended by the Clinical and Laboratory Standards Institute (CLSI) 29th Edition guidelines [19]. Briefly, bacterial suspensions in sterile normal saline equivalent to 0.5 McFarland standard solution were prepared and inoculated on the entire surfaces of Mueller Hinton agar (MHA; HiMedia, India). Then, disks of ceftazidime 30 µg (CAZ 30 µg; HiMedia, India) with and without clavulanic acid 10 µg (CAZ/CA 30/10 µg; HiMedia, India) were seeded on inoculated MHA plates and incubated aerobically at 37 °C for 16–18 h. Isolate exhibiting a difference of ≥5 mm zone of inhibition between CAZ 30 µg and CAZ/CA 30/10 µg were phenotypically confirmed as ESBL-EC.

2.5.5. Quality Control

E. coli ATCC 25,922 and *E. coli* ATCC 35,218 were used as control strains to control the performance of culture media, incubation conditions, and antibiotic disks.

2.6. Data Management and Analysis

Data were entered into Microsoft Excel for cleaning and coding, then into STATA version 13.0 for analysis and NRI computer software, where calculations were performed to define wild type populations by establishing ECVs. All isolates that showed resistance to one or more antibiotic agents in at least three classes were considered multidrug-resistant (MDR) strains. Continuous data were presented as mean (±standard deviation: SD)/median (interquartile range: IQR), whereby categorical data were presented as percentages. Chi square analysis was used to show the association between outcome (i.e., MDR colonization) and variables (i.e., antimicrobial exposure, breed of livestock, and species). A p-value of <0.05 was considered statistically significant. Epidemiological cut-off values were determined by computer software called Normalized resistance interpretation (NRI; Bioscand AB, Täby, Sweden, International Patent Application WO 02/083935 A1). This method analyzes inhibition zone diameters produced from the disk diffusion technique of antimicrobial susceptibility testing. The NRI software produced a histogram that showed the ECVs and distribution of wild type (sensitive isolates) and non-wild type (resistant isolates) bacteria and the number of SD from the mean [21,22]. In some circumstances where the obtained ECVs were very low, i.e., a zero or negative number, due to the high resistance of *E. coli* to a particular antibiotic, mean was used as a tentative ECV estimate as reported elsewhere [11,23].

3. Results

3.1. Characteristics of and Antimicrobials Use among Livestock Enrolled in the Study

A total of 691 livestock, including 27.6% (191/691) broilers, 29.4% (203/691) layers, and 42.9% (297/691) pigs, were sampled from 29 livestock keepers. The majority of livestock were exotic—71.4% (493/691). The recent date of antimicrobial use (AMU) was not known 62.8% (434/691); however, the majority of livestock were administered antimicrobials for therapeutic reasons 86.4% (597/691). About 41.1% (284/691) of livestock were on antimicrobials prescribed by a Veterinary officer (Table 1)

Table 1. Characteristics and AMU of livestock.

Variables		Frequency (*n*)	Percentage (%)
Livestock	Broiler	191	27.6
	Layers	203	29.4
	Pigs	297	42.9
Breed	Exotic	493	71.4
	Local	198	28.7
Recent antimicrobial date	2 months ago	48	6.9
	1 month ago	30	4.3
	2 weeks ago	100	14.5
	1 week ago	79	11.4
	Not known	434	62.8
Purpose of antimicrobial use	Prophylaxis and Therapeutic	94	13.6
	Therapeutic	597	86.4
Antimicrobial prescription	Agro vet shop/Vet shop	60	8.7
	Another farmer	45	6.5
	Myself/family member	142	20.6
	Paraveterinarian	115	16.6
	Paraveterinarian/Vet shop	25	3.6
	Veterinary officer	284	41.1
	Vet officer/myself/family member	20	2.9

3.2. Commonly Used Classes of Antibiotics among Livestock Enrolled in this Study

It was observed from this study that antibiotic agents such as tetracycline, sulfonamides, and quinolones were commonly used in poultry keeping. However, in pigs, antibiotic agents in tetracycline and sulfonamides were common, although classes of quinolones, macrolides, and aminoglycosides were not reported to be used in pigs in this study (Table 2).

Table 2. Antimicrobial classes commonly used in livestock keeping.

Antimicrobial Class	Poultry		Pigs	
	Frequency (*n*)	Percentage (%)	Frequency (*n*)	Percentage (%)
Tetracycline, sulfonamides	121	30.72	161	54.2
Tetracycline, quinolones	202	51.29		
Tetracycline, sulfonamides, macrolides	20	5.08	-	-
Tetracycline, macrolides	10	2.54	-	-
Tetracycline, aminoglycosides	21	5.33	-	-
Quinolones, aminoglycosides	20	5.08	-	-
Not known	-	-	136	45.8

3.3. Culture Results

A total of 95.1% (657/691) *E. coli* were isolated from pigs and poultry, of which a total of 63.2% (415/657) were MDR *E. coli*. The MDR proportions in broilers, layers, and pigs were 86.3% (164/190), 87.5% (161/184), and 31.8% (90/283), respectively. A total of 17.8% (117/657) screened positive for potential ESBL production by MCA-C plates. Furthermore, all presumptive ESBL producing *E. coli* 100% (117/117) were phenotypically confirmed to be ESBL producers. ESBL production was significantly high among *E. coli* from layers (30.9%, 57/184) compared to pigs (17.7%, 50/283) and broilers (5.3%, 10/190) *p* < 0.001 (Figure 1).

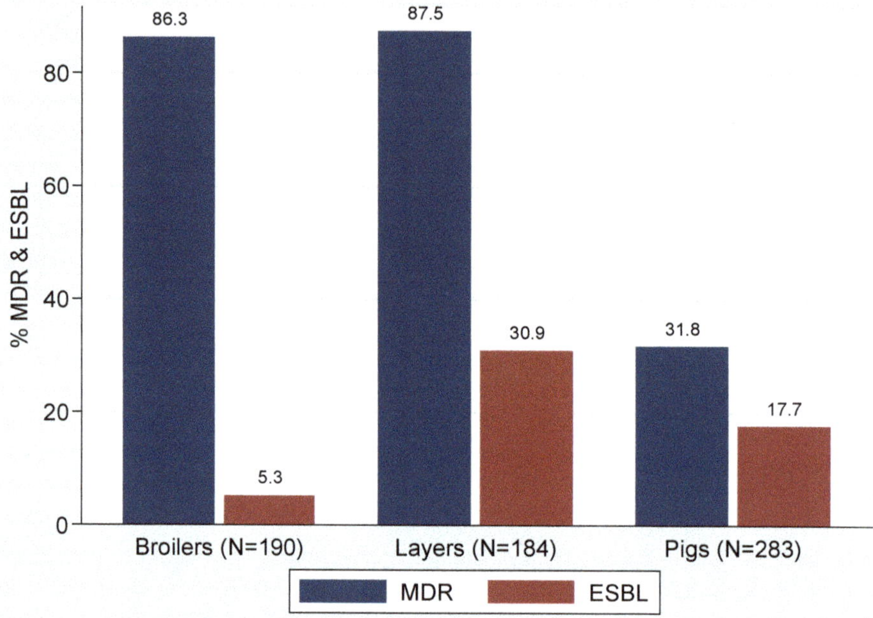

Figure 1. Percentages of MDR and ESBL *E. coli* from broilers, layers, and pigs.

3.4. Resistance Patterns of E. coli to Antibiotics Tested

Percentage resistance for ciprofloxacin, tetracycline, ampicillin, and trimethoprim/sulfamethoxazole, was high among *E. coli* isolated from poultry compared to pigs. However, tetracycline was the most resistant antibiotic among *E. coli* from pigs compared to other antibiotics tested (Table 3).

Table 3. Percentage susceptibility of *E. coli* to antibiotic agents tested.

Antibiotics	Broilers (n = 190)	Layers (n = 184)	Pigs (n = 283)	p Value
	R	R	R	
CIP	180 (94.7%)	165 (89.7%)	67 (23.7%)	0.001
AMP	134 (70.5%)	115 (62.5%)	80 (28.2%)	0.001
MEM	6 (3.2%)	11 (6%)	37 (13.1%)	0.001
TET	166 (87.4%)	165 (89.7%)	140 (49.5%)	0.001
CAZ	36 (18.9%)	80 (43.5%)	76 (26.9%)	0.001
SXT	131 (68.9%)	161 (87.5%)	63 (22.3%)	0.001
CN	44 (23.2%)	41 (22.3%)	30 (10.6%)	0.001
FEP	35 (18.4%)	71 (38.6%)	69 (24.4%)	0.001

Key: CIP, ciprofloxacin; AMP, ampicillin; MEM, meropenem; TET, tetracycline; CAZ, ceftazidime; SXT, trimethoprim/sulfamethoxazole; CN, gentamicin; FEP, cefepime.

3.5. ECVs of Tested Antibiotics against E. coli

The ECV of antibiotics tested against *E. coli* isolated from broilers ranged from 11 mm for TET to 18 mm for FEP and CIP; in layers, it ranges from 9 mm for TET to 18 mm for MEM; and from 9 mm for TET to 30 mm for SXT in pigs. In general, *E. coli* from poultry and pigs exhibited different resistance patterns to antibiotics tested and, therefore, ECVs. This observation indicates that *E. coli* from poultry and pigs belong to different populations, whereby *E. coli* from pigs were more susceptible to antibiotics tested than *E. coli* from poultry (Table 4 & Figure 2).

Table 4. ECVs of tested antibiotics against *E. coli*.

Antimicrobials	Disk Content	Broiler ECVs	SD	Layer ECVs	SD	Pigs ECVs	SD
CIP	5 µg	18 *	5.55	17	2.29	22	4.18
TET	30 µg	11 *	4.13	9 *	9.14	9	4.65
AMP	10 µg	17	2.00	15	3.62	24	1.85
MEM	10 µg	16	3.94	18	4.29	18	5.25
CAZ	30 µg	15	3.15	10	4.39	12	4.08
FEP	30 µg	18	3.26	11	5.19	18	4.36
CN	30 µg	15 *	3.04	10	2.95	16	2.84
SXT	1.25/23.75 µg	15 *	6.41	12 *	4.63	30	2.97

Key: CIP, ciprofloxacin; AMP, ampicillin; MEM, meropenem; TET, tetracycline; CAZ, ceftazidime; SXT, trimethoprim/sulfamethoxazole; CN, gentamicin; FEP, cefepime. * Mean used as ECV tentative estimate. SD = standard deviation.

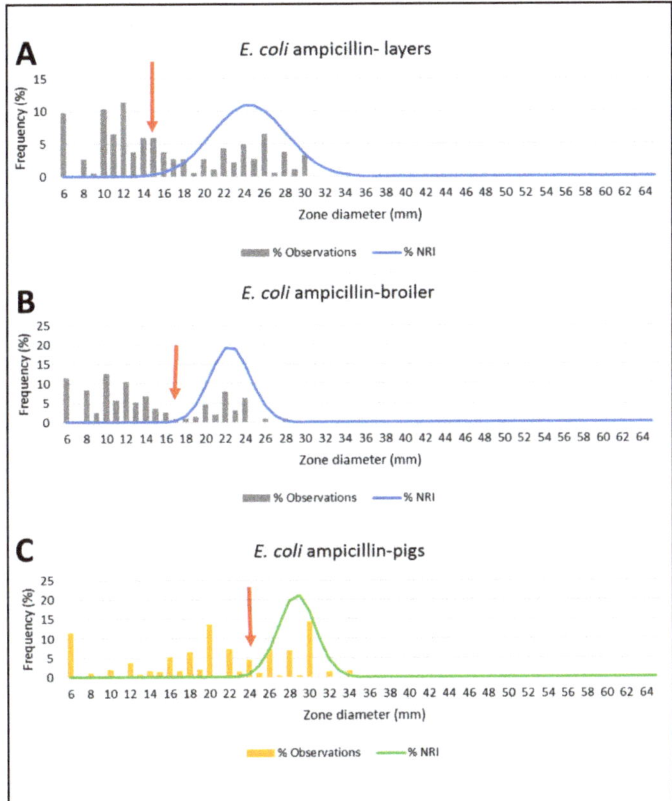

Figure 2. The ECVs of AMP tested against *E. coli* isolated from layers, broilers, and pigs. The arrow indicates where the ECV is located in a histogram, and distribution of WT and non-WT *E. coli* as determined by NRI. Subfigure (**A**) shows ECV of AMP tested against *E. coli* from layers which is 15 mm; subfigure (**B**) shows ECV of AMP tested against *E. coli* from broilers which is 17 mm; and subfigure (**C**) shows ECV of AMP tested against *E. coli* from pigs which is 24 mm.

There is a varying proportion of wild type (WT) distribution when a human clinical breakpoint is used compared to when ECVs are used. Notably, clinical breakpoints under-report percentages of WT-*E. coli* for CIP, TE, MEM, FEP, and CAZ in poultry, whereas clinical breakpoints over-report percentages of WT *E. coli* for AMP, CN, and SXT in pigs (Table 5).

Table 5. Comparison of ECVs and CLSI clinical breakpoints of antibiotics tested against *E. coli* isolated from poultry and pigs.

Antimicrobial Agents	Broilers: ECVs (%WT)	Broilers: CBs (%S)	Layers: ECVs (%WT)	Layers: CBs (%S)	Pigs: ECVs(%WT)	Pigs: CBs (%S)
CIP	31.1	5.3	51.1	10.3	84.5	76.3
TE	51.6	12.6	48.4	10.3	70.3	50.5
AMP	29.5	29.5	57.6	37.5	38.5	71.7
MEM	100	96.8	96.2	94.02	98.6	86.9
CAZ	99.5	81.1	95.7	56.5	97.2	73.1
FEP	100	81.6	97.3	61.4	96.8	75.3
CN	77.4	76.8	95.1	77.7	83.7	89.4
SXT	36.8	31.1	41.3	12.5	63.3	77.7

Key: CIP = ciprofloxacin; TE = tetracycline; AMP = ampicillin; MEM = meropenem; CAZ = ceftazidime; FEP = cefepime; CN = gentamicin; SXT = trimethoprim-sulfamethoxazole; ECVs = epidemiological cutoff values; WT = wild type as per ECVs; CBs = clinical breakpoints; and S = susceptible as per clinical breakpoints.

3.6. Factors Associated with MDR Colonization

The chi square test showed poultry (broilers and layers) are significantly colonized with MDR *E. coli* ($p < 0.001$), furthermore, exotic breeds were significantly colonized with MDR *E. coli* ($p < 0.001$), and the recency of the antimicrobial use predicts the colonization of MDR ($p < 0.001$) (Table 6).

Table 6. Factors associated with MDR colonization.

Variables		MDR Positive *n* (%)	MDR Negative *n* (%)	CHI ANALYSIS χ^2	*p* Value
Livestock	Broiler	164 (86.3)	26 (13.7)		
	Layers	161 (87.5)	23 (12.5)	210.24	<0.001
	Pigs	90 (31.8)	193 (68.2)		
Breed	Exotic	353 (75.6)	114 (24.4)	107.11	<0.001
	Local	62 (32.6)	128 (67.4)		
Recent antimicrobial use	2 months ago	22 (52.4)	20 (47.6)		
	1 month ago	22 (78.6)	6 (21.4)		
	2 weeks ago	89 (91.8)	8 (8.3)	153.51	<0.001
	1 week ago	68 (86.1)	11 (13.9)		
	Not known	214 (52.1)	197 (47.9)		

4. Discussion

Our findings showed different levels of resistance and different ECVs to commonly used antimicrobials in livestock among *E. coli* isolated from poultry and pigs. Most of the animals enrolled in this study (95.1%) were colonized with *E. coli,* similar to a study by Kimera et al., which observed colonization in 80.5% of animals [10] The majority of livestock enrolled were exotic breeds (71.4%) compared to a study done by Seni et al., who enrolled more local breeds (66.7%) [8]. Furthermore, exotic breeds were exposed to antimicrobial use more than local breeds. Generally, about two-thirds (63.2%) of *E. coli* isolated from livestock in our study were MDR, similar to a study done in Morogoro, which reported a prevalence of 65.1% [24]. A study done in Dar es Salaam reported low MDR prevalence (51.6%), which is lower than in the current study [10] The high prevalence of MDR in our study might be due to variations in farming conditions and antibiotic use. Furthermore, the level of MDR in the present study is lower than what was observed in China (83%) [25]. This might be explained by the fact that there is different antimicrobial exposure time and frequency, and China is one of the largest users of livestock antimicrobials in the world, increasing the antimicrobials selection pressure as a result of high MDR prevalence [25,26].

It was observed that over three-quarters and one-third of *E. coli* isolated from poultry and pigs were MDR, respectively. As documented previously [10] the level of MDR was significantly higher in poultry (in broilers and layers) than in pigs because there is a higher antimicrobial use in poultry keeping than in keeping pigs, as observed in this and other studies [24,27,28].

We observed more resistance to antibiotics of class quinolones, penicillin, tetracycline, and sulfonamides. High resistance to antibiotics of these classes was not surprising because these were the antibiotics reported to be commonly used by the livestock keepers in this study. MDR patterns observed in this study are in line with what was previously reported in Tanzania (quinolones, penicillin, tetracycline, and sulfonamides), China (tetracycline, sulfonamides, penicillin, quinolones), and Nigeria (tetracycline, sulfonamides, penicillin) [7,10,25,29].

ESBL production is one of the commonest MDR phenotypes. In this study, it was observed that about one-fifth of *E. coli* isolates from poultry and pigs were ESBL producers. Similar to a previous study done in the same region, Mwanza, Tanzania, reported a prevalence of 21.7% [8]. However, the proportion of ESBL producing *E. coli* in our study is lower than in a study done in another region in Tanzania, which reported a prevalence of 65.3% [10]. This might be due to different exposure times to beta-lactam and cephalosporin in livestock keeping, and different *E. coli* populations in different geographic locations.

ECVs determined from NRI are based on the assumption that *E. coli* isolated from broilers, layers, and pigs carrying resistance mechanism/non-wild type exhibit smaller inhibition zone diameters (IZD) than *E. coli* without resistance mechanism/wild type. To the best of our knowledge, this is the first study in Mwanza, Tanzania, to report the ECVs to the commonly used antibiotics tested against *E. coli* isolated from poultry and pigs.

According to this study, the majority of *E. coli* from poultry and pigs were wild type meropenem, ceftazidime, cefepime, and gentamicin. This could be explained by the fact that these antibiotics are not readily available over the counter; that they are expensive (meropenem), while the gentamicin, ceftazidime, and cefepime mode of administration is intravenous (IV), which is not easy for the livestock keepers. This is further supported by the fact that livestock keepers did not report using these antibiotics.

The majority of *E. coli* from broilers were non-wild type to ciprofloxacin, trimethoprim-sulfamethoxazole, and ampicillin, while the majority of *E. coli* from non-wild type layers to tetracycline and trimethoprim-sulfamethoxazole. The majority of *E. coli* from pigs were non-wild type to ampicillin. This is not surprising since these antibiotics are commonly used, readily available over the counter, and the cheapest [9]. Furthermore, most livestock keepers from this study reported commonly using antibiotics from tetracycline, quinolones, sulfonamides categories.

A study in China reported 92.3% of *E. coli* from chicken respiratory tract infections were non-wild type to danofloxacin (quinolones), while 22.3% were non-wild type to apramycin (aminoglycosides). Similar observations were seen in this study in *E. coli* from broilers, where the majority were non-wild type to ciprofloxacin (quinolones), and the majority of *E. coli* from poultry were wild type to gentamicin (aminoglycosides) [27]. A study done by Yang et al. reported ECV of danofloxacin (quinolones) tested against *E. coli* from pigs to be 8 µg/mL using MIC [28].

However, contrary to poultry, most *E. coli* from pigs were wild type to most antibiotics tested. This can be explained by the fact that pigs are not exposed to high antimicrobial use compared to poultry. Furthermore, some antibiotics reported to be used in poultry were not reported to be used in pigs in this study, such as quinolones, macrolides, and aminoglycosides.

This study observed different ECVs to the same antibiotics tested to *E. coli* from poultry and pigs, meaning different *E. coli* population, this is not astonishing since poultry and pigs are exposed to different antibiotics with different frequencies of use. Contrary to what was observed in Sweden, the wild type distribution of *E. coli* isolated from human and wild birds in the same area was identical [23].

This study observed that human clinical breakpoints (CLSI) could lead to both over and under-reporting antimicrobial resistance burdens. Similar observations were made by Doidge et al. [11] in the UK, although it was in sheep and beef.

MDR colonization can be attributed to different factors. The most common is antimicrobial pressure which causes the selection of resistant bacteria. In this study, it was observed that exotic breeds were found to predict MDR colonization among poultry and pigs, and there was a significant association of MDR colonization to broilers and layers. This is not surprising as it might be contributed by the fact that there is high antimicrobial use in exotic breeds and poultry keeping. A study done by Seni et al. observed that exotic breeds were more at risk of being colonized with ESBL bacteria than local breeds [8]. ESBL is one of the common MDR phenotypes. Similar to what was observed in the current study, it was observed in a study done by Nonga et al., Katakweba et al., and Kimera et al. that poultry farming is associated with uncontrolled use of both veterinary and human antimicrobials [5,24,28]. A systematic review by Mshana et al. also reported that poultry production accounts for high antimicrobial use in Africa [7]. Recent antimicrobial use was found to predict MDR colonization in the present study. This can be explained by the fact that antimicrobial use creates selection pressure allowing resistant bacteria to multiply and propagate. It was reported that uncontrolled use of antimicrobials in livestock keepings as a growth factor, prophylaxis, and/or therapeutics had been associated with the emergence of MDR bacteria [29].

5. Conclusions

E. coli from layers and broilers are more resistant to commonly used antibiotics than *E. coli* isolates from pigs. Distinct populations of *E. coli* were circulating in layers, broilers, and pigs based on ECVs of different antibiotics tested, which was likely due to differences in antibiotic exposure and breeding type. Furthermore, recent antimicrobial use and exotic breeds predicted MDR colonization which might result from high antimicrobial use. Further studies involving other laboratories are needed to establish the ECVs for commonly used antibiotics and the data used to monitor resistance and further research to establish veterinary breakpoints. Veterinary officers should implement regulations that prohibit the inappropriate use of antimicrobial agents in livestock keeping. More studies to establish the genotypes of *E. coli* circulating in these species are warranted to provide data to monitor the emergence of new *E. coli* strains/genotypes.

6. Study Limitation

Limited knowledge of livestock keepers on antimicrobial used in their livestock may impact statistical analysis, particularly associated with ESBL carriage. In addition, the ECVs data are from a single laboratory, and, therefore, should be carefully interpreted to reflect the local settings. The resistance frequency may be underestimated, as only one colony per sample was analyzed.

Author Contributions: C.I.M., H.N., M.M.M. and S.E.M. conceptualized this study; C.I.M., R.J.B. and J.M.G. worked on the process of sample collection; C.I.M., H.N. and V.S. performed sample processing; C.I.M., V.S., H.N., M.F.M., J.M., E.K., M.M.M. and S.E.M. performed data analysis, C.I.M. wrote the first draft of the manuscript, M.F.M., E.K., M.M.M., J.M. and S.E.M. critically reviewed the manuscript and all authors approved the final manuscript. All authors have read and agreed to the published version of the manuscript.

Funding: This study was in-parts funded by SNAP-AMR (MR/S004815/1) and HATUA (MR/S004785/1) projects at CUHAS.

Institutional Review Board Statement: The study was conducted in accordance with the Declaration of Helsinki, and approved by the Institutional Review Board of the joint CUHAS/BMC research ethics and review committee (CREC/474/2021).

Informed Consent Statement: Not applicable.

Data Availability Statement: The data presented in this study are available on request from the corresponding author.

Acknowledgments: We would like to thank the livestock keepers (farmers) for granting permission for their livestock to participate in the study.

Conflicts of Interest: The authors declare no conflict of interest.

References

1. Van Boeckel, T.P.; Brower, C.; Gilbert, M.; Grenfell, B.T.; Levin, S.A.; Robinson, T.P.; Teillant, A.; Laxminarayan, R. Global trends in antimicrobial use in food animals. *Proc. Natl. Acad. Sci. USA* **2015**, *112*, 5649–5654. [CrossRef] [PubMed]
2. Kolář, M.; Urbánek, K.; Látal, T. Antibiotic selective pressure and development of bacterial resistance. *Int. J. Antimicrob. Agents* **2001**, *17*, 357–363. [CrossRef]
3. Marshall, B.M.; Levy, S.B. Food animals and antimicrobials: Impacts on human health. *Clin. Microbiol. Rev.* **2011**, *24*, 718–733. [CrossRef]
4. Millner, P.D. Manure management. In *The Produce Contamination Problem*; Elsevier: Amsterdam, The Netherlands, 2009; pp. 79–104.
5. Nonga, H.; Mariki, M.; Karimuribo, E.; Mdegela, R. Assessment of antimicrobial usage and antimicrobial residues in broiler chickens in Morogoro Municipality, Tanzania. *Pak. J. Nutr.* **2009**, *8*, 203–207. [CrossRef]
6. Caudell, M.A.; Quinlan, M.B.; Subbiah, M.; Call, D.R.; Roulette, C.J.; Roulette, J.W.; Roth, A.; Matthews, L.; Quinlan, R.J. Antimicrobial use and veterinary care among agro-pastoralists in Northern Tanzania. *PLoS ONE* **2017**, *12*, e0170328. [CrossRef]
7. Mshana, S.E.; Sindato, C.; Matee, M.I.; Mboera, L.E. Antimicrobial Use and Resistance in Agriculture and Food Production Systems in Africa: A Systematic Review. *Antibiotics* **2021**, *10*, 976. [CrossRef]
8. Seni, J.; Falgenhauer, L.; Simeo, N.; Mirambo, M.M.; Imirzalioglu, C.; Matee, M.; Roth, A.; Matthews, L.; Quinlan, R.J. Multiple ESBL-producing *Escherichia coli* sequence types carrying quinolone and aminoglycoside resistance genes circulating in companion and domestic farm animals in Mwanza, Tanzania, harbor commonly occurring plasmids. *Front. Microbiol.* **2016**, *7*, 142. [CrossRef]
9. Kimera, Z.I.; Mshana, S.E.; Rweyemamu, M.M.; Mboera, L.E.; Matee, M.I. Antimicrobial use and resistance in food-producing animals and the environment: An African perspective. *Antimicrob. Resist. Infect. Control.* **2020**, *9*, 37. [CrossRef]
10. Kimera, Z.I.; Mgaya, F.X.; Misinzo, G.; Mshana, S.E.; Moremi, N.; Matee, M.I.N. Multidrug-Resistant, Including Extended-Spectrum Beta Lactamase-Producing and Quinolone-Resistant, *Escherichia coli* Isolated from Poultry and Domestic Pigs in Dar es Salaam, Tanzania. *Antibiotics* **2021**, *10*, 406. [CrossRef]
11. Doidge, C.; West, H.; Kaler, J. Antimicrobial Resistance Patterns of *Escherichia coli* Isolated from Sheep and Beef Farms in England and Wales: A Comparison of Disk Diffusion Interpretation Methods. *Antibiotics* **2021**, *10*, 453. [CrossRef]
12. Varga, C.; Rajić, A.; McFall, M.E.; Reid-Smith, R.J.; Deckert, A.E.; Pearl, D.L.; Avery, B.P.; Checkley, S.L.; McEwen, S.A. Comparison of antimicrobial resistance in generic *Escherichia coli* and *Salmonella* spp. cultured from identical fecal samples in finishing swine. *Can. J. Vet. Res.* **2008**, *72*, 181. [PubMed]
13. Organization WHO. *WHO Global Principles for the Containment of Antimicrobial Resistance in Animals Intended for Food: Report of a WHO Consultation with the Participation of the Food and Agriculture Organization of the United Nations and the Office International Des Epizooties*; World Health Organization: Geneva, Switzerland, 2000.
14. Murray, B.E. Antibiotic resistance. *Adv. Intern. Med.* **1997**, *42*, 339–367. [PubMed]
15. van den Bogaard, A.E.; Stobberingh, E.E. Epidemiology of resistance to antibiotics: Links between animals and humans. *Int. J. Antimicrob. Agents* **2000**, *14*, 327–335. [CrossRef]
16. Aarestrup, F.M.; Bager, F.; Jensen, N.; Madsen, M.; Meyling, A.; Wegener, H.C. Resistance to antimicrobial agents used for animal therapy in pathogenic-, zoonotic-and indicator bacteria isolated from different food animals in Denmark: A baseline study for the Danish Integrated Antimicrobial Resistance Monitoring Programme (DANMAP). *Apmis* **1998**, *106*, 745–770. [CrossRef] [PubMed]
17. Procop, G.W.; Church, D.L.; Hall, G.S.; Janda, W.M. *Koneman's Color Atlas and Textbook of Diagnostic Microbiology*; Jones & Bartlett Publishers: Burlington, MA, USA, 2020.
18. Biemer, J.J. Antimicrobial susceptibility testing by the Kirby-Bauer disc diffusion method. *Ann. Clin. Lab. Sci.* **1973**, *3*, 135–140.
19. CLSI. *Performance Standards for Antimicrobial Susceptibility Testing*, 30th ed.; CLSI: Wayne, PA, USA, 2020.
20. Silago, V.; Kovacs, D.; Samson, H.; Seni, J.; Matthews, L.; Oravcová, K.; Lupindu, A.M.; Hoza, A.S.; Mshana, S.E. Existence of Multiple ESBL Genes among Phenotypically Confirmed ESBL Producing *Klebsiella pneumoniae* and *Escherichia coli* Concurrently Isolated from Clinical, Colonization and Contamination Samples from Neonatal Units at Bugando Medical Center, Mwanza, Tanzania. *Antibiotics* **2021**, *10*, 476.
21. Turnidge, J.; Kahlmeter, G.; Kronvall, G. Statistical characterisation of bacterial wild-type MIC value distributions and the determination of epidemiological cut-off values. *Clin. Microbiol. Infect.* **2006**, *12*, 418–425. [CrossRef]

22. Kronvall, G.; Karlsson, I.; Walder, M.; Sörberg, M.; Nilsson, L.E. Epidemiological MIC cut-off values for tigecycline calculated from Etest MIC values using normalized resistance interpretation. *J. Antimicrob. Chemother.* **2006**, *57*, 498–505. [CrossRef]

23. Sjölund, M.; Bengtsson, S.; Bonnedahl, J.; Hernandez, J.; Olsen, B.; Kahlmeter, G. Antimicrobial susceptibility in *Escherichia coli* of human and avian origin—A comparison of wild-type distributions. *Clin. Microbiol. Infect.* **2009**, *15*, 461–465. [CrossRef]

24. Katakweba, A.A.; Muhairwa, A.P.; Lupindu, A.M.; Damborg, P.; Rosenkrantz, J.T.; Minga, U.M.; Mtambo, M.M.A.; Olsen, J.E. First report on a randomized investigation of antimicrobial resistance in fecal indicator bacteria from livestock, poultry, and humans in Tanzania. *Microb. Drug Resist.* **2018**, *24*, 260–268. [CrossRef]

25. Yassin, A.K.; Gong, J.; Kelly, P.; Lu, G.; Guardabassi, L.; Wei, L.; Han, X.; Qiu, H.; Price, S.; Cheng, D.; et al. Antimicrobial resistance in clinical *Escherichia coli* isolates from poultry and livestock, China. *PLoS ONE* **2017**, *12*, e0185326. [CrossRef] [PubMed]

26. Hvistendahl, M. *China Takes Aim at Rampant Antibiotic Resistance*; American Association for the Advancement of Science: Washington, DC, USA, 2012.

27. Zhang, H.-L.; Wu, S.-L.; Fu, J.-L.; Jiang, H.-X.; Ding, H.-Z. Research Note: Epidemiological cutoff values and acquired resistance mechanisms of three veterinary antibiotics against *Escherichia coli* from chicken respiratory tract infections. *Poult. Sci.* **2021**, *100*, 1093–1097. [CrossRef] [PubMed]

28. Yang, Y.; Zhang, Y.; Li, J.; Cheng, P.; Xiao, T.; Muhammad, I.; Yu, H.; Liu, R.; Zhang, X. Susceptibility breakpoint for Danofloxacin against swine *Escherichia coli*. *BMC Vet. Res.* **2019**, *15*, 51. [CrossRef] [PubMed]

29. Nguyen, V.T.; Carrique-Mas, J.J.; Ngo, T.H.; Ho, H.M.; Ha, T.T.; Campbell, J.I.; Nguyen, T.N.; Hoang, N.N.; Pham, V.M.; Wagenaar, J.A.; et al. Prevalence and risk factors for carriage of antimicrobial-resistant *Escherichia coli* on household and small-scale chicken farms in the Mekong Delta of Vietnam. *J. Antimicrob. Chemother.* **2015**, *70*, 2144–2152. [CrossRef] [PubMed]

Article

Metabolomic Profiles of Multidrug-Resistant *Salmonella* Typhimurium from Humans, Bovine, and Porcine Hosts

Jessie M. Overton [1], Lyndsey Linke [1], Roberta Magnuson [1], Corey D. Broeckling [2] and Sangeeta Rao [1,*]

[1] Department of Clinical Sciences, College of Veterinary Medicine and Biomedical Sciences, Colorado State University, Fort Collins, CO 80523, USA; jessie.earhart@colostate.edu (J.M.O.); lyndsey.linke@colostate.edu (L.L.); magnusonroberta@gmail.com (R.M.)

[2] Analytical Resources Core: Bioanalysis and Omics, Colorado State University, Fort Collins, CO 80523, USA; corey.broeckling@colostate.edu

* Correspondence: sangeeta.rao@colostate.edu

Simple Summary: The global threat that is imposed by the resistance the pathogens develop to antimicrobial drugs is escalating. Tools to detect the resistance (with evidence on molecular and cellular outcomes) would reveal intricate mechanisms through which novel drugs could be developed. Approaches such as metabolomics, which involve metabolite detection, provide scientific evidence of metabolite expression of antimicrobial-resistant pathogens. The current study involved metabolomics of antimicrobial-resistant *Salmonella* Typhimurium collected from various hosts (human, porcine, bovine) and were exposed to antimicrobial drugs—ampicillin, chloramphenicol, streptomycin, sulfisoxazole, and tetracycline—as one set of the experiment. The same isolates were also cultured with no drug exposure as a comparison. There are certain pathways of metabolite expression that are impacted by drug exposure when compared to no drug exposure, meaning that the expressed metabolites could be potential targets for drug companies for the treatment of antimicrobial-resistant pathogens.

Citation: Overton, J.M.; Linke, L.; Magnuson, R.; Broeckling, C.D.; Rao, S. Metabolomic Profiles of Multidrug-Resistant *Salmonella* Typhimurium from Humans, Bovine, and Porcine Hosts. *Animals* **2022**, *12*, 1518. https://doi.org/10.3390/ani12121518

Academic Editors: Paola Roncada and Bruno Tilocca

Received: 1 April 2022
Accepted: 7 June 2022
Published: 10 June 2022

Publisher's Note: MDPI stays neutral with regard to jurisdictional claims in published maps and institutional affiliations.

Abstract: Antimicrobial resistance (AMR) is a global public health threat, yet tools for detecting resistance patterns are limited and require advanced molecular methods. Metabolomic approaches produce metabolite profiles and help provide scientific evidence of differences in metabolite expressions between *Salmonella* Typhimurium from various hosts. This research aimed to evaluate the metabolomic profiles of *S.* Typhimurium associated with AMR and it compares profiles across various hosts. Three samples, each from bovine, porcine, and humans (total *n* = 9), were selectively chosen from an existing library to compare these nine isolates cultured under no drug exposure to the same isolates cultured in the presence of the antimicrobial drug panel ACSSuT (ampicillin, chloramphenicol, streptomycin, sulfisoxazole, tetracycline). This was followed by metabolomic profiling using UPLC and GC–mass spectrometry. The results indicated that the metabolite regulation was affected by antibiotic exposure, irrespective of the host species. When exposed to antibiotics, 59.69% and 40.31% of metabolites had increased and decreased expressions, respectively. The most significantly regulated metabolic pathway was aminoacyl-tRNA biosynthesis, which demonstrated increased expressions of serine, aspartate, alanine, and citric acid. Metabolites that showed decreased expressions included glutamate and pyruvate. This pathway and associated metabolites have known AMR associations and could be targeted for new drug discoveries and diagnostic methods.

Keywords: antimicrobial resistance; metabolomics; metabolites; *Salmonella* Typhimurium; resistance markers

1. Introduction

Antimicrobial resistance (AMR) in bacteria isolated from animal hosts is a major global public health threat. The Centers for Disease Control and Prevention have determined that AMR is "one of the greatest public health challenges of our time" [1]. One of the key goals for slowing or decreasing AMR identified by the White House in the National Strategy for Combatting Antibiotic-Resistant Bacteria is to "accelerate basic and applied research and development for new antibiotics … " [2]. However, current tools for detecting phenotypic resistance patterns are limited and require advanced molecular methods to reveal associations with AMR patterns [3,4].

Metabolomics is a relatively new tool that can be used to construct metabolite profiles and these metabolite patterns provide evidence of metabolite regulation at the cellular level. When bacteria are exposed to antibiotics, this exposure can trigger cellular changes within the bacterial cell that results in specific metabolic patterns that can help predict antimicrobial drug resistance profiles. Such predictability can have an immediate impact on human and animal health by leading to advancements in drug discoveries by targeting the expression of certain metabolites, as well as diagnostic tools to screen large numbers of samples for AMR. Identifying possible new drug targets would help pharmaceutical companies develop more specific and effective antibiotics to combat AMR bacterial infections [4].

Recent studies have shown that bacteria produce specific metabolic fingerprints when exposed to different classes of antibiotics. These fingerprints can help predict the mode of action used by antibiotics [4] to help develop any novel therapies. For many years, the development of antibiotic resistance could be partly explained by the synthesis of novel analogues of existing compounds [4]. However, such chemical modifications are finite, to keep pace with the remarkable adaptability of the bacteria when exposed to these selective drug pressures in the environment. To combat the prevalence of multidrug-resistant (MDR) pathogens, novel antibiotics that target distinct cellular functions are needed [5]. Better understanding the metabolic patterns in AMR bacteria to expose new cellular functions associated with drug resistance and susceptibility is one way to identify new drug targets. One of these studies also suggested that a core metabolic profile for each bacterium is identifiable regardless of the environmental condition, suggesting bacteria could be identified using in vitro metabolic profiles whether in a wound, on surgical equipment, or in the environment [6].

Salmonella is a rod-shaped, Gram-negative bacillus that belongs to the Enterobacteriaceae family. This organism is of high public health importance due to its ability to cause several syndromes in both animals and humans. Enteritis, septicemia, abortion, and asymptomatic miscarriages are the major syndromes that present in animals. Enteric fever, gastroenteritis, septicemia, and focal infections are the major syndromes that present in humans infected with *Salmonella* bacteria [7]. *Salmonella enterica* serovar Typhimurium is of utmost importance to public health due to its ability to infect human hosts via contaminated foods. It is one of the most identified serovars in cattle, humans, and pigs and has displayed resistance to ampicillin, chloramphenicol, sulfamethoxazole, and tetracycline [8].

The aim of this project was to identify cellular biomarkers (metabolites) associated with mechanisms of AMR in *Salmonella* Typhimurium using metabolomics and investigate the diversity of those markers among established genetic patterns of resistance in *S.* Typhimurium isolated from humans, bovine, and porcine samples. Metabolomics can be used as a tool to identify the cellular effects of AMR in this pathogen of public health importance. This research has two specific aims (1) to establish metabolomic profiles of *Salmonella* Typhimurium isolated from humans, porcine, and bovine and cultured in the presence and absence of an ACSSuT panel of drugs, and (2) to evaluate similarities and differences in these metabolomic profiles in *Salmonella* Typhimurium across isolates originating from humans, porcine, and bovine hosts. The hypothesis of the study is that non-targeted metabolite profiling will identify biomarker profiles distinctive of AMR in *S.* Typhimurium and, more specifically, the metabolite patterns will differ across various host species.

2. Materials and Methods

2.1. Isolate Collection, Screening, Identification, and Growth

Salmonella isolates from various institutes (*Salmonella* Typhimurium isolates were contributed by the Colorado Department of Public Health and Environment, CSU-Veterinary Diagnostic Laboratory, Ohio State University, University of Illinois, University of Pennsylvania, and Washington State University) in the US were shipped directly to the Animal Population Health Institute laboratory. A total of 88 human, 33 bovine, and 36 porcine isolates were screened to verify proper serovar typing belonging to *Salmonella* Typhimurium. Briefly, samples were streaked for isolation onto blood agar plates containing 5% sheep blood and incubated overnight at 37 °C. A single colony was first tested with the *Salmonella* O Antiserum group Poly A-I, & Vi, and then *Salmonella* O Antiserum Group B, factors 1, 4, 5, 12 (BD Diagnostic Systems, Fisher Scientific, Hampton, NH, USA). After antibody confirmation, the *Salmonella* Typhimurium isolates were grown in 1 mL of trypticase soy broth (TSB), and generated stocks were frozen at −80 °C in 10% sterile glycerol.

2.2. Integron and AMR Testing

A portion of each *Salmonella* Typhimurium stock isolate was scraped into a separate microcentrifuge tube, thawed, and centrifuged for 5 min at $5000 \times g$. The supernatant was removed, and each pellet was resuspended in molecular grade water in a 1:3 ratio (10 μL cell pellet suspended in 30 μL of water). A total of 5 μL of each washed, resuspended isolate was used as a template and added to the following PCR mastermix for a 25-μL total reaction volume: 2.5 μL $1\times$ Amplitaq Gold Buffer II and 1.5 mM $MgCl_2$ (Applied Biosystems, Foster City, CA, USA), 0.8 mM dNTPs (0.2 mM each) (Roche Applied Sciences, Indianapolis, IN, USA), 0.4 μM of each primer (Int forward primer sequence: 5′-GGC ATC CAA GCA GCA AGC-3′; Int reverse primer sequence: 5′-AAG CAG ACT TGA CCT GAT-3′), 1.875 U Amplitaq Gold polymerase (Applied Biosystems, Foster City, CA, USA), and 2.5 μL $5\times$ Q-Solution (Qiagen, Valencia, CA, USA).

The primers amplify the variable region between the 5′CS to 3′CS region of class 1 integrons [9]. Each reaction was overlaid with 30 μL of Chill Out wax (Bio-Rad, Hercules, CA, USA) to prevent evaporation and placed into an MJ Research 60 place thermal cycler (Bio-Rad). Thermal cycling conditions consisted of an initial incubation at 94 °C for 10 min to activate the polymerase and lyse cells, followed by 35 cycles of 94 °C for 30 s, 54 °C for 1 min, 72 °C for 1.5 min, and a final extension incubation at 72 °C for 10 min.

PCR products were analyzed by agarose gel electrophoresis using the FlashGel® DNA System (Lonza Group, Ltd., Basel, Switzerland) and visualized by UV light transillumination. A 100 bp–4 kb molecular weight marker (Lonza Group, Ltd., Basel, Switzerland) was concordantly run on the gel as a ladder to aid in the calculation of the size of the amplified DNA fragments. A positive control sample generated from purified DNA from two isolates previously analyzed [9] for class 1 integrons and containing integron sizes of 1000, 1200, and 1600 was included (5 pg total) with each PCR and gel. Samples containing integron sizes of 1000, 1200, 1600, 1800, or both 1000 + 1200 bp were recorded and subsequently re-run on a 1% agarose gel containing a marker and a positive control for proper band size identification. Integron bands were excised from the gel and submitted for DNA purification using the QIAquick PCR Purification kit (Qiagen, Hilden, Germany).

All *Salmonella* Typhimurium isolates used in this study were tested for susceptibility to 16 antimicrobial agents by the disk diffusion assay according to CLSI standard procedures. The AMR testing panel consisted of the following sixteen antimicrobial drugs—amoxicillin–clavulanate (AMC-30), cephalothin (CF-30), chloramphenicol (C-30), ampicillin (AM-10), ceftiofur (CTO-30), enrofloxacin (ERF-5), streptomycin (S-10), triple sulfa (SSS-0.25), tetracycline (TE-30) sulfamethoxazole/trimethoprim (SXT 23.75–1.25), cefoxitin (FOX-30), ciprofloxacin (CIP-5), florfenicol (FFC-30), gentamicin (GM-10), kanamycin (K-30), and nalidixic acid (NA-30). *Escherichia coli* (*E. coli*) ATCC 25922 and *Staphylococcus aureus* ATCC 25923 were used as quality controls.

2.3. Isolate Growth and Extraction for Proteomic and Metabolomics Profiling

Nine *S.* Typhimurium isolates (three human, three porcine, and three bovine) were selected to undergo an antimicrobial drug growth challenge followed by a non-targeted metabolomics analysis. Criteria for selection were the presence of both 1000 and 1200 base pair integrons, and matching susceptibility/resistance profiles across the 16 drugs tested. Five drugs, ampicillin, chloramphenicol, streptomycin, sulfisoxazole, and tetracycline (ACSSuT panel; Sigma Aldrich, St. Louis, MO, USA), were selected for the *S.* Typhimurium antimicrobial drug challenge.

Ampicillin, chloramphenicol, tetracycline, and streptomycin were each dissolved in water to the desired stock concentration. Sulfisoxazole was added to 10% HCl and heated at 80 °C until dissolved. The sulfisoxazole–acid mix was added to TSB, the broth was neutralized to pH 7.0 using NaOH, and the other antibiotics were subsequently added. The final concentration of each antibiotic was based on the recommended minimum inhibitory concentration (MIC) recommended by the Clinical Laboratory and Standards Institute, as shown in Table 1.

Table 1. Recommended MIC values for *Salmonella enterica* serotype Typhimurium for the ACSSuT pattern according to 2014 CLSI standards.

Drug Panel	MIC
Ampicillin	32 µg/mL
Chloramphenicol	32 µg/mL
Streptomycin	64 µg/mL
Sulfisoxazole	512 µg/mL
Tetracycline	16 µg/mL

Isolates were processed using standard laboratory procedures. They were thawed and streaked for isolation on sheep blood agar plates. One resulting colony from each selected isolate was suspended in 0.5 mL of TSB; 100 µL was inoculated into 20 mL of normal TSB (no drug = ND) and 100 µL was inoculated into 20 mL of ACSSuT TSB (Drug = D). The only difference between the ND and D was that the ND group of cultures were without antimicrobials. Cultures were then incubated with shaking at 37 °C for 24 h. After pelleting at $4300 \times g$ for 10 min at 4 °C and supernatant removal, the wet weight of each culture pellet was recorded and adjusted to 20 mg. Pellets were washed with phosphate-buffered saline (PBS) and centrifuged again as above; after discarding PBS supernatant, the pellets were frozen at −20 °C. Each sample pellet was thawed at 4 °C, suspended in methyl tert-butyl ether (MTBE), and sonicated for 30 s intervals for a total of 6 cycles, with a 30 s cooling on ice between cycles. The sonicated lysates were then centrifuged at $2500 \times g$ for 5 min at 4 °C, and 150 µL of LC-MS grade water and an additional 100 µL MTBE was added to the cleared supernatants. After sealing with Parafilm, sample tubes were vortexed at room temperature for 15 min, incubated at −80 °C for 15 min, and centrifuged at $15,890 \times g$ for 15 min at 4 °C. Samples were then divided by a non-polar supernatant, a polar supernatant, and protein lysates. Each layer was dried via nitrogen gas and stored at −80 °C for metabolomics analysis.

2.4. Metabolomic Profiling by UPLC- and GC–MS

An ultra-performance liquid chromatography–mass spectrometry (UPLC-MS) analysis was performed on a Waters Xevo G2-TOF MS coupled with a Waters Acquity UPLC [10]. Separation was performed on a UPLC T3 reverse phase column and data were collected in MSE mode (alternating low and high collision energy) [11]. For the gas chromatography–mass spectrometry (GC–MS) analysis, cell extracts were dried and derivatized using a standard protocol. Briefly, GC–MS data were acquired on a Thermo Scientific Trace-ISQ GC–MS system (Waltham, MA, USA) with separation using a 30 m TG-5MS column. Data from both UPLC-MS and GC–MS acquisitions were processed using XCMS (https://www.bioconductor.org/packages/release/bioc/html/xcms.html, accessed on 31 March

2022) for peak detection, retention time alignment, and normalization [12]. Metabolite annotation of GC–MS data was performed by grouping molecular features into peak groups using AMDIS software (http://www.amdis.net/, accessed on 31 March 2022) and screening spectra against the CSU in-house spectral library, NIST GC–MS spectral library, and the Golm Metabolite Database (http://gmd.mpimp-golm.mpg.de/, accessed on 31 March 2022). Annotations of UPLC-MS data were performed by an unbiased grouping of molecular features into spectra based on correlational clustering across the dataset [10] and screening spectra against the CSU in-house spectral library (consisting of approximately 1100 compounds), NIST LC-MS spectral library, and MassBank spectral library [12].

2.5. Statistical Analysis (MetaboAnalyst 4.0)

Data analysis of the biomarkers was completed using MetaboAnalyst 4.0 (MetaboAnalyst 4.0 is available at https://www.metaboanalyst.ca/ (accessed on 26 April 2021) and its R packages are available at https://github.com/xia-lab/MetaboAnalystR, (accessed on 26 April 2021)). The UPLC and GC–MS spectra were combined, normalized, and scaled. To determine the statistically significant (S.S.) metabolites, a pairwise analysis was conducted, including a non-parametric Wilcoxon rank-sum test and fold-change analysis. A two-way analysis of variance (ANOVA) followed by a principal component analysis (PCA) and heatmapping were used to determine and visualize the species and drug effects and interactions. A pathway analysis was then conducted to match S.S. metabolites to known metabolic pathways and determine the biological significance of those pathways.

Multiple features of this program were used, including "Two-factor", "Statistical Analysis", and "Pathway Analysis", to conduct multiple statistical tests, including Wilcoxon rank-sum, fold-change, two-way ANOVA, PCA, and heatmapping. Conducting the "Pathway Analysis" in MetaboAnalyst required all metabolites to have an HMDB identifier. The Human Metabolome Database (HMDB) is a website that compiles detailed information about metabolites and their roles in human metabolic pathways and assigns HMDB identifiers or numbers.

3. Results

3.1. AMR Patterns and Integrons

The most common AMR pattern among all resistant samples (23/126 = 18.3%) was ampicillin, amoxicillin–clavulanate, streptomycin, sulfonamides, tetracycline, chloramphenicol, and florfenicol (coded as AMC-AM-S10-SSS-TE-C-FFC). All isolates with this AMR pattern carried both the 1000 and 1200 bp integrons.

3.2. Metabolite Expression by Drug Treatment and Host Species

Visualization by the principal component analysis (Figure 1) and the two-way ANOVA heatmap (Figure 2) showed that a greater effect on metabolite production was apparent when the samples were exposed to the full drug (ACSSuT panel) treatment, irrespective of species.

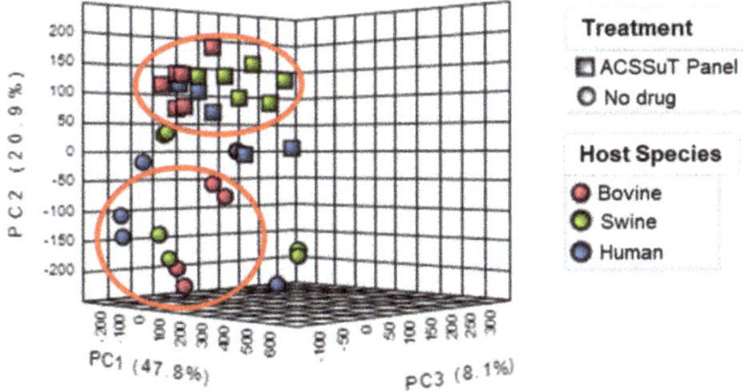

Figure 1. PCA chart derived from two–way ANOVA showing clustering of samples by drug treatment.

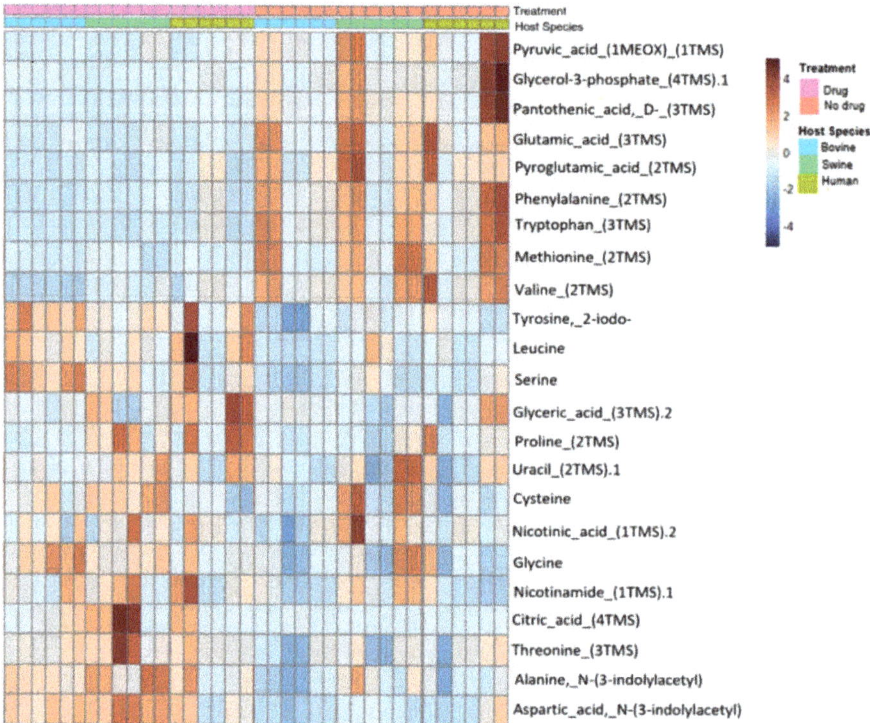

Figure 2. Heatmap derived from two–way ANOVA showing clustering of metabolite concentrations based on drug treatment.

3.3. Metabolite Expression and Matched Metabolic Pathways

Wilcoxon rank-sum showed 653 metabolites that had an S.S. concentration difference (59.69% increased and 40.31% decreased expressions) when the sample was exposed to the ACSSuT antibiotic panel versus when it was not. Of those 653 metabolites, 23 unique metabolites were annotated by the PMF, identifiable by HMDB, and matched to one or multiple of the 9 statistically significant metabolic pathways in MetaboAnalyst. Of these, 60.87% of metabolites had an increased expression when exposed to antibiotics and 39.13% had a decreased expression (Figure 2).

Methionine, nicotinamide, nicotinate, pantothenate, phenylalanine, proline, pyroglutamic acid, pyruvate, serine, threonine, tryptophan, tyrosine, uracil, and valine significantly increased with full drug treatment. Conversely, alanine, aspartate, citrate, cysteine, glutamate, glycerate, glycerone phosphate, glycine, and leucine decreased with full drug treatment.

The metabolic pathways matched to the significantly different metabolites include glycine, serine, and threonine metabolism; alanine, aspartate, and glutamate metabolism; aminoacyl-tRNA biosynthesis; pantothenate and CoA biosynthesis; glutathione metabolism; valine, leucine, and isoleucine biosynthesis; nicotinate and nicotinamide metabolism; glyoxylate and dicarboxylate metabolism; and beta-Alanine metabolism, in order of descending pathway impact scores (Table 2).

Table 2. Significant metabolites from the Wilcoxon rank-sum test with the metabolic pathway and the associated fold changes observed across host species.

Metabolic Pathway	Impact	p-Value	FDR	Metabolite	All Hosts Fold Change (FD/ND)	All Hosts p-Value	All Hosts FDR	Bovine Fold Change (FD/ND)	Bovine p-Value	Bovine FDR	Swine Fold Change (FD/ND)	Swine p-Value	Swine FDR	Human Fold Change (FD/ND)	Human p-Value	Human FDR
Aminoacyl-tRNA biosynthesis	0.2	1.93×10^{-7}	1.66×10^{-5}	Phenylalanine	0.565	0.0000581	0.00043499	0.433	0.0021645	0.0096991	0.737			0.535		
				Cysteine	1.515	0.0096309	0.023444	0.118	0.0021645	0.0096991	0.228	0.0021645	0.022605	1.400		
				Glycine	1.858	0.00000394	0.0000655	2.003	0.0021645	0.0096991	1.812			1.723		
				Aspartate	3.077	3.06×10^{-8}	0.00000283	3.174	0.0021645	0.0096991	3.883	0.0021645	0.022605	2.178		
				Serine	2.837	8.22×10^{-8}	0.00000426	4.469	0.0021645	0.0096991	1.755	0.0021645	0.022605	2.240		
				Methionine	0.594	0.020464	0.042605	0.498	0.0021645	0.0096991	0.588			0.708		
				Valine	0.675	0.0064022	0.017165	0.391	0.0021645	0.0096991	0.925			0.782		
				Alanine	2.065	3.06×10^{-8}	0.00000283	2.149	0.0021645	0.0096991	2.240	0.0021645	0.022605	1.910		
				Leucine	0.087	0.0023492	0.0080271	0.106	0.015152	0.041482	0.063	0.0021645	0.022605	0.116		
				Threonine	2.233	2.2×10^{-10}	0.000000285	2.065	0.0021645	0.0096991	3.156	0.0021645	0.022605	1.834		
				Tryptophan	2.143	0.0000164	0.00017127	1.789	0.0021645	0.0096991	2.671			1.647		
				Tyrosine	2.183	1.48×10^{-8}	0.00000212	2.932	0.0021645	0.0096991	2.054	0.0021645	0.022605	1.959		
				Proline	2.150	0.00075777	0.0033152	1.153			3.432			2.041		
				Glutamate	0.321	0.0000239	0.00021944	0.353			0.497			0.243		
Pantothenate and CoA biosynthesis	0.144	0.00304	0.10686	Pantothenate	0.414	0.000000	0.000005	0.440	0.0021645	0.0096991	0.389	0.0021645	0.022605	0.429		
				Valine	0.675	0.006402	0.017165	0.391	0.0021645	0.0096991	0.925			0.782		
				Aspartate	3.077	0.000000	0.000003	3.174	0.0021645	0.0096991	3.883	0.0021645	0.022605	1.300		
				Cysteine	1.515	0.009631	0.023444	0.118	0.0021645	0.0096991	0.228	0.0021645	0.022605	1.400		
				Pyruvate	0.458	0.002642	0.008841	0.382	0.008658	0.026381	1.570	0.0021645	0.022605	0.394		
				Uracil	1.477	0.020464	0.042605	0.977			1.848			1.474		
Glycine, serine, and threonine metabolism	0.456	0.00373	0.10686	Serine	2.837	0.000000	0.000004	4.469	0.0021645	0.0096991	1.755	0.0021645	0.022605	2.240		
				Glycine	1.858	0.000004	0.000066	2.003	0.0021645	0.0096991	1.812			1.723		
				Aspartate	3.077	0.000000	0.000003	3.174	0.0021645	0.0096991	3.883	0.0021645	0.022605	1.300		
				Glycerate	1.691	0.000503	0.002423	0.631	0.008658	0.026381	2.225			2.473		
				Threonine	2.233	0.000000	0.000000	2.065	0.0021645	0.0096991	3.156	0.0021645	0.022605	1.834		
				Pyruvate	0.458	0.002642	0.008841	0.382	0.008658	0.026381	1.570	0.0021645	0.022605	0.394		
				Tryptophan	2.143	0.000016	0.000171	1.789	0.0021645	0.0096991	2.671			1.647		
Glutathione metabolism	0.118	0.01061	0.2144	Glycine	1.858	0.000004	0.000066	2.003	0.0021645	0.0096991	1.812			1.723		
				Cysteine	1.515	0.009631	0.023444	0.118	0.0021645	0.0096991	0.228	0.0021645	0.022605	1.400		
				Pyroglutamic Acid	0.679	0.005177	0.014605	0.613			0.748			0.693		
				Glutamate	0.321	0.000024	0.000219	0.353			0.497			0.243		

Table 2. *Cont.*

Metabolic Pathway	Impact	p-Value	FDR	Metabolite	All Hosts			Bovine			Swine			Human		
					Fold Change (FD/ND)	p-Value	FDR	Fold Change (FD/ND)	p-Value	FDR	Fold Change (FD/ND)	p-Value	FDR	Fold Change (FD/ND)	p-Value	FDR
Nicotinate and Nicotinamide metabolism	0.066	0.0125	0.2144	Aspartate	3.077	0.000000	0.000003	3.174	0.0021645	0.0096991	3.883	0.0021645	0.022605	1.300		
				Glycerone phosphate	0.582	0.000001	0.000027	0.644	0.004329	0.016886	0.579	0.0021645	0.022605	0.536		
				Nicotinamide	1.858	0.000245	0.001366	1.505			1.666			2.495		
				Nicotinate	1.435	0.000007	0.000098	1.587	0.008658	0.026381	1.455			1.832		
Glyoxylate and dicarboxylate metabolism	0.055	0.0269	0.38631	Citrate	7.934	0.000024	0.000219	2.799			20.017			6.503		
				Glycerate	1.691	0.000503	0.002423	0.631	0.008658	0.026381	2.225			2.473		
				Glycine	1.858	0.000004	0.000066	2.003	0.0021645	0.0096991	1.812			1.723		
				Glutamate	0.321	0.000024	0.000219	0.353			0.497			0.243		
				Serine	2.837	0.000000	0.000004	4.469	0.0021645	0.0096991	1.755	0.0021645	0.022605	2.240		
				Pyruvate	0.458	0.002642	0.008841	0.382	0.008658	0.026381	1.570	0.0021645	0.022605	0.394		
beta-Alanine Metabolism	0	0.0458	0.43434	Aspartate	3.077	0.000000	0.000003	3.174	0.0021645	0.0096991	3.883	0.0021645	0.022605	1.300		
				Pantothenate	0.414	0.000000	0.000005	0.440	0.0021645	0.0096991	0.389	0.0021645	0.022605	0.429		
				Uracil	1.477	0.020464	0.042605	0.977			1.848			1.474		
Valine, leucine, and isoleucine biosynthesis	0.107	0.0475	0.43434	Threonine	2.233	0.000000	0.000000	2.065	0.0021645	0.0096991	3.156	0.0021645	0.022605	1.834		
				Leucine	0.087	0.002349	0.008027	0.106	0.015152	0.041482	0.063	0.0021645	0.022605	0.116		
				Pyruvate	0.458	0.002642	0.008841	0.382	0.008658	0.026381	1.570	0.0021645	0.022605	0.394		
				Valine	0.675	0.006402	0.017165	0.391	0.0021645	0.0096991	0.925			0.782		
Alanine, aspartate and glutamate metabolism	0.45	0.0475	0.43434	Aspartate	3.077	0.000000	0.000003	3.174	0.0021645	0.0096991	3.883	0.0021645	0.022605	1.300		
				Alanine	2.065	0.000000	0.000003	2.149	0.0021645	0.0096991	2.240	0.0021645	0.022605	1.910		
				Glutamate	0.321	0.000024	0.000219	0.353			0.497			0.243		
				Pyruvate	0.458	0.002642	0.008841	0.382	0.008658	0.026381	1.570	0.0021645	0.022605	0.394		

FDR = false discovery rate; FDR helps control for falsely positive significant features; FDR < 0.05 has less than a 5% probability of being a falsely significant feature. Fold change of > 1 indicates an increase in metabolite expression when exposed to full drug treatment and a fold change of <1 indicates a decrease in metabolite expression when exposed to the full drug treatment. Non-significant metabolites.

3.4. Univariate Analysis

A between-subject, two-way ANOVA identified 297 metabolites (Table 3) that were statistically significant only for the treatment factor. No metabolites were found to be significant for the host species factor or the interaction between host species and treatment.

Table 3. Significant identifiable metabolites found via univariate, between-subject, two-way ANOVA.

Metabolite	p-Value	FDR
2-Piperidinecarboxylic_acid_1MEOX_2TMS	0.00958	0.04464
Adenine_1TMS	0.00275	0.02104
Alanine, *N*-3-indolylacetyl	0.00002	0.00135
Aspartic acid, *N*-3-indolylacetyl	0.00000	0.00032
Butanoic acid,_3-hydroxy-0.2	0.00064	0.00847
Butanoic_acid, 4-hydroxy-_2TMS	0.00035	0.00576
Cinnamic_acid, 2-hydroxy-, trans-	0.000003	0.00040
Cohibin_A.1	0.00374	0.02508
Coixenolide_2	0.00869	0.04154
Coixenolide_4	0.00561	0.03144
Cysteamine_3TMS	0.00531	0.03071
Cysteine_3TMS	0.01144	0.04989
Glycerol-3-phosphate_4TMS.2	0.00118	0.01276
Glycine	0.00004	0.00173
Guanosine	0.00001	0.00088
Guanosine,_2′-deoxy-_4TMS.1	0.00027	0.00504
Guanosine_4TMS coeluting_with_Guanosine_5TMS	0.00261	0.02063
Iminodiacetic_acid_3TMS	0.00185	0.01715
Isoleucine_2TMS	0.00071	0.00914
Lactose	0.00001	0.00065
Leucine	0.00032	0.00541
Leucine,_cyclo-	0.00352	0.02428
Leucine_2TMS	0.00051	0.00704
Levulinic_acid	0.00057	0.00760
Luteolin	0.00474	0.02871
Naringenin	0.00066	0.00867
Oxamide_3TMS	0.00017	0.00383
Pantothenic_acid,_D-_3TMS	0.00799	0.04009
Phenylalanine_2TMS	0.00003	0.00147
Phosphomycin	0.000000008	0.00001
Pinitol,_D-_5TMS	0.00363	0.02488
Putrescine_4TMS	0.00224	0.01912
Pyridine	0.00443	0.02729
Pyridoxamine	0.00859	0.04136
Pyroglutamic_acid_2TMS	0.00976	0.04528
Quercetin	0.00003	0.00147
Serine	0.00000003	0.00001
Thiamine	0.00004	0.00173
Threitol,_dithio-	0.00166	0.01589
Thymidine-5′-monophosphoric-acid-3TMS	0.00020	0.00426
Tryptophan_3TMS	0.00007	0.00249
Tyrosine,_2-iodo-	0.000000017	0.00001
Tyrosine_3TMS	0.00001	0.00094
Uric_acid	0.00017	0.00383
Valine_2TMS	0.00008	0.00280
Xanthine_3TMS	0.00772	0.03903

FDR = false discovery rate; FDR helps control for falsely positive significant features; FDR < 0.05 has less than a 5% probability of being a falsely significant feature.

4. Discussion

In this study, we investigated metabolite expression patterns in AMR *Salmonella* Typhimurium isolated from human, bovine, and swine when exposed to antibiotics. We were able to demonstrate a greater difference in metabolite expression when the isolates were exposed to the full drug challenge compared to no drug exposure, irrespective of host species. The univariate analysis further confirmed that metabolite expression changes were significant only according to the treatment factor, not according to the host species or interaction of the host species and treatment. Metabolite expression being non-host specific suggests that AMR *Salmonella* Typhimurium drug targets are consistent across human, bovine, and swine hosts. This finding has great significance when considering that future drug testing on AMR *Salmonella* Typhimurium in swine and bovine could be translated to human treatments.

While the expression of 23 specific metabolites significantly changed when exposed to the full drug treatment and these upregulated metabolites each matched significant metabolic pathways, a specific resistance mechanism remains unclear. These isolates were exposed to multiple antimicrobial drugs and each drug has a different mechanism of action. Therefore, there are potentially many mechanisms of resistance that have developed in these isolates [4]. As per Hoerr et al. (2016), the metabolic profiles could be separated in a fingerprint, and based on the specific fingerprints obtained for different classes of antibiotics, the mode of action of several antibiotics could be predicted. The profiles could also be used as potential drug targets for pharmaceutical companies. Over the past few decades, there has been a decline in approvals of new antibiotic drugs in the market by the US Food and Drug Administration (FDA) [13,14]. The number of new antibiotics being developed every year decreases due to the challenges of effectively dispatching both antibiotic-resistant bacteria and novel infectious bacteria [15]. If we want to reverse these trends and facilitate new approaches to overcoming resistance, we must first understand the microbial forces responsible for developing resistance [16]. Metabolomics in particular offers a unique strategy to detect metabolic changes that occur in an organism in response to drugs and the outcomes of such studies can provide insights into their corresponding modes of action [17,18].

The significant changes observed in our study include increases in methionine, nicotinamide, nicotinate, pantothenate, phenylalanine, proline, pyroglutamic acid, pyruvate, serine, threonine, tryptophan, tyrosine, uracil, and valine, and decreases in alanine, aspartate, citrate, cysteine, glutamate, glycerate, glycerone phosphate, glycine, and leucine. These metabolites were matched to nine significant metabolic pathways, including glycine, serine, and threonine metabolism; alanine, aspartate, and glutamate metabolism; aminoacyl-tRNA biosynthesis; pantothenate and CoA biosynthesis; glutathione metabolism; valine, leucine, and isoleucine biosynthesis; nicotinate and nicotinamide metabolism; glyoxylate and dicarboxylate metabolism; and beta-Alanine metabolism.

Lin et al. (2019) [19] found biosynthesis of amino acids, biosynthesis of phenylpropanoids, and purine metabolism were commonly enriched in MDR strains of *E. coli*, and the results concurred that antibiotic resistance affects the metabolite profiles of MDR bacteria. Several related metabolites, such as glycerol, were increased in MDR strains, while citric acid and succinic acid were decreased in MDR strains [19].

An established resistance mechanism against β-lactams, such as ampicillin, includes the production of metallo-β-lactamases, which inactivate the drug through a cleavage process. The metallo-β-lactamases are especially threatening due to their ability to inactivate multiple β-lactams and their insensitivity to β-lactamase inhibitors that target the acyl serine transferases. This resistance mechanism has been identified in extended-spectrum β-lactamases where two amino acid substitutions are critical, a serine-for-arginine and a lysine-for-glutamate [20]. This substitution may explain the increased expression of serine and the decreased expression of glutamate observed when isolates are exposed to the ACSSuT drug panel in our study. Aspartate has also been identified as a critical component of the metallo-β-lactamases; thus, the increased expression of aspartate may support this mechanism [20].

Perhaps one of the largest resistance mechanisms is through decreasing TCA cycle flux. Previous studies have shown that exogenous alanine and/or glucose increase susceptibility to antibiotic treatment by increasing TCA flux and thereby increasing drug uptake by the cell [21]. Therefore, it is possible that decreased TCA flux could contribute to decreased drug susceptibility. Decreased concentrations of pyruvate and glutamate in our study support this conclusion, as pyruvate directly feeds the TCA cycle and glutamate is converted to pyruvate by α-ketoglutarate [22].

These data from our study suggest that another resistance mechanism utilized by these AMR isolates may be initiated from the aminoacyl-tRNA pathway. Aminoacyl-tRNA biosynthesis is responsible for changing cell membrane properties and increasing a pathogen's resistance. It has previously been identified as an attractive drug target [22]. This pathway likely acts by decreasing cell permeability and, thus, inhibiting drug entrance into the cell. The aminoacyl-tRNA biosynthesis pathway in our study is significantly altered when isolates are exposed to the ACSSuT antibiotic panel.

Alanine is a required component of cell wall peptidoglycan and it has been demonstrated that inhibition of alanine transport results in increased susceptibility to drugs [23]. Increased concentrations of alanine may indicate that the cell wall has undergone peptidoglycan remodeling, resulting in decreased susceptibility.

In our study, citrate (citric acid) has the highest fold change of any of the metabolites matched to a significant pathway, but its possible role in antimicrobial resistance is less clear. Citrate has previously been described as having a role in the regulation of cell division and gene expression and is known to be a chelator, which may allow bacteria to manage intracellular concentrations of cations. Previous research has shown an increase in citrate concentrations when *Salmonella* aureus is exposed to cold temperatures, as well as upregulated cell division proteins [24]. Therefore, increased citrate concentrations may suggest that *S.* Typhimurium depends on this metabolite to maintain intracellular Ca++ concentrations and increases the rate of cell division. An increased rate of cell division would also increase the chances of DNA mutation occurring and antibiotic resistance developing. Further examining the role of citrate in bacterial survival and AMR is warranted.

In future research, exposing isolates to only one antibiotic or one class of antibiotics would allow for a more specific interpretation of the expressed metabolites and potentially provide more robust evidence on resistance mechanisms. Interpretation of these data is limited due to the multiple mechanisms by which the ACSSuT panel targets bacteria. Resistance mechanisms against one class of antibiotics differ from those against another class, hence why bacteria resistant to one class may be susceptible to a different one [25]. This explains why a distinct resistance mechanism was not identifiable in this project. Exposure to a single antibiotic class may create a more easily identifiable profile of metabolites attributable to a specific resistance mechanism.

5. Conclusions

The findings of this study suggest that exposing AMR *Salmonella* Typhimurium to an ACSSuT panel significantly alters metabolic pathways and, thus, metabolite expression by the bacteria. This research supports the continuation of using metabolomics to study AMR and identify resistance mechanisms, which could become future drug or testing targets. However, further studies are necessary to identify specific resistance mechanisms for different classes of antibiotics.

Author Contributions: Conceptualization, S.R., C.D.B., L.L. and R.M.; Methodology, C.D.B., L.L., S.R. and R.M.; software, J.M.O. and S.R.; validation, J.M.O. and S.R.; formal analysis, J.M.O. and S.R.; resources, C.D.B.; data curation, J.M.O.; writing—original draft preparation, J.M.O.; writing—review and editing, S.R., C.D.B., L.L. and R.M.; visualization, J.M.O.; supervision, S.R.; project administration, S.R.; funding acquisition, S.R., C.D.B., L.L. and R.M. All authors have read and agreed to the published version of the manuscript.

Funding: This study was funded by the USDA Animal Health & Disease (USDA COLV-2015-36100-06008) through the College of Veterinary Medicine and Biomedical Sciences College Research Council at Colorado State University.

Institutional Review Board Statement: Not applicable.

Informed Consent Statement: Not applicable.

Data Availability Statement: All data generated or analyzed during this study are included in this published article.

Acknowledgments: We are thankful to all the institutes who have contributed the *Salmonella* Typhimurium isolates for this study: Colorado Department of Public Health and Environment, Washington State University Paul G Allen School for Global Animal Health, Ohio State University College of Veterinary Medicine, University of Illinois Urbana-Champaign Veterinary Diagnostic Laboratory, University of Pennsylvania Veterinary Diagnostic Services, and Colorado State University Veterinary Diagnostic Laboratory. Metabolomic profiling was performed at the Proteomics and Metabolomics Facility (currently known as Analytical Resource Core) at Colorado State University.

Conflicts of Interest: The authors declare no conflict of interest.

References

1. CDC. What Exactly Is Antibiotic Resistance? Centers for Disease Control and Prevention. 2020. Available online: https://www.cdc.gov/drugresistance/about.html (accessed on 21 January 2022).
2. CDC. National Action Plan for Combating Antibiotic-Resistant Bacteria. Centers for Disease Control and Prevention. 2015. Available online: https://www.cdc.gov/drugresistance/pdf/national_action_plan_for_combating_antibotic-resistant_bacteria.pdf (accessed on 21 January 2022).
3. Van Belkum, A.; Burnham, C.D.; Rossen, J.W.A.; Mallard, F.; Rochas, O.; Dunne, W.M. Innovative and rapid antimicrobial susceptibility testing systems. *Nat. Rev. Microbiol.* **2020**, *18*, 299–311. [CrossRef] [PubMed]
4. Hoerr, V.; Duggan, G.; Zbytnuik, L.; Poon, K.H.; Grobe, C.; Neugebauer, U.; Methling, K.; Loffler, B.; Vogel, H. Characterization and prediction of the mechanism of action of antibiotics through NMR metabolomics. *BMC Microbiol.* **2016**, *16*, 82. [CrossRef] [PubMed]
5. Walsh, C. Where will new antibiotics come from? *Nat. Rev. Microbiol.* **2003**, *1*, 65–70. [CrossRef] [PubMed]
6. Dailey, A.; Saha, J.; Zaidi, F.; Abdirahman, H.; Haymond, A.; Alem, F.; Hakami, R.; Couch, R. VOC fingerprints: Metabolomic signatures of biothreat agents with and without antibiotic resistance. *Sci. Rep.* **2020**, *10*, 11746. [CrossRef]
7. Agbaje, M.; Begum, R.H.; Oyekunle, M.A.; Ojo, O.E.; Adenubi, O.T. Evolution of *Salmonella* nomenclature: A critical note. *Folia Microbiol.* **2011**, *56*, 497–503. [CrossRef]
8. European Food Safety Authority & European Centre for Disease Prevention and Control. The European Union Summary Report on Antimicrobial Resistance in zoonotic and indicator bacteria from humans, animals and food in 2017/2018. *EFSA J.* **2020**, *18*, 6007.
9. Rao, S.; Maddox, C.; Hoien-Dalen, P.; Lanka, S.; Weigel, R. Diagnostic accuracy of class 1 integron PCR method in detection of antibiotic resistance in *Salmonella* isolates from swine production systems. *J. Clin. Microbiol.* **2008**, *46*, 916–920. [CrossRef]
10. Broeckling, C.; Afsar, F.; Neumann, S.; Ben-Hur, A.; Prenni, J. RAMClust: A Novel Feature Clustering Method Enables Spectral-Matching-Based Annotation for Metabolomics Data. *Anal. Chem.* **2014**, *86*, 6812–6817. [CrossRef]
11. Plumb, R.S.; Johnson, K.A.; Rainville, P.; Smith, B.W.; Wilson, I.D.; Castro-Perez, J.M.; Nicholson, J.K. UPLC/MS^E; A new approach for generating molecular fragment information for biomarker structure elucidation. *Rapid Commun. Mass Spectrom.* **2006**, *20*, 1989–1994. [CrossRef]
12. Smith, C.; Want, E.; Elizabeth, J.; O'Maille, G.; Abagyan, R.; Siuzdak, G. XCMS: Processing Mass Spectrometry Data for Metabolite Profiling Using Nonlinear PSeak Alignment, Matching, and Identification. *Anal. Chem.* **2006**, *78*, 779–787. [CrossRef]
13. Spellberg, B.; Guidos, R.; Gilbert, D.; Bradley, J.; Boucher, H.W.; Scheld, W.M.; Bartlett, J.G.; Edwards, J., Jr. The epidemic of antibiotic-resistant infections: A call to action for the medical community from the Infectious Diseases Society of America. *Clin. Infect. Dis.* **2008**, *46*, 155–164. [CrossRef] [PubMed]
14. Shlaes, D.M.; Sahm, D.; Opiela, C.; Spellberg, B. Commentary: The FDA reboot of antibiotic development. *Antimicrob. Agents Chemother.* **2013**, *57*, 4605–4607. [CrossRef] [PubMed]
15. Aries, M.L.; Cloninger, M.J. NMR Hydrophilic Metabolomic Analysis of Bacterial Resistance Pathways Using Multivalent Antimicrobials with Challenged and Unchallenged Wild Type and Mutated Gram-Positive Bacteria. *Int. J. Mol. Sci.* **2021**, *22*, 13606. [CrossRef] [PubMed]
16. Spellberg, B. The future of antibiotics. *Crit. Care* **2014**, *18*, 228. [CrossRef]
17. Lindon, J.C.; Holmes, E.; Nicholson, J.K. Metabonomics Techniques and Applications to Pharmaceutical Research & Development. *Pharm. Res.* **2006**, *23*, 1076–1088. [CrossRef]

18. Dörries, K.; Schlueter, R.; Lalk, M. Impact of antibiotics with various target sites on the metabolome of *Staphylococcus aureus*. *Antimicrob. Agents Chemother.* **2014**, *58*, 7151–7163. [CrossRef]
19. Lin, Y.; Li, W.; Sun, L.; Lin, Z.; Jiang, Y.; Ling, Y.; Lin, X. Comparative metabolomics shows the metabolic profiles fluctuate in multi-drug resistant *Escherichia coli* strains. *J. Proteom.* **2019**, *207*, 103468. [CrossRef]
20. Franklin, T.J.; Snow, G.A. Biochemical Mechanisms of Resistance to Antimicrobial Drugs. In *Biochemistry and Molecular Biology of Antimicrobial Drug Action*, 5th ed.; Springer: Berlin/Heidelberg, Germany, 2005; pp. 149–174.
21. Su, Y.-B.; Peng, B.; Li, H.; Cheng, Z.-X.; Zhang, T.-T.; Zhu, J.-X.; Li, D.; Li, M.-Y.; Ye, J.-Z.; Du, C.-C.; et al. Pyruvate Cycle Increases Aminoglycoside Efficacy and Provides Respiratory Energy in Bacteria. *Proc. Natl. Acad. Sci. USA* **2018**, *115*, 1578–1587. [CrossRef]
22. Grube, C.; Hervé, R. A Quantitative Spectrophotometric Assay to Monitor the TRNA-Dependent Pathway for Lipid Aminoacylation In Vitro. *J. Biomol. Screen.* **2016**, *21*, 722–728. [CrossRef]
23. Gallagher, L.; Shears, R.; Fingleton, C.; Alvarez, L.; Waters, E.; Clarke, J.; Bricio-Moreno, L.; Campbell, C.; Yadav, A.K.; Razvi, F.; et al. Impaired Alanine Transport or Exposure to D-Cycloserine Increases the Susceptibility of MRSA to β-Lactam Antibiotics. *J. Infect. Dis.* **2020**, *221*, 1000–1016. [CrossRef]
24. Alreshidi, M.; Dunstan, R.; Macdonald, M.; Smith, N.; Gottfries, J.; Roberts, T. Metabolomic and Proteomic Responses of *Staphylococcus aureus* to Prolonged Cold Stress. *J. Proteom.* **2015**, *121*, 44–55. [CrossRef] [PubMed]
25. De Sousa Oliveria, K.; de Lima, L.A.; Cobacho, N.B.; Dias, S.C.; Franco, O.L. Mechanisms of Antibacterial Resistance: Shedding Some Light on These Obscure Processes? In *Antibiotic Resistance: Mechanisms and New Antimicrobial Approaches*; Kon, K., Rai, M., Eds.; Academic Press: Cambridge, MA, USA, 2016; Chapter 2; pp. 19–35.

MDPI

St. Alban-Anlage 66

4052 Basel

Switzerland

Tel. +41 61 683 77 34

Fax +41 61 302 89 18

www.mdpi.com

Animals Editorial Office

E-mail: animals@mdpi.com

www.mdpi.com/journal/animals